Y0-BSU-851

Introductory

Physical

Chemistry

PRENTICE-HALL INTERNATIONAL, INC. London
PRENTICE-HALL OF AUSTRALIA, PTY. LTD. Sydney
PRENTICE-HALL OF CANADA, LTD. Toronto
PRENTICE-HALL OF INDIA PRIVATE LIMITED New Delhi
PRENTICE-HALL OF JAPAN, INC. Tokyo

Introductory
Physical
Chemistry

A. R. KNIGHT

Associate Professor of Chemistry
University of Saskatchewan

PRENTICE-HALL, INC., ENGLEWOOD CLIFFS, N.J.

© 1970 by Prentice-Hall, Inc., Englewood Cliffs, N.J.
All rights reserved. No part of this book may be reproduced in any form or by any means without permission in writing from the publisher.

Library of Congress Catalog Card Number 74-100586

Printed in the United States of America

13-502203-7

Current Printing (last digit):

10   9   8   7   6   5   4   3   2

Preface

This book is based on a one-semester course for pre-medical students and other life science majors, given by the author at the University of Saskatchewan, Saskatoon.

The approach is one in which the basic ideas of thermodynamics and kinetics are developed at the beginning in a way which attempts to introduce the necessary mathematical concepts in simple terms as a preparation for their use both immediately and in later discussions. Previous knowledge of the calculus is not necessary. The relatively small amount of integration necessary to deal effectively with a number of phenomena is introduced, in context, in the discussion. Hopefully, the subject of physical chemistry is more satisfying and meaningful to the student with a limited mathematical background if the fundamental relationships are presented in at least a "semi-derived" manner with emphasis on their use in real problems, and if the significance and utility of the functions encountered are emphasized.

Many of the subject areas traditionally covered in a first course in physical chemistry are not included, principally because of the time limitations of a one semester course, but also because of the growing trend to treat many of these areas in greater detail in introductory general chemistry courses. It is assumed that the student has had such a course, or at least a good high school chemistry background and has retained a knowledge of basic chemical ideas and facts.

An effort has been made to indicate the relevance of the fundamental principles of physical chemistry to systems of biological interest. These systems are not examined in any real detail, but rather they are discussed from the point of view of illustrating how the general ideas developed for chemical and physical processes can be applied to some specific cases. In this way the approach which can be taken in more complex situations in real biochemical systems can be demonstrated in an introductory manner.

I am happy to acknowledge the encouragement and assistance of the members of the Department of Chemistry and Chemical Engineering, University of Saskatchewan, Saskatoon. In particular, I am grateful to Dr. K. J. McCallum who offered very helpful comments on the draft manuscript. The constructive comments of the pre-publication reviewers is also very much appreciated. It is, of course, incumbent on the author to accept responsibility for residual errors.

Finally, I would express my gratitude to the students in my physical chemistry course who inspired this book in the first instance, and to my family for their patience during its preparation.

A. R. KNIGHT

Saskatoon, Saskatchewan

Contents

Thermodynamics: The First Law and Thermochemistry, 1   **1**

1-1 INTRODUCTION, 1

   1-1-1 Thermodynamic Concepts, 1

1-2 THE FIRST LAW, 4

   1-2-1 Energy, 4
   1-2-2 Work, 5
   1-2-3 Heat, 11
   1-2-4 Statement of the First Law, 13

1-3 APPLICATIONS OF THE FIRST LAW, 14

   1-3-1 Constant Volume Processes, 15
   1-3-2 Constant Pressure Processes—Enthalpy, 16
   1-3-3 Types of Thermodynamic Process, 18
   1-3-4 Isothermal Expansion of an Ideal Gas, 20
   1-3-5 Vaporization, 24

1-4 THERMOCHEMISTRY, 26

    1-4-1   Heat Capacity, 27
    1-4-2   Thermochemical Equations and Hess' Law, 29
    1-4-3   Standard Enthalpies of Formation, 32
    1-4-4   Enthalpy Changes in Physical and
            Chemical Processes, 35
    1-4-5   Variation of ΔH with Temperature, 38
    1-4-6   Experimental Determination of Heat Effects, 40
    1-4-7   Biological Applications of Thermochemistry, 42

Thermodynamics: Entropy and Free Energy, 46   **2**

2-1 THE SECOND LAW, 46

    2-1-1   The Concept of Entropy, 47
    2-1-2   Statement of the Second Law, 49
    2-1-3   Calculation of ΔS in Physical Processes, 52

2-2 THE THIRD LAW—ABSOLUTE ENTROPIES, 55

2-3 ENTROPY CHANGES IN CHEMICAL REACTIONS, 58

2-4 THE FREE ENERGY, 60

    2-4-1   Definition of Free Energy, 60
    2-4-2   Significance of ΔG, 61
    2-4-3   Standard Free Energy Changes, 64
    2-4-4   Free Energy Changes in Physical Processes, 66
    2-4-5   Calculation of ΔG for a Chemical Reaction, 70
    2-4-6   Temperature Dependence of ΔG, 72

2-5 APPLICATIONS OF THE FREE ENERGY CONCEPT TO BIOLOGICAL SYSTEMS, 74

Chemical Equilibrium in Molecular Systems, 81   **3**

3-1 INTRODUCTION, 81

3-2 CALCULATIONS INVOLVING THE EQUILIBRIUM CONSTANT, 83

3-3 EQUILIBRIUM AND FREE ENERGY, 85

    3-3-1 Relation between $\Delta G°$ and K, 86

    3-3-2 Calculations Involving $\Delta G°$ and K, 91

    3-3-3 Temperature Dependence of the Equilibrium Constant, 93

3-4 EQUILIBRIUM IN NONIDEAL SYSTEMS: THE
ACTIVITY, 96

    3-4-1 Real Gases, 97

    3-4-2 Nonideal Solutions, 99

Ionic Equilibrium I: Acid-Base Systems, 103   **4**

4-1 FUNDAMENTAL CONCEPTS, 103

    4-1-1 Acids and Bases, 103

    4-1-2 Proton Transfer in Pure Water, 106

    4-1-3 The pH Scale, 107

    4-1-4 Principle of Electroneutrality, 108

4-2 AQUEOUS SOLUTIONS OF STRONG ACIDS
AND BASES, 109

4-3 SOLUTIONS OF WEAK ACIDS AND BASES, 111

    4-3-1 Equilibrium and pH Calculations, 111

    4-3-2 Conjugate Acid and Base Pairs, 115

    4-3-3 Hydrolysis, 116

4-4 MULTIPLE ACID-BASE EQUILIBRIA, 117

    4-4-1 Amphoteric Substances, 117

    4-4-2 Polyprotic Acids and Bases, 119

4-5 BUFFER SOLUTIONS, 122

    4-5-1 The Henderson-Hasselbalch Equation, 123

    4-5-2 Properties of Buffer Solutions, 124

    4-5-3 Preparation of Buffer Solutions, 128

4-6 ACID-BASE TITRATIONS, 129

    4-6-1 Colorimetric Determination of pH, 130

    4-6-2 Change in pH in Acid-Base Titrations, 133

4-7 ACID-BASE EQUILIBRIA IN BIOLOGICAL
      SYSTEMS, 142

      4-7-1 The Bicarbonate System as a Physiological Buffer, 142
      4-7-2 Equilibria Involving Amino Acids and Proteins, 145

Ionic Equilibrium II: Ion Activities, 152   **5**

5-1 CONCENTRATION EFFECTS IN IONIC SOLUTIONS, 152

      5-1-1 Determination of Equivalent Conductance, 153
      5-1-2 Concentration Dependence of Λ, 155

5-2 ACTIVITY COEFFICIENTS OF IONS, 157

5-3 pH CALCULATIONS USING THE ACTIVITY, 161

      5-3-1 pH° of Strong Acids and Bases, 161
      5-3-2 pH° of Weak Acid and Buffer Solutions, 163

5-4 SLIGHTLY SOLUBLE SALTS, 166

      5-4-1 Solubility Products, 166
      5-4-2 Thermodynamic Solubility Product, 169

Chemical Kinetics, 174   **6**

6-1 RATES OF CHEMICAL REACTIONS, 175

      6-1-1 Concept of Reaction Rate, 175
      6-1-2 Rate Constant, Order, and Molecularity, 180
      6-1-3 First Order Reactions, 183
      6-1-4 Second and Zero Order Reactions, 189

6-2 DETERMINATION OF REACTION ORDER, 191

      6-2-1 Mechanical Slope Determination, 192
      6-2-2 Integrated Rate Law Method, 192
      6-2-3 Empirical Fit Method, 193
      6-2-4 Half-life Method, 193

6-3 COMPLEX REACTIONS, 195

    6-3-1 Types of Reaction, 195
    6-3-2 Reaction Mechanisms, 198
    6-3-3 Kinetic Treatment of Complex Reactions, 200

6-4 TEMPERATURE DEPENDENCE OF REACTION RATES, 202

6-5 REACTION-RATE THEORIES, 205

    6-5-1 Collision Theory, 205
    6-5-2 Transition State Theory, 209

6-6 CATALYSIS: ENZYME CATALYZED REACTIONS, 212

Electrochemistry, 222   **7**

7-1 INTRODUCTION, 222

7-2 BASIC ELECTROCHEMICAL CONCEPTS, 225

    7-2-1 Units and Definitions, 225
    7-2-2 Cells and Electrodes, 227

7-3 THERMODYNAMICS OF GALVANIC CELLS, 230

    7-3-1 Measurement of Reversible emf, 230
    7-3-2 Relation of $\mathscr{E}$ to ΔG, 233
    7-3-3 Concentration Dependence of $\mathscr{E}$, 235
    7-3-4 Calculation of $\mathscr{E}°$ from Half-cell emf Values, 240
    7-3-5 Experimental Determination of Standard Oxidation Potentials, 247

7-4 APPLICATIONS OF EMF MEASUREMENTS, 248

    7-4-1 Potentiometric Titrations, 248
    7-4-2 Determination of pH, 249

7-5 ELECTROCHEMISTRY OF BIOLOGICAL OXIDATION-REDUCTION PROCESSES, 251

Phase Equilibria, 255   **8**

8-1 THE VAPOR PHASE, 255

8-2 EQUILIBRIA BETWEEN TWO PHASES, 259

    8-2-1   Liquid-Vapor Equilibrium, 259
    8-2-2   Solid-Vapor Equilibrium, 265
    8-2-3   Solid-Liquid Equilibrium, 265
    8-2-4   Properties of Gases and Liquids near the Critical Point, 266

8-3 EQUILIBRIA BETWEEN THREE PHASES, 268

Properties of Solutions, 273   **9**

9-1 RAOULT'S LAW, 274

9-2 BINARY LIQUID MIXTURES, 275

    9-2-1   Ideal Solutions, 275
    9-2-2   Deviations from Raoult's Law, 278
    9-2-3   Fractional Distillation, 280
    9-2-4   Partially Miscible Liquids, 282
    9-2-5   Immiscible Liquids, 284

9-3 SOLUTIONS OF GASES IN LIQUIDS, 286

9-4 COLLIGATIVE PROPERTIES I—NONELECTROLYTE SOLUTIONS, 288

    9-4-1   Vapor Pressure Lowering, 289
    9-4-2   Boiling Point Elevation, 290
    9-4-3   Freezing Point Depression, 292
    9-4-4   Osmosis, 294

9-5 COLLIGATIVE PROPERTIES II—ELECTROLYTE SOLUTES, 299

9-6 DISTRIBUTION BETWEEN TWO SOLVENTS, 300

***Physical Chemistry of Macromolecular
Systems, 304***   **10**

10-1 INTRODUCTION, 304

10-2 GENERAL PROPERTIES OF MACROMOLECULES, 305

　　10-2-1　Nature of Macromolecules, 305
　　10-2-2　Macromolecular Weights, 307
　　10-2-3　Surface Effects, 308

10-3 PROPERTIES OF SOLUTIONS OF
　　MACROMOLECULES, 314

　　10-3-1　Light Scattering, 314
　　10-3-2　Sedimentation, 317
　　10-3-3　Osmotic Pressure—Donnan Membrane
　　　　　　Equilibrium, 318
　　10-3-4　Other Properties Used in Determination of
　　　　　　Macromolecular Weights, 322

Appendix **1: Use of Logarithms, 324**

Appendix **2: Graphical Analysis, 328**

Bibliography, 335

Index, 337

Introductory

Physical

Chemistry

Thermodynamics: The First Law and Thermochemistry

1-1 INTRODUCTION

Physical chemistry is concerned with a wide range of chemical endeavor, but the fundamental ideas of thermodynamics, kinetics, and molecular structure and the topics related to them remain the core of the subject. The approach here is based on thermodynamics and kinetics and how the principles developed in these areas can be applied to physical chemistry problems.

We begin our study with thermodynamics. A large proportion of the subsequent discussion is centered around thermodynamic principles. In addition, even areas not directly related to this topic are characterized by the same kind of exact application of fundamental principles and equations.

1-1-1 Thermodynamic Concepts

Thermodynamics is often a great enigma to beginning students in physical chemistry. This need not be the case if the quantitative relationships developed are seen as part of an overall development of a useful method of treating physical and chemical changes.

It is the generality of thermodynamics that makes it such a worthwhile tool. It is not necessary to deal with any particular kind of system—the same rules apply to all. There is a basic distinction between the chemical and physical laws derived for and applied to specific types of system and the fundamental relations applicable to any system. The general principles of thermodynamics are of the latter type. For example, the rules governing the theoretical maximum efficiency of a heat engine apply equally well to the conversion of hydrogen and oxygen to water and to the vaporization of the liquid water formed in the process. A large part of thermodynamics was developed in terms of heat engines, but our concern is with applications to systems of direct interest to the chemist.

The initial discussions, however, are centered around systems that at first sight appear to have very little to do with chemical reactions. This background is nevertheless necessary to carry out thermodynamic calculations for processes of interest.

The approach in thermodynamics can be summarized as follows:

THERMODYNAMICS DEALS WITH THE VALUE OF CERTAIN FUNCTIONS OF A SYSTEM IN A GIVEN STATE AND THE IMPLICATIONS OF THE CHANGE IN THESE VALUES AS THE SYSTEM GOES FROM ONE STATE TO ANOTHER.

There are three important terms here that must be thoroughly understood. A frequent error in the study of thermodynamics and physical chemistry in general is lack of appreciation of the fact that a given term may have various meanings in different contexts but, when we use it in a thermodynamic sense, the term takes on a very specific and precisely defined meaning.

The first of these three terms, the *system*, may be defined as follows:

SYSTEM: THE PART OF THE UNIVERSE WE WISH TO CONSIDER

Whatever is not included in the system is referred to as the *surroundings*. Usually a fixed amount of material is taken as the system, although we may choose any convenient dividing line between system and surroundings. For example, in considering a reaction taking place in a reaction vessel in the laboratory, the system could be the reactants and products or these materials along with the vessel. We could include the entire laboratory if we wished to consider the effect on the temperature of the room. In any treatment we also must determine what, if anything, can be transferred from system to surroundings. We consider whether the system is *closed* (no material transfer) or *open* (material transfer across the system boundary) and whether the system is *isolated* (no energy transfer). Generally, closed systems that are not isolated are dealt with here, for example, the reaction vessel that confines products and reactants but is subject to heat exchange with the surroundings.

Once we have a satisfactory description of the system it is necessary to define the *state* of the system as follows:

STATE: THE CONDITION OF THE SYSTEM AS DEFINED BY SPECIFYING VALUES OF SUFFICIENT PARAMETERS SUCH THAT ALL REMAINING PARAMETERS ARE FIXED.

Ordinarily the temperature T and pressure P and composition of the system would be specified. For example, if the system is 1 mole of ideal gas, if T and P are known, then the state has been specified. The value of the only remaining parameter, the volume V, is fixed by the ideal gas law $PV = nRT$ since n, the number of moles of gas, is unity, and R, the gas constant, has a fixed numerical value appropriate to the P, V, and T units employed.

The functions calculated for the states of a system and for the changes between these states are of two kinds:

STATE FUNCTIONS: FUNCTIONS WHOSE VALUE IS INDEPEND-ENT OF THE PAST HISTORY OF THE SYSTEM AND DETER-MINED SOLELY BY THE THERMODYNAMIC PROPERTIES OF THE SYSTEM IN A GIVEN STATE.

PATH-DEPENDENT FUNCTIONS: FUNCTIONS WHOSE VALUE DEPENDS ON THE WAY IN WHICH THE STATE-TO-STATE TRAN-SITION IS MADE.

Both types of function are important in thermodynamics, but it is the state functions that are the pivot for the fundamental laws of the subject. The considerable advantage of dealing with state functions is that the system can move from initial to final state by *any* necessary or convenient path, but we can *calculate* the initial and final value of the required function by choosing the specific path that allows direct computation of the change in the state function. This important technique is used very often in thermodynamics. As will be seen, the change in the value of a state function is often numerically equal to the change in some path-dependent function for a specific pathway. We therefore imagine that the state-to-state transition occurs via this route. The calculation made on this basis gives the change in the state function. This does *not* restrict us in practice from carrying out the process by any real path, because the change in the state function invariably will have the same value so long as the process involves the same initial state and the same final state. The specific path is invoked solely for the purpose of calculation.

In a way the name thermo*dynamics* is a misnomer, since the areas of the subject that concern us here are not *primarily* involved with the way in which the system moves from state to state. However, we often specify certain definite routes to calculate path-dependent functions. Furthermore, we are not at all concerned with the time interval involved in the transition.

In thermodynamics we attempt, from our accumulated experience, to deduce rather than to derive a number of perfectly general statements. What is fundamentally important and valuable about thermodynamics is that it allows us to make predictions. Some predictions turn out to be correct only in an academic sense. For example, the transition between two states may be possible, but the rate of the process may be too small to be observed. Thermodynamics cannot predict the rate. Despite such eventualities, however, the ability to predict is the essential value of the subject. This is the central theme in the development of thermodynamic principles here. Many laws of physical chemistry that can be used to describe quantitatively processes of interest have their foundation in thermodynamics. The basic relationships can be derived or deduced directly from the laws of thermodynamics.

The change in the *free energy*, a state function defined later, is the basis of any prediction, however, and this is the essence of our approach. A thorough understanding of the concept of free energy and its applications is vital.

I-2 THE FIRST LAW

The First Law of Thermodynamics deals with three fundamental concepts—energy, work, and heat—and the relations among them. Before stating the law, then, we must examine each of these terms.

I-2-I Energy

The formal definition of energy as the "ability to do work" is of little value, since work itself is defined later in terms of energy. Perhaps a more satisfactory approach at this stage is to rely on our intuitive idea of what energy is, coupled with what can be appreciated about specific kinds of energy. For example, we can appreciate kinetic energy if we stand in the path of a moving object.

Chemical thermodynamics deals with the *internal energy E*, energy possessed by the system by virtue of the mass and motion of the molecules, intermolecular forces, chemical composition, and similar factors. Any energy the system possesses because of other considerations—its position in the earth's magnetic field, for example—are ignored.

The second feature of energy considerations is concern with the *change in energy* rather than its absolute value. The following expresses this concept mathematically:

$$\Delta E = E_2 - E_1 \qquad (1\text{-}1)$$

This indicates that the finite change in total internal energy ΔE is the difference between the energy in the final state and that in the initial state. Note that we invariably subtract the initial (E_1) from the final (E_2) value in calculating the change in any thermodynamic function. The most significant aspect of this kind of relation is that the energy change depends only on the initial and final states and is independent of the path linking these states; that is,

INTERNAL ENERGY IS A STATE FUNCTION.

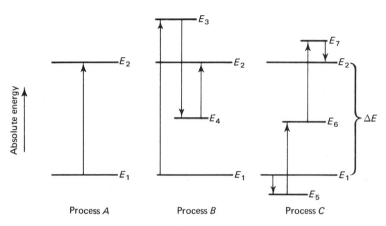

FIG. 1-1 Schematic representation of three possible routes between initial state (E_1) and final state (E_2) that have the same ΔE regardless of the intermediate stages through which the system passes.

This fundamentally important idea is illustrated schematically in Fig. 1-1, in which the absolute internal energy of the system (with unspecified origin) is plotted for the initial and final states. The three processes shown begin and end in the same energy states, and therefore the net change in energy is identical regardless of the intermediate states through which the system passes.

There are many forms that energy can assume, but it is convenient to classify all energy entering or leaving the system as either heat or work. As a preliminary to stating the principle governing a change in energy, these two modes of energy alteration must be investigated. These ideas are important not only in this context but also because they illustrate a number of generally useful points of thermodynamic technique. Thus the discussion here is perhaps more detailed than is mechanically necessary for later work, but the insight into thermodynamic procedures that results is well worth the effort.

1-2-2   Work

On the basis of the twofold classification of energy, work may be defined in the following manner:

WORK: ENERGY LOST FROM THE SYSTEM BY MECHANISMS OTHER THAN HEAT TRANSFER.

There are several varieties of work, but initially only one type is considered: *pressure–volume* or *PV* work—work done by the system in pushing back the surrounding atmosphere—for example, the case of a gas formed in a chemical reaction. If a chemical process occurs in an electrochemical cell, it is characterized by an amount of electrical work. This type of energy transfer is discussed in detail in Chapter 7.

Note that the definition is stated in terms of energy *lost*, and on this basis work done *by the system* on the surroundings is a positive quantity. The implications of this convention become apparent in the way in which the first law is stated, but we should note it carefully since it involves a commitment to a particular sign convention.

To obtain a working expression for calculating the change in the energy of a system because of PV work done on or by it, we must begin by considering mechanical work in general, as defined by the relation

$$\text{Mechanical work} = \text{force} \times \text{distance}$$

Consider the specific case of PV work shown in Fig. 1-2, where an amount of gas in a cylinder expands against a constant applied pressure. Writing the work equation in more formal terms we have $w = fl$, where f is the force and l is the distance through which the piston moves. The force can be written as a product of the external pressure and area of the piston, $f = PA$, and therefore work can be represented by $w = PAl$. The term Al is the change in volume when the piston moves through the distance l, and this volume change can be represented by ΔV. Using this relation we obtain the desired equation

$$w = P\,\Delta V \qquad\qquad (1\text{-}2)$$

The pressure used in the application of this equation must be the *external* pressure. No statement has been made about the internal pressure of the gas. In this type of process it is the constant external pressure and the change in the volume that determine the amount of work involved. Whatever the nature of the process, no PV work will be done if the volume does not change.

Although Eq. (1-2) has been derived for the case of gas expansion, it is equally applicable to a compression process. In an expansion, $\Delta V\;(= V_2 - V_1)$ will be positive since $V_2 > V_1$ and the numerical value of w, defined as the work done by the system, will be a positive quantity. In a compression, $V_2 < V_1$, and therefore w will be negative as required since the compression

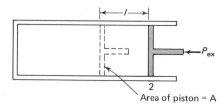

FIG. I-2 Expansion of a gas in a cylinder of cross-sectional area against a constant external pressure P_{ex}.

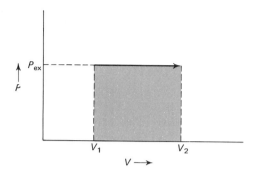

FIG. 1-3 *Pressure against volume for the expansion of a gas from* V_1 *to* V_2, *as shown in Fig. 1-2.*

work is done on the system by the surroundings. An important point is illustrated here. In thermodynamics we set up a general equation and allow the sign of the function to be determined by the numerical variables in the equation and not by a sign that we arbitrarily impose on the answer, because we "know" whether the value should be positive or negative. Considerable difficulty is avoided if the general equation for any situation is set up completely in the first instance; then, in making use of this relation, we let the signs "take care of themselves."

For the simple case of constant external pressure to which Eq. (1-2) can be applied, the situation can be analyzed graphically,* as shown in Fig. 1-3. Here, since $w = P \Delta V = P(V_2 - V_1)$, we are effectively calculating the area of the rectangular, shaded portion in the graph—what is referred to as the *area under the curve*. The "curve" in this case is the straight line representing the constant value of the external pressure as the volume changes.

Although this type of calculation is elementary, we encounter a difficulty, which presents itself several times in subsequent discussions in general, when the pressure does not remain constant. This situation is shown in Fig. 1-4. We will see presently that the work is again equal to the area under the curve, but we cannot use Eq. (1-2), which now becomes the inequality $w \neq P \Delta V$, simply because we do not know what value of P should be substituted—it changes continuously throughout the process.

How can we circumvent this difficulty? One possible device would be to take an average pressure, but this evidently would yield an acceptable result only if the pressure varied in exactly the same manner above and below the average or if the pressure changed only very little during the process. The latter alternative, however, provides the key to the solution of the problem. We could consider the overall change as a *series* of processes, in each of which $w = P_{av} \Delta V$ and where each ΔV is rather small and P_{av} is the

* See Appendix 2 for a discussion of the general techniques associated with graphical presentation of data.

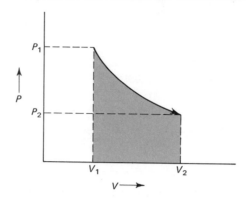

FIG. 1-4 *Pressure against volume for the expansion of a gas against a variable external pressure.*

average external pressure during each "subprocess." Summation of the work terms for all steps in the sequence would then give the work done in the total volume change. This procedure is illustrated in Fig. 1-5, which considers a volume change from V_1 to V_2 as a sequence of three steps. For this case, then, we can write Eq. (1-2) for each step, assuming the error involved in taking a constant average external pressure in each is acceptable, and then sum the terms to obtain the total work:

$$w_1 = P_a(V_a - V_1) \qquad w_2 = P_b(V_b - V_a)$$
$$w_3 = P_c(V_2 - V_b) \qquad w_{tot} = w_1 + w_2 + w_3$$

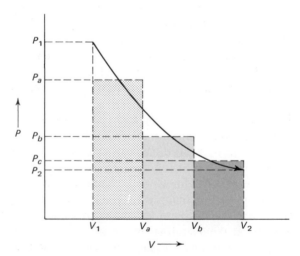

FIG. 1-5 *Pressure against volume for the expansion of a gas against a variable external pressure. The process is considered to occur in three steps for which average external pressures of P_a, P_b, and P_c are taken.*

To carry this technique one step further we take an infinitely large number of segmental processes in which the pressure during each infinitesimal volume change therefore remains constant within an infinitely small error, and we obtain an answer as accurate as when P is constant during a finite volume change. Expressing this operation in a word equation,

Work$_{tot}$ = sum of work terms for each infinitesimal process

Work$_{tot}$ = sum of infinitely small volume change times the pressure during this change for each component process

It is apparent from Fig. 1-5 that the overall process can be divided into a number of smaller steps, but how is it possible to accomplish this for an infinitely large number of steps? Here we can make use of the mathematical techniques of calculus, although it is not necessary to become deeply involved in the theory of the procedure.* The technique suggested above, which amounts to summing an infinite number of area segments under a curve, is termed *definite integration*. The conventionally employed mathematical symbols represent the procedure indicated in the word equation for the total work. Expressed in this notation the word equation becomes

$$\int_1^2 dw = w = \int_{V_1}^{V_2} P \, dV \tag{1-3}$$

where

$\int_1^2$ = the sum of the infinitely large number of work terms as the system goes from the initial to final state

w = the total work done as the volume changes from V_1 to V_2

$P \, dV$ = the product of the applicable pressure and the infinitesimal volume change at this pressure†

$\int_{V_1}^{V_2}$ = the sum of all the $P \, dV$ terms between the limits of V_1 and V_2

Thus we can consider the integral sign with *limits of integration* V_1 and V_2 as an instruction to plot a graph of external pressure as a function of volume and measure the area under the curve between these two values of the volume, for example, the shaded area in Fig. 1-4.

In the simplest case where the pressure is constant, it can be shown directly that Eq. (1-3) becomes identical with Eq. (1-2). In this situation the

---

* Sufficient details of the methods of calculus as it applies to this and subsequently encountered cases are given in context in the general discussion. The reader who wishes to review the very limited amount of calculus employed is referred to one of the standard texts in introductory calculus. A particularly pertinent and useful book is J. N. Butler and D. G. Bobrow, *The Calculus of Chemistry* (New York: W. A. Benjamin, Inc., 1965).

† The d indicates an infinitely small change in the quantity indicated, as opposed to Δ for finite changes.

value of P in each $P\,dV$ term in the summation is the same. We therefore can add the dV terms first and then multiply the result by P. This procedure is analogous to taking a common factor "outside the brackets" in algebra. However, the sum of dV terms is just ΔV, and the expression therefore is $P\,\Delta V$.

$$w = \int_{V_1}^{V_2} P\,dV = P\int_{V_1}^{V_2} dV = P\,\Delta V \quad \text{(pressure constant)}$$

For any case where P is not constant, Eq. (1-3) must be used. Mathematically we require an equation relating P and V that allows us to substitute in the integral for P in terms of V. We then will have to integrate a function of a single variable only. Such integrals can be evaluated directly using the standard rules of integration. The result is usually a straightforward relation that can be solved for w by substituting the appropriate numerical values of V.

Although the approach to calculus here is rather naïve, it is sufficient for our purposes. Integration is regarded as a method of evaluating a thermodynamic function—in the case discussed above, the PV work—by computing the area under a curve representing the behavior of the relevant variables. In any situation where it does become necessary to evaluate an integral, one always has the option of simply plotting the function and measuring the area mechanically. Alternatively, the integrated form of the function can be obtained from the rules of calculus. We will encounter both types of procedure here.

For example, it is possible to evaluate the integral

$$\int_{x_1}^{x_2} y\,dx$$

where $y = x^2$, either by plotting y vs. x and measuring the area under the curve between x_1 and x_2 or by evaluating the integral using calculus. For this function the rules of integration yield the following expression for the value of the integral:

$$\int_{x_1}^{x_2} x^2\,dx = \frac{(x_2)^3}{3} - \frac{(x_1)^3}{3}$$

Substitution of the numerical values of x_2 and x_1 then gives the numerical value of the integral.

Unlike energy, work is not a state function. The amount of work involved in a process depends on the path along which the transition from state to state occurs. Consider a change in state where 1 mole of ideal gas is expanded from state 1—where the internal pressure is P_1 and the volume is V_1—to state 2, with P_2 and V_2. Two possible routes are illustrated in Fig. 1-6. In (a) the pressure is suddenly decreased from $P_{\text{external}} = P_1$ (initial conditions, system in equilibrium) to the final value $P_{\text{external}} = P_2$

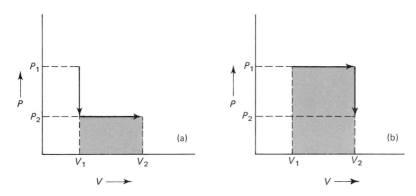

FIG. 1-6 *Expansion of a gas by two different routes for which the amount of work obtained is different.*

and $V = V_2$. In process (b) the expansion is carried out at an external pressure equal to P_1. When the final volume V_2 is attained, the pressure is lowered to P_2 and the system is again in equilibrium. It is apparent that in (b) there will have to be suitable adjustments to the temperature. We are concerned here, however, with the change in energy due to PV work. Comparison of the areas under the curves in the figure shows that the amount of work involved in these two processes that occur between the same initial state and final state is quite different.

This simple example illustrates the important principle that the energy alteration owing to work depends on the way in which the process occurs.

1-2-3 Heat

Experience indicates that when and if there is a difference in temperature between two bodies in contact there occurs what is loosely referred to as a flow of heat between them until the temperature differential disappears. Nothing of substance, of course, actually flows. We are simply describing a process involving a transfer of thermal energy. This form of energy is unique. All energy is perfectly convertible into heat although, as we will see the converse is not true. This convertibility gives rise to the convention of measuring energies and expressing amounts of energy in terms of heat units, calories or joules. The calorie is defined as exactly 4.1840 joules.

We can define this form of energy as follows:

HEAT: AN ENERGY ALTERATION MANIFESTED IN A CHANGE IN THE TEMPERATURE OF THE SYSTEM OR ITS SURROUNDINGS.

Like work, heat is not a state function. The heat effect in a process depends on the way in which it is carried out. In the two processes illustrated in Fig. 1-6, for example, it was noted that in process (b) suitable adjustments to the temperature would have to be made. In other words, processes (a) and (b) would involve differing amounts of heat transfer because the paths from the same initial to the same final state are not the same.

A quantity of heat q is a product of two terms, the difference in temperature—a potential factor in which the greater the ΔT the greater the possibility of thermal energy transfer—and the *heat capacity* C, which is a measure of the responsiveness of the system to the ΔT. Thus,

$$q = C \, \Delta T \qquad (1\text{-}4)$$

Equation (1-4) can be recast to yield the defining equation for a general heat capacity:

$$C \, (\text{cal/deg}) = q \, (\text{cal})/\Delta T \, (\text{deg}) \qquad (1\text{-}5)$$

This is a perfectly general definition, and it is apparent that the heat capacity of a system depends not only on the heat-absorbing ability but also in a simple way on the amount of substance present. It is necessary, then, to define heat capacities in terms of the amount of material in the system, and two notations are used.

> **SPECIFIC HEAT CAPACITY: HEAT ABSORBED BY ONE GRAM OF MATERIAL THAT UNDERGOES A RISE IN TEMPERATURE OF ONE CELCIUS DEGREE.**

> **MOLAR HEAT CAPACITY: HEAT ABSORBED BY ONE GRAM MOLE OF MATERIAL THAT UNDERGOES A RISE IN TEMPERATURE OF ONE CELCIUS DEGREE.**

The latter is just a product of the specific heat capacity and the gram molecular weight. The heat capacities used in this book, unless stated otherwise, are molar heat capacities* with units of calories per degree mole.

We will see later that there is a further distinction to be made between heat capacities, depending on whether the volume or pressure remains constant during the heating process.

The heat effect in a process q can be calculated from

$$q = C \, \Delta T = C(T_2 - T_1) \qquad (1\text{-}6)$$

if it is assumed that the heat capacity C, which generally varies with temperature, remains essentially constant between T_1 and T_2. The situation here is analogous to the calculation of PV work in a constant pressure process, and q can be represented graphically by the area under the curve, as shown in Fig. 1-7a. Again, however, integration must be used if C varies significantly with temperature. This case is illustrated in Fig. 1-7b. If we use the same

---

* The device of designating molar quantities in general as $\bar{X}$ has been avoided because of possible confusion concerning the use of the bar for molar or partial molar quantities—a distinction related to a more detailed treatment of thermodynamics.

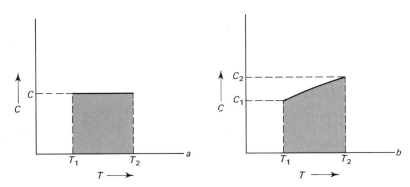

FIG. I-7 Heat capacity vs. temperature for (a) C invariant with T and (b) C a function of T.

arguments as before, the heat absorbed is the area under the curve,

$$q = \int_{T_1}^{T_2} (C \text{ as a function of } T) \, dT \qquad (1\text{-}7)$$

If the heat capacity cannot be expressed conveniently as a function of temperature, q can be evaluated by mechanical area determination in a graph of heat capacity vs. temperature over the range of interest. If a suitable function is available, the integration can be performed numerically by the use of integration rules, but this will not be done here.

1-2-4 Statement of the First Law

We have now examined in detail three thermodynamic properties of the system: the energy associated with particular states and the two interrelated mechanisms by which this energy can be changed—heat and work. The first law expresses the fundamental relationship between these three quantities. The law can be stated in a number of equivalent ways, perhaps the most familiar of which is the *Law of Conservation of Energy: Energy cannot be created or destroyed.**

Since the energy that leaves or enters the system must do so, by definition, as either heat or work, and furthermore since energy cannot be dissipated by some extraneous process, the net change in energy must be the difference between the heat energy gained and the work energy lost. This gives us the statement of the First Law of Thermodynamics in the form

$$\Delta E = q - w \qquad (1\text{-}8)$$

* Since mass and energy are interconvertible, this principle is more correctly stated as the *Conservation of Mass–Energy.* In practice, however, the change in mass of a system undergoing a normal chemical reaction is negligible, and the statement given in the text is adequate if not strictly correct.

The law governs the transition of the system from initial to final state where

ΔE = change in internal energy of the system
q = heat absorbed by the system
w = work done by the system

Note carefully the sign convention* referred to earlier in Sec. 1-2-2. Since the first law is stated here in terms of the *difference* between heat and work, both the heat absorbed by the system and the work done by it are positive quantities. In utilizing the first law, then, we evaluate the change in energy by adding the heat energy gained by the system and subtracting the energy lost as the system does work on its surroundings. The following examples will serve to clarify the procedure:

The system absorbs 10 cal of heat and does 5 cal of work.

$$\Delta E = q - w = 10 - 5 = +5 \text{ cal (net gain in energy)}$$

The system evolves 10 cal of heat and does 5 cal of work.

$$\Delta E = q - w = -10 - 5 = -15 \text{ cal (net loss of energy)}$$

The system absorbs 10 cal of heat and has 5 cal of work done on it.

$$\Delta E = q - w = 10 - (-5) = +15 \text{ cal (net gain in energy)}$$

The central feature of the first law is that although both heat and work depend on the nature of the process, ΔE as defined by Eq. (1-8) is path independent, a state function. This arises as a direct result of the way in which the terms have been defined and the restriction that energy cannot be created or destroyed.

1-3 APPLICATIONS OF THE FIRST LAW

If the first law is to be applied to a chemical process in order to calculate the change in energy of the system as the reactants are converted to products, data for q and w in the process are required. It is seen later, in the discussion of thermochemistry, how the heat effect can be measured. For the moment we will simply note that q can be determined experimentally in a straightforward manner. Depending on whether or not the pressure is constant, we can use Eq. (1-2) or (1-3) to calculate the PV work done.

Frequently only PV work is possible in a chemical reaction. The only work done by the system is in pushing back the surrounding atmosphere as a

---

* The alternative convention, perhaps more self-consistent, is to consider work done *on* the system as a positive quantity since it adds to the energy. This leads to the first law in the form $\Delta E = q + w$, which is used in a number of standard thermodynamic references. Equation (1-8) will be used in this book.

gas is formed. If this is the case Eq. (1-3) for w can be used in the first law, which becomes

$$\Delta E = q - \int_{V_1}^{V_2} P \, dV \qquad (1\text{-}9)$$

In any process there are a number of parameters that can be allowed to vary or can be maintained constant. There is often considerable simplification if the process is arranged so that one or more variables are kept constant. Although the equations are available to deal with situations in which these parameters vary, it does simplify the elementary approach at this stage to set up these restrictions. This technique is being used, not only for ease of calculation but also because in practice one or more variables remain unchanged during the process. For example, if a reaction is carried out in a sealed vessel, then the volume of the system is constant. This is, in fact, the first type of process to which the first law is applied here.

1-3-1 Constant Volume Processes

The restriction of constant volume can be stated mathematically as $dV = \Delta V = 0$. Therefore, the integral in Eq.(1-9) will be zero whether or not the pressure is constant—again the basic idea that no work is done unless V changes—and the energy alteration will be $\Delta E = q$ (volume constant). The convention in thermodynamics is to indicate quantities that do not vary in a process as right subscripts, and thus the conventional form of this expression is

$$\Delta E = q_v \qquad (1\text{-}10)$$

We can interpret this equation as an alternative definition of ΔE as *the heat absorbed by the system in a constant volume process.*

It is important to appreciate the fact that this statement is not inconsistent with energy being a state function and q being path dependent. The system moving from state 1 to state 2 will have the same ΔE regardless of path. If we carry out the process along a specific path, that of constant volume where no work is done, the only possible energy alteration is through heat transfer. The thermal energy input and the total energy change therefore are identical.

If the particular process of interest is a chemical reaction, then the significance of the energy change can be indicated as follows:

ΔE: *TIIE HEAT ABSORBED BY THE REACTION CARRIED OUT AT CONSTANT VOLUME.*

ΔE thus can be regarded as the *heat of reaction at constant volume.*

If we perform a particular process in a sealed vessel so that the volume is invariant, a measurement of the heat absorbed gives directly the energy change in the reaction. This energy change is a characteristic of the reaction under *any* conditions—we use the constant volume conditions only to calculate the numerical value of ΔE.

I-3-2 Constant Pressure Processes—Enthalpy

Although ΔE is a fundamentally important quantity, from a practical point of view it is not a quantity we normally determine directly. Most chemical reactions are not carried out under constant volume conditions in a sealed reactor but rather at constant pressure in a reaction vessel open to the atmosphere. We must remember, however, that dealing with constant pressure processes is a matter of choice and convenience and is not indicative of any fundamental limitation on the applicability of thermodynamics.

To deal conveniently with constant pressure processes a new thermodynamic function, *enthalpy*, H, is used. It is defined by the equation

$$H = E + PV \qquad (1\text{-}11)$$

As with energy, it is the *change in enthalpy* that is of interest rather that its absolute value. Consequently the defining equation is more frequently written in the form

$$\Delta H = \Delta E + \Delta(PV) \qquad (1\text{-}12)$$

H is frequently referred to as the *heat content*. This is a somewhat misleading term for reasons that will become apparent shortly, although it does indicate in a way the origin of the function. There is probably less confusion if one accepts in a simple way enthalpy as a thermodynamic function of the system, a function that can be used to deal effectively with processes occurring at constant pressure.

We can attach a physical meaning to enthalpy through the way in which the change in its value is most often measured experimentally. For a constant pressure process, Eq. (1-12) becomes

$$\Delta H = \Delta E + P\,\Delta V \,(p \text{ constant}) \qquad (1\text{-}13)$$

When only PV work is possible and at constant P, we have from Eq. (1-2) that $w = P\,\Delta V$. Making this substitution into the first law, Eq. (1-8) gives

$$\Delta E = q_p - P\,\Delta V \qquad (1\text{-}14)$$

The use of this relation for the energy change in Eq. (1-13) yields the desired equation for ΔH,

$$\Delta H = q_p - P\,\Delta V + P\,\Delta V$$
$$\Delta H = q_p \qquad (1\text{-}15)$$

This relation indicates the most convenient way to measure the enthalpy change in a process and is a definition of ΔH as the *heat absorbed in a constant pressure process*. Again, as with the equivalence of ΔE and q_v, there is no conflict between enthalpy being a state function and a change in its value being the same as the path-dependent heat effect along the specific path of constant pressure. It is nonetheless perfectly valid to compute an enthalpy change for *any* process—we are not restricted to constant pressure. Every change of state will have an associated ΔH, which we can determine, however, by measuring the quantity of heat absorbed when the change of state occurs at constant pressure. Under these conditions, although ΔH does correspond to the change in heat content, this term is misleading because for the same initial state to final state transition by a variable pressure path, the heat effect may be vastly different although ΔH is the same.

We are also concerned, here as with energy, with chemical processes, and from the general definition written earlier we can now state the following:

ΔH: *THE HEAT ABSORBED BY A REACTION WHEN CARRIED OUT AT CONSTANT PRESSURE.*

Note the sign convention. The change in enthalpy or *heat of reaction* is defined in terms of heat absorbed. When heat is evolved—an exothermic reaction—negative quantity of heat is said to be absorbed and ΔH will be negative.

Having defined the change in two important thermodynamic quantities, ΔH and ΔE, it is interesting to see how different they are in some processes of interest.

Recall that when $\Delta P = 0$ we have Eq. (1-13). If the pressure remains constant, the difference between the energy and enthalpy change evidently depends solely on the $P \Delta V$ term. For moderate pressure, if *all* reactants and products are liquids or solids, the volume change will be quite small, if the temperature remains unchanged as well, and the $P \Delta V$ term will be insignificant. Equation (1-13) reduces under these conditions to

$$\Delta H = \Delta E \quad \text{(solids and liquids only)} \qquad (1\text{-}16)$$

For a reaction involving one or more gases there may be a significant contribution from change in the pressure–volume product. Here we must make use of the more general enthalpy change equation, Eq. (1-12). If we assume ideal behavior for the gases involved,

$$PV = n(RT) \qquad (1\text{-}17)$$

and if the temperature is constant,

$$\Delta(PV) = \Delta n(RT) \qquad (1\text{-}18)$$

Δn is the number of moles of gaseous products less the number of moles of gaseous reactant in the reaction equation as written. Making this substitution in Eq. (1-12) gives

$$\Delta H = \Delta E + \Delta n(RT) \qquad (1\text{-}19)$$

For calculations using this equation care must be exercised to use the numerical value of the gas constant appropriate to the applicable units. With $R = 0.0821$ liter atm/deg mole, ΔH and ΔE would be given in liter atmospheres, whereas for the more usual case of these quantities in units of calories, $R = 1.987$ cal/deg mole.

EXAMPLE 1-1.   The energy change in the reaction

$$FeS(s) + 2\,HCl(aq) = FeCl_2(aq) + H_2S(g)$$

is $-3{,}646$ cal at 27 °C. Calculate ΔH at this temperature.

$\Delta n = 1$ (1 mole of gaseous product formed)

$\Delta H = \Delta E + \Delta n(RT) = -3{,}646 + 1 \times 1.987 \times 300 = -3{,}646 + 596$

$\underline{\Delta H = -3{,}050\ \text{cal}}$

Note that the units of ΔH are simply calories, with "for the reaction as written" being understood. For any chemical reaction we always calculate thermodynamic functions for the reaction as written, and we reserve such units as calories per mole for processes involving a single substance.

Viewing this example reaction from the first law, there is a difference of 3,646 cal in the energy of the initial state and final state, which would be the heat evolved were the process conducted at constant volume. At constant pressure, however, 3,050 cal are produced. Thus when the reaction is carried out in a vessel open to atmospheric pressure, the system has to do 596 cal of work pushing back the atmosphere as H_2S is formed.

I-3-3 Types of Thermodynamic Process

The final discussion necessary before calculating energy and enthalpy changes in some processes of interest involves a careful examination of the distinction made in thermodynamics between various kinds of processes.

The most significant specification that can be made about the nature of a process is to characterize it as a *reversible process*, which can be defined in the following manner:

> **REVERSIBLE PROCESS: A PROCESS IN WHICH THE OPPOSING FORCE IS ONLY INFINITESIMALLY SMALLER THAN THE DRIVING FORCE AND WHICH CONSEQUENTLY MAY BE REVERSED BY AN INFINITESIMAL INCREASE IN THE OPPOSING FORCE.**

The principal importance of the reversible process is in its association with maximum efficiency, which will be discussed in detail later.

One way in which a reversible process can be envisioned is in terms of the series of equilibrium states. The system moves from initial to final state by a series of infinitely small steps, which are so small that the change in the system during any one of them is imperceptible, and thus it is in continuous equilibrium. Furthermore, as the steps are infinitely small, it follows that a truly reversible process would take an infinitely long time to carry out. If such a situation appears to be unrealistic, it is only because a truly reversible reaction is an abstraction—an idealized process to which we can compare real processes that occur in nature. On the other hand, certain chemical processes, with oxidation-reduction reactions in electrochemical cells as the classic example (see Chapter 8), can be made to proceed in a manner that approaches reversibility very closely. In addition, any system in equilibrium —either chemical or physical—can legitimately be thought of as undergoing a reversible (in the strict thermodynamic sense) transformation.

When a chemical reaction such as $A + B \rightleftharpoons AB$ is referred to as being thermodynamically reversible, the implication is far more than the superficial technical meaning that the reaction can proceed to a measurable extent in both directions, forward and reverse. Most chemical reactions are reversible in this sense, but they approach thermodynamic reversibility either not at all or only under very carefully controlled and defined conditions.

There is another point that can be made profitably here concerning the way in which this simple example reaction is written. By convention only, we designate the species on the left as reactants and those opposite as products. When reactions proceed from right to left, the products become reactants, although the conventional nomenclature continues to be used. This may appear to be a trivial point, but it becomes significant in later discussions because we invariably define the thermodynamic functions for the reaction (which may be chemically reversible) in terms of the reaction proceeding from left to right. If on the basis of thermodynamic predictions we conclude that the reaction will not proceed, we mean that the reactants will not be converted to products. This does *not* imply that "nothing will happen"— indeed, in such a case it is probable that we would observe a conversion of the products to reactants.

The simplest illustration of a reversible process is the case of gas expansion, which can be so classified if at all times

$$P_{\text{internal}} = P_{\text{external}} + dP \qquad (1\text{-}20)$$

At each value of the external pressure the system will be in equilibrium since there is only an infinitely small difference between opposing and driving forces, in this case internal and external pressures. If this difference is allowed, or made, to become greater than the infinitesimal increment indicated by the dP term in Eq. (1-20), the system is no longer in equilibrium and the process becomes a thermodynamically *irreversible* one. On this basis—the criterion of whether the system is in what we can think of as continuous equilibrium— all thermodynamic processes can be classified as reversible or irreversible.

Within the framework of this classification there are several other terms used to describe processes. The general basis of this subclassification is the parameter maintained constant. The most frequently encountered terms are as follows:

Isothermal—temperature constant
Isobaric   —pressure constant
Isometric —volume constant
Adiabatic— no heat enters or leaves the system

In this book we encounter primarily isothermal and isobaric processes. The term isometric, although convenient, is not in general use.

I-3-4 Isothermal Expansion of an Ideal Gas

We are now in a position to apply the discussion to some real processes of interest. It might appear that the volume change of an ideal gas is not very closely related to the stated central theme of examining the thermodynamics of chemical reactions to the exclusion of irrelevant systems. However, many reactions involve the production or consumption of gases whose contribution to the volume change in the system must be considered and whose behavior under the most frequently encountered conditions is ideal within experimental error. In addition, the restriction of constant temperature means only that reactants and products must be at the same temperature.

One property of an ideal gas that gives rise to a considerable simplification is the fact—a postulate of the Kinetic Molecular Theory—that the internal energy of the gas is dependent solely on the absolute temperature. Thus in an isothermal process involving only an ideal gas the energy change is zero. Similarly, from Eq. (1-19) it follows that if Δn is zero (that is, there is no change in the amount of gas present), $\Delta H = 0$ when there is no energy change. For this situation the first law becomes

$$\Delta E = 0 = q - w = q - \int_{V_1}^{V_2} P_{external} \, dV \qquad (1\text{-}21)$$

P is explicitly identified as the *external* pressure because we presently will need to make a careful distinction between internal and external pressure. Given that the process is isothermal, the expansion could be conducted in any one of the following ways:

Expansion in a vacuum. Here $P_{external} = 0$, and Eq. (1-21) gives simply $q = w = 0$. This is just the classic experiment where, given ideal gas behavior, no temperature change can be observed on free expansion of a gas.

Expansion at a fixed pressure. If we again consider the example of the gas expansion illustrated in Fig. 1-2, we find that if the pressure remains constant, Eq. (1-21) takes the form

$$q = w = P_{external} \, \Delta V \qquad (1\text{-}22)$$

Reversible expansion. Although this is an idealized process, it might be possible in practice to approach reversibility, if not attain it, by allowing the gas to expand very slowly by gradual reduction of the external pressure. The criterion of reversibility here is Eq. (1-20). The internal and external pressure must be the same within an infinitely small error. The system is in continuous equilibrium during the expansion. Consequently, the ideal gas law applies throughout. We can write for the internal pressure of the system $P_{internal} = nRT/V$, and since Eq. (1-20) applies, we can substitute for $P_{external}$ in Eq. (1-21) as follows:

$$q = w = \int_{V_1}^{V_2} P_{external}\, dV = \int_{V_1}^{V_2} P_{internal}\, dV = \int_{V_1}^{V_2} nRT\, \frac{dV}{V} \qquad (1\text{-}23)$$

The meaning of the integral can be elucidated in terms of the graph in Fig. 1-8. At every point along the P vs. V curve we take an infinitely small rectangle (one side being dV and the other $P_{internal}$ expressed as nRT/V) and find its area ($dV \times nRT/V$), and finally sum all of these area segments under the curve from V_1 to V_2. The relation of one such area segment to the overall pressure–volume graph is shown schematically in the figure. The only reason the substitution of internal for external pressure is legitimate is that the system is always in equilibrium in this *reversible* process. As before, we can evaluate an integral such as that in Eq. (1-23) either by plotting the curve and determining the area mechanically or by using the rules of integration. In this particular case the latter procedure is considerably more convenient. The first simplification is to take the constant nRT term outside the integral

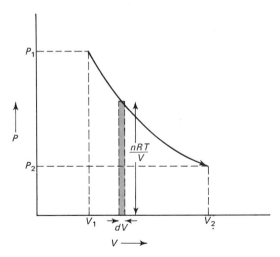

FIG. I-8 Pressure against volume for the isothermal reversible expansion of an ideal gas. The shaded rectangle represents a $P\, dV$ term in Eq. (I-23).

sign. This is permissible since we are in effect multiplying each dV/V term in the summation by the same factor. It remains, then, only to make use of the rule of integration for the general function dx/x. When this is done integration of Eq. (1-23) leads to the equation*

$$q_{\text{rev}} = w = nRT \int_{V_1}^{V_2} dV/V = nRT(2.303 \log V_2 - 2.303 \log V_1)$$

This equation is readily transformed to a form more convenient for calculations:

$$q_{\text{rev}} = w = 2.303\, nRT \log \frac{V_2}{V_1} \qquad (1\text{-}24)$$

Occasionally it is convenient to use the initial and final pressures, and for this we invoke Boyle's law, $P_1 V_1 = P_2 V_2$ (T constant) and substitute from the volume ratio in Eq. (1-24).

$$q_{\text{rev}} = w = 2.303\, nRT \log \frac{P_1}{P_2} \qquad (1\text{-}25)$$

There are a number of extremely useful applications of these equations encountered later in other contexts. For the moment the discussion is confined to the most significant aspect of this process, described by Eq. (1-24): *The work derived through a reversible expansion is the maximum obtainable in an isothermal volume change.* This fact can be demonstrated by the following numerical example, which illustrates how the various equations derived for changes in gas volume are applied to actual problems.

We consider the expansion of 1 mole of ideal gas from a volume of 10 liters at 4 atm pressure to 20 liters at 2 atm, the temperaure remaining constant at 487 °K.† The initial and final states can be represented in a PV diagram, such as Fig. 1-8. Applying that graph to the case in question, the initial state is represented by $V_1 = 10$ liters and $P_1 = 4$ atm and the final state is at the point (P_2, V_2). The curve between these two points can be interpreted as representing the internal pressure of the gas *at equilibrium*, at any volume between V_1 and V_2.

Figure 1-9 illustrates two possible paths between the initial and final states. Initially, the system is in equilibrium with $P_{\text{internal}} = P_{\text{external}} = 4$ atm. Applying Eq. (1-17) we have

$$PV = 4 \times 10 = nRT = 1 \times 0.0821 \times 487 = 40 \text{ liter atm}$$

---

* Calculus frequently yields an integrated function in terms of *natural logarithms*. Since tables of common logarithms are more readily available, the equations are usually stated in terms of common rather than natural logs, in which case the factor 2.303 is the conversion from the base e to the base 10. Common and natural logarithms are discussed in detail in Appendix 1.

† With the specified values of P, V, and n, the ideal gas law indicates that $T = PV/nR = (10 \times 40)/(0.0821 \times 1) = 487$ °K.

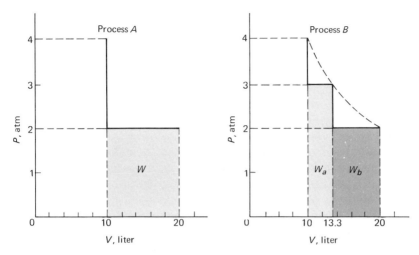

FIG. 1-9 *Isothermal expansion of one mole of ideal gas from 10 liter and 4 atm to 20 liter and 2 atm by two different routes.*

Since we are dealing with an ideal gas at constant temperature, the PV product will have a value of 40 liter atm under any conditions where the gas is in equilibrium. Thus in the final state the conditions must be $P = 2$ atm and $V = 20$ liters.

We can now make use of the gas expansion equations to calculate the work performed in the two processes represented in Fig. 1-9.

Process A. The external pressure is reduced to 2 atm, and the expansion occurs at this constant pressure. The system comes to final equilibrium when $P_{internal} = P_{external} = 2$ atm, where $V_2 = 20$ liters. Using Eq. (1-2),

$$w = P\,\Delta V = 2(20 - 10) = \underline{20 \text{ liter atm}}$$

This amount of work is the area of the rectangle indicated in Fig. 1-9.

Process B. The gas is expanded in two steps, (1) against an external pressure of 3 atm to an intermediate volume and then (2) against an external pressure of 2 atm to the final volume. In the first step equilibrium will be attained when $P_{internal} = 3$ atm. At this point, according to the required constancy of the PV product at 40 liter-atm, the volume will be 13.3 liters. That is, the point representing the PV product of the gas at equilibrium after the first step of the expansion must fall on the ideal gas PV curve shown as the dashed curve in the figure. The work obtained in step (1) is

$$w_{(1)} = P\,\Delta V = 3(13.3 - 10) = 9.9 \text{ liter atm}$$

In step (2) the gas expands against the constant pressure to its final volume, and the work is

$$w_{(2)} = P \, \Delta V = 2(20 - 13.3) = 13.4 \text{ liter atm}$$

Finally, the total work can be computed as follows:

$$w = w_{(1)} + w_{(2)} = 9.9 + 13.3 = \underline{23.3 \text{ liter atm}}$$

Comparison of these two processes indicates that as the number of steps is increased, the amount of work obtained also is increased. So long as the temperature remains constant, the PV product cannot exceed 40 liter atm when equilibrium is reached at the end of each intermediate step. The PV point representing the termination of each small expansion must lie along the curve obtained by plotting nRT/V as a function of V, such as the one shown in Fig. 1-8. To obtain the *maximum work* we should carry out the process in as many steps as possible. The gas would come to equilibrium at the end of each "substep," and both the internal *and* external pressures are given by the PV curve in Fig. 1-8. If this is done, employing an infinite number of steps, we will have a continuous equilibrium during the expansion—*the process producing the maximum amount of work is therefore an expansion under reversible conditions.* This process gives in the expansion from P_1V_1 to P_2V_2 an amount of work that can now be specified explicitly as w_{max} and can be calculated from Eq. (1-24) as follows:

$$q_{rev} = w_{max} = 2.303 \, nRT \, \log \frac{V_2}{V_1} = 2.303 \times 1 \times 0.0821 \times \log 2 \times 487$$

$$q_{rev} = w_{max} = \underline{27.7 \text{ liter atm}}$$

This amount of work is represented by the area under the curve in a graph, such as shown in Fig. 1-8, between the limits V_1 and V_2.

I-3-5 Vaporization

This is another type of process of considerable interest in physical chemistry to which we can apply the first law. It is discussed here, not only because of interest in vaporization as such, but also because we can see how the first law can be applied to any process by writing the general expressions and then making the simplifications appropriate to that particular case.

In the vaporization of a pure substance we are involved with a phase change at constant temperature. If the external pressure acting on the condensed phase is constant and equal to the vapor pressure at that temperature, there is true equilibrium between the condensed phase and the vapor. The system is undergoing a reversible process. Under these conditions, as heat is added to the system to maintain constant temperature, more and more of the material vaporizes. Only after conversion to the vapor phase is complete will the temperature rise. The importance of the fact that the process is reversible will become apparent later, but relevant to the present discussion is the constancy of the external pressure. This allows the enthalpy change to be calculated from Eq. (1-13), which is written here specifically for the vaporization process.

$$\Delta H_{vap} = \Delta E_{vap} + P \Delta V \qquad (1\text{-}26)$$

where P is the vapor pressure of the condensed phase at the temperature in question and ΔV is the change in volume per mole as the substance is vaporized. The treatment is specifically for vaporization of a liquid, but identical equations using ΔH and ΔE for the sublimation process can be written for the solid phase.

Usually the molar volume of the liquid is negligible with respect to that of the vapor, and the latter can be substituted for ΔV, that is,

$$\Delta V = V_{vap} - V_{liquid} \approx V_{vap}$$

For the vaporization process the first law can be written as

$$\Delta E_{vap} = q_{vap} - w = q_{vap} - P \Delta V \qquad (1\text{-}27)$$

The $P \Delta V$ term is the work done as the vaporizing material expands against the atmosphere. Combination of Eqs. (1-26) and (1-27) gives

$$\Delta H_{vap} = q_{vap} - P \Delta V + P \Delta V$$

$$\Delta H_{vap} = q_{vap} \qquad (1\text{-}28)$$

This relation thus represents a specific case of Eq. (1-15), which expresses the equality of the enthalpy change and the heat absorbed in a constant pressure process and also gives a definition of a specific ΔH:

ΔH_{vap}: *HEAT OF VAPORIZATION—THE HEAT ABSORBED IN THE CONVERSION OF ONE MOLE OF A SUBSTANCE FROM THE LIQUID TO THE VAPOR PHASE AT CONSTANT PRESSURE.*

The energy change attending the vaporization can be obtained from a rearrangement of Eq. (1-26),

$$\Delta E_{vap} = \Delta H_{vap} - P \Delta V \qquad (1\text{-}29)$$

Calculations using this equation require care with units. Both ΔH and ΔE normally will be in calories per mole, and thus the $P \Delta V$ term must be converted to these units from the usual liter atmospheres. The conversion factor from liter atmospheres to calories is 24.22 cal/liter atm.

EXAMPLE 1-2. In the vaporization of 10 gm cyclohexane, C_6H_{12}, at an external pressure of 400 torr* and at $T = 60.8$ °C, 854 cal were required. The vapor pressure of cyclohexane at this temperature is 400 torr. Calculate the energy and enthalpy change in the vaporization process.

Molecular weight $C_6H_{12} = 84.2$; $P = 400/760 = 0.526$ atm. Vaporization of 10.0/84.2 moles requires 854 cal.

$$\Delta H_{vap} = 854 \times \frac{84.2}{10.0} = 7{,}190 \text{ cal/mole}$$

$$\Delta V \approx V_{vapor} = nRT/P = \frac{1 \times 0.0821 \times 334}{0.526} = 52.1 \text{ liters/mole}$$

$$\Delta E_{vap} = \Delta H_{vap} - P\,\Delta V = 7{,}190 - 0.526 \times 52.1 \times 24.22$$
$$= 7{,}190 - 663$$

$$\Delta E_{vap} = 6{,}527 \text{ cal/mole}$$

Although in this example the numerical value of V_{vap} was calculated for purposes of illustration, the solution of the problem is more simply obtained by making the substitution from the ideal gas law directly into the equation for ΔE_{vap}. This procedure also eliminates the necessity of having a value of P available (but it is still required that $P_{applied} = P_{vap\ pressure}$ whatever the value). This approach leads to the following solution for ΔE_{vap}:

$$\Delta E_{vap} = \Delta H_{vap} - P(V_{vap}) = \Delta H_{vap} - P\left(\frac{RT}{P}\right) = \Delta H_{vap} - RT$$

$$\Delta E_{vap} = 7{,}190 - 1.987 \times 334 = 7{,}190 - 663 = 6{,}527 \text{ cal/mole}$$

Note that by using the value of the gas constant R of 1.987 cal/deg mole the RT term has units of calories and that no conversion factor need be used subsequently.

It is also instructive to consider how the first law applies to this process. Since $\Delta E = q - w$ takes the form $6{,}527 = 7{,}190 - 663$, it is apparent that in converting cyclohexane from liquid to vapor we have increased its energy by only 6,527 cal/mole. Of the 7,190 cal/mole heat energy put into the system, 663 cal were used in performing the PV work of expansion of the vapor phase against the applied pressure.

I-4 THERMOCHEMISTRY

This aspect of thermodynamics is concerned with enthalpy and energy changes in chemical and physical processes and how they can be calculated from available data or measured as experimentallly determined heat effects.

---

* Current usage has replaced the pressure units millimeters mercury by torr. The two units are of identical size, but the latter is preferred since pressures actually in millimeters mercury are designated frequently as pressure in millimeters, an ambiguous designation.

Perhaps one of the most useful quantities involved with thermochemical measurements is heat capacity. It therefore is necessary to reexamine the concept of heat capacity and to determine how this quantity is intimately related to measurement and calculation of enthalpy changes.

I-4-I Heat Capacity

Heat capacity was previously defined by Eq. (1-5) as $C = q/\Delta T$. This relation now can be used in conjunction with two specific equations, Eqs. (1-10) and (1-15), involving heat, to write defining equations for two specific types of heat capacity:

$$\text{Heat capacity at constant pressure} = C_p = \Delta H/\Delta T \qquad (1\text{-}30)$$

$$\text{Heat capacity at constant volume} = C_v = \Delta E/\Delta T \qquad (1\text{-}31)$$

An increase in temperature is almost invariably accompanied by an increase in volume. Thus under constant pressure conditions the enthalpy change will be larger than ΔE, and for a given compound $C_p > C_v$. Heat capacity data are most often tabulated in the form of C_p values, but conversion can readily be made to C_v if required. As indicated previously, the units are calories per degree mole.

As already noted in conjunction with Eq. (1-6), the heat capacities calculated from Eqs. (1-30) and (1-31) are valid only for rather small temperature ranges. Thus, for example, if Eq. (1-31) is to be used to calculate an enthalpy change owing to some ΔT, there is the inherent assumption that C_p remains constant between the two temperatures in question.

As we proceed it will become clear that the heat capacity of a substance is a datum of considerable versatility. It will be involved in many subsequent calculations. Historically, the prediction of C values from first principles has been of considerable interest. Here only some of the results that have been confirmed experimentally for a few types of system of particular interest at room temperature are discussed.

Gases. The relevant equation can be obtained from Eqs. (1-30) and (1-31) as follows:

$$C_p - C_v = \frac{\Delta H - \Delta E}{\Delta T}$$

Substitution for ΔH from the defining equation for this quantity, Eq. (1-12), gives

$$C_p - C_v = \frac{\Delta E + \Delta(PV) - \Delta E}{\Delta T} = \frac{\Delta(PV)}{\Delta T}$$

For 1 mole of ideal gas $\Delta(PV) = \Delta(RT) = R\,\Delta T$, and thus the difference between the two heat capacities is simply

$$C_p - C_v = R \qquad (1\text{-}32)$$

For a *perfect monatomic gas* the only mode of thermal excitation is translational motion, and the calculated and observed values are in good agreement at $C_p \approx 5$ cal/deg mole and $C_v \approx 3$ cal/deg mole. As the complexity of the gas molecules increases, so does the number of ways in which the molecule can take up thermal energy and, consequently, the more difficult is the prediction.

Liquids and solids. Although some progress has been made with prediction of the heat capacities of solids, for the liquid phase we rely almost exclusively on experimentally determined values. There are, however, a few useful general rules that can be helpful. For elemental solids, particularly those of higher atomic mass, $C_p \approx 6$ cal/deg mole. If we refer again to Eqs. (1-30) and (1-31), since for processes involving liquids we have seen that $\Delta H \approx \Delta E$, we find that C_p and C_v are usually the same within experimental error.

The heat capacities at constant pressure at 25 °C are given for a number of substances in Table 1-1.

TABLE 1-1.   HEAT CAPACITIES AT CONSTANT PRESSURE AT $298\,°K^a$

Substance	$C_p^{\circ},{}^b$ cal/deg mole	Substance	C_p°, cal/deg mole
$CO_2(g)$	8.87	$CH_4(g)$	8.54
$CO(g)$	6.97	$H_2O(g)$	8.03
$CO(NH_2)_2(s)$	22.3	$H_2O(l)$	18.0
$CH_3OH(l)$	19.5	$HCl(g)$	6.96
$CH_3COOH(l)$	29.5	$N_2(g)$	6.96
$CCl_4(l)$	31.5	$NH_3(g)$	8.52
$CaO(s)$	10.2	$NH_4Cl(s)$	20.1
$H_2(g)$	6.89	$O_2(g)$	7.02

[a] From *Selected Values of Chemical Thermodynamic Properties* (Circular 500, National Bureau of Standards, 1952), and *Selected Values of Properties of Hydrocarbons* (Circular C461, National Bureau of Standards, 1947).

[b] The superscript ∘ indicates that the data apply to the substance in its standard state, as defined in Sec. 1-4-3.

In general we will not be concerned with the variation of heat capacity with temperature, except for one very important case where C_p values down to absolute zero will be required. It is useful to note, however, that except at the lowest temperatures the heat capacity at constant pressure of most substances can be expressed as a function of temperature with an equation of the form $C_p = a + bT + cT^{-2}$, where a, b, and c are constants characteristic of the given material.

1-4-2   Thermochemical Equations and Hess' Law

We can represent any chemical or physical change as a thermochemical equation that embodies not only a mass balance but also the change in energy or enthalpy of the system as it is altered from the initial (reactants) to final (products) state.

The equation

$$H_2(g, 1 \text{ atm}) + \tfrac{1}{2} O_2(g, 1 \text{ atm}) \rightarrow H_2O(l) \qquad \Delta H_{298} = -68,300 \text{ cal}$$

illustrates a number of important points associated with the correct presentation of thermochemical information in equation form. The complete relation is based primarily on the usual balanced chemical equation. It is necessary to specify completely the state of the system by indicating the phase, concentration, or pressure and the nature of the solvent if the species is in solution. Thus we designate a gas as $(g, x \text{ atm})$ and a solute as $(aq, x \text{ M})$, indicating for the latter that it is present in aqueous solution at a concentration of x moles/liter. Unless stated otherwise it can be assumed that the pressure is 1 atm; thus we would write just $H_2(g)$ and $H_2O(l)$, indicating gaseous H_2 at 1 atm and liquid water under an applied pressure of 1 atm. The ΔH for the reaction has units of calories, but note that this implies calories "for the reaction as written." In the example the ΔH is thus calories per mole of water formed. Since by definition ΔH is the heat *absorbed* at constant pressure, the minus sign indicates that 68,300 cal are evolved in this process. The subscript 298 is the temperature in °K. If no temperature is stated we assume $T = 298$ °K. In using sources of thermodynamic information care must be taken to establish the conditions to which the data apply.

The process represented by the example equation can be envisaged in two ways. From the physical point of view we consider that $\tfrac{1}{2}$ mole of oxygen gas at 1 atm combines with 1 mole of hydrogen gas at 1 atm to give 1 mole of liquid water, evolving 68,300 cal of heat in the process. Alternatively, we can think of a hydrogen–oxygen system in which there is a decrease in enthalpy when the system moves from the initial state (H_2 and O_2 in the gas phase) to the final state (liquid water). The first approach is more useful if we are primarily interested in the heat effect of the reaction. The second is more informative when we wish to carry out thermodynamic calculations and interpret what is happening to the thermodynamic state of the system. Since enthalpy is a state function, ΔH will be the same for given initial and final states in any type of process. ΔH gives the *experimentally observed heat effect*, however, only when the experiment is performed under isobaric conditions.

The same information about this hydrogen–oxygen system can be given in alternative ways:

$$H_2O(l) \rightarrow H_2(g) + \tfrac{1}{2} O_2(g) \qquad \Delta H_{298} = +68,300 \text{ cal}$$

$$2 H_2(g) + O_2(g) \rightarrow 2 H_2O(l) \qquad \Delta H_{298} = -136,600 \text{ cal}$$

In the first equation note that since the direction of the process is reversed there is an increase in enthalpy in going from the initial to final state, but the magnitude of ΔH is unaltered. The physical interpretation is that 68,300 cal are absorbed in the indicated decomposition of H_2O. The second equation demonstrates that enthalpy is an *extensive* property, depending on the amount of material reacting.

Considering the extremely large number of possible chemical reactions for which a ΔH value might be required, a complete set of data would be rather formidable. A great simplification, however, arises from the fact that H is a state function. Since the path by which we proceed from reactants to products will not alter the magnitude of ΔH, thermochemical equations can be treated by simple algebra. We can calculate the enthalpy change in reactions of interest by combining ΔH values for other reactions involving the same species. These ideas are embodied in a versatile thermodynamic principle that can be illustrated by the following example involving the formation of the biologically important compound urea, $CO(NH_2)_2$.

Consider first the process in which urea is formed, along with water, from ammonia gas, elemental oxygen, and carbon:

I. $C(s) + O_2(g) + 2\,NH_3(g) \rightarrow CO(NH_2)_2(s) + H_2O(l)$

$$\Delta H_{\mathrm{I}} = -288{,}540 \text{ cal}$$

The equation indicates the enthalpy difference between the C, N, and H system in the two states, as reactants and products. Thus the same *total* ΔH should result when the same process occurs in a sequence of two steps:

II. $C(s) + O_2(g) \rightarrow CO_2(g)$

III. $CO_2(g) + 2\,NH_3(g) \rightarrow CO(NH_2)_2(s) + H_2O(l)$

The fact that we have proceeded along this alternative route does not alter the enthalpy of the initial reactants and final products in the sequence and, therefore,

$$\Delta H_{\mathrm{I}} = \Delta H_{\mathrm{II}} + \Delta H_{\mathrm{III}}$$

This equality is illustrated in Fig. 1-10.

This simple example illustrates the general principle of additivity of chemical equations and their associated state function changes, which is equally applicable to any more complex system or other state function. The principle is known as

HESS' LAW: THE ENTHALPY CHANGE IN A REACTION IS THE SAME WHETHER IT OCCURS IN ONE OR SEVERAL STEPS.

The usefulness of this law is that it permits us to obtain a ΔH value for a particular reaction of interest without having to measure the heat effect experimentally, from other known ΔH values. Not only is this a convenient

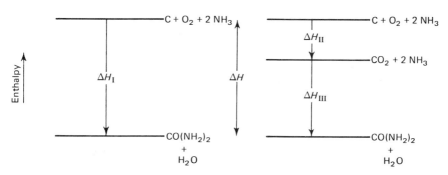

FIG. 1-10 Enthalpy change in the formation of urea. The direct and two-step processes have the same ΔH.

procedure, but it also often allows computation of ΔH for a reaction that cannot be studied experimentally without considerable effort.

Consideration can now be given to the application of Hess' law to a series of thermochemical equations to extract a particular ΔH value. The next section examines one particularly convenient form of the law that can be used when one specific type of thermochemical information is available.

Suppose we wished to calculate the enthalpy change at 298 °K in the following process:

$$2\,C(s) + 2\,H_2(g) + O_2(g) \rightarrow CH_3COOH(l) \tag{a}$$

Although the ΔH in question is readily available, it is instructive to see how it is possible to calculate this value by choosing just one set, of the very large number available, of thermochemical data for the components of this system. We take as our basis data the following three reactions and their associated enthalpy changes:

$$CH_3COOH(l) + 2\,O_2(g) \rightarrow 2\,CO_2(g) + 2\,H_2O(l) \quad \Delta H = -209{,}400 \text{ cal} \tag{b}$$
$$C(s) + O_2(g) \rightarrow CO_2(g) \quad \Delta H = -94{,}100 \text{ cal} \tag{c}$$
$$H_2(g) + \tfrac{1}{2}\,O_2(g) \rightarrow H_2O(l) \quad \Delta H = -68{,}300 \text{ cal} \tag{d}$$

The object of the manipulations will be to alter these equations and their ΔH so that they will be in a form suitable for simple algebraic summation yielding equation (a) and its enthalpy change. Although there are no specific methods of procedure, in general each equation must be rearranged, where necessary, in the following way: By writing it in reverse or by multiplying by a suitable factor, any particular equation is altered so that all species that are products in the final equation appear exclusively on the right and species not involved in the final equation cancel in the summation. It is important that the same operation be performed on *both* the equation and its ΔH. For the three equations in this example, Eq. (b) must be reversed so that CH_3COOH appears as a product. Equation (c) is to be multiplied by a factor of 2 to permit cancellation of the CO_2, and Eq. (d) requires multiplication by 2 so that the 2 moles of water in this and Eq. (b) will cancel. With these alterations the three equations can be written and their sum taken as follows:

$$2\ CO_2(g) + 2\ H_2O(l) \rightarrow CH_3COOH(l) + 2\ O_2(g) \qquad \Delta H = \\ +209{,}400\ \text{cal}$$

$$2\ C(s) + 2\ O_2(g) \rightarrow 2\ CO_2(g) \qquad \Delta H = -188{,}200\ \text{cal}$$

$$2\ H_2(g) + O_2(g) \rightarrow 2\ H_2O(l) \qquad \Delta H = -136{,}600\ \text{cal}$$

$$2\ H_2(g) + O_2(g) + 2\ C(s) \rightarrow CH_3COOH(l) \qquad \Delta H = -115{,}400\ \text{cal}$$

The result is the required equation of interest and its enthalpy change.

I-4-3 Standard Enthalpies of Formation

We have seen that specifying the state of each species in a thermochemical equation is a necessity since the enthalpy of any substance—and therefore the contribution it makes to ΔH in a reaction—depends on its state. To simplify the tabulation of thermodynamic data, a reference state is assumed and normally available data apply to the substance in that state. If the substance of interest in a particular process is not in this reference state, then we must calculate the difference in the state function between the reference and other state.

This reference set of conditions is known as the *standard state*, established essentially arbitrarily but conventionally adhered to, which can be defined as

> *STANDARD STATE: THE STABLE FORM OF THE SUBSTANCE AT 1 ATM PRESSURE AT A SPECIFIED TEMPERATURE, USUALLY 298 °K.*

One of the most common sources of misunderstanding in thermodynamics is the specification of the standard state temperature. Although values are almost invariably tabulated at 298 °K, it must be realized that we can have a standard state *at any temperature*.

That we are dealing with materials in their standard state is indicated by a right superscript zero. In a reaction in which all products and reactants are in their standard state, the ordinarily measured ΔH is the *standard enthalpy change*, written $\Delta H°$.

Within the framework of the general definition of the standard state and the requirement that the substance be in its most stable form at that temperature, some standard states can be specified as follows:

Substance	Standard state
Ideal gas	1 atm pressure
Liquid and solids	Pure material under 1 atm pressure
Solution components	Pure substance (mole fraction $= 1$)
	or as a solute at a concentration of
	1 mole/liter

These conditions must be modified somewhat, particularly for solutes as we will see later, to take nonideal behavior into account, but for the moment they can be taken as good approximations. The discussion of thermochemistry will be concerned only with pure liquids and solids. Standard states in general are examined at the outset, however, since the same considerations apply to subsequent discussions of all other state functions. It is worth commenting, finally, that there is nothing mysterious about standard states! They are nothing more than a set of reference conditions that have been chosen rather arbitrarily to simplify the tabulation and calculation of thermodynamic data.

It is apparent that the effort in general here is directed toward reducing the amount of data that must be directly available to calculate a particular ΔH or $\Delta H°$—the primary goal of thermochemistry. One of the most useful types of information, which permits many possible calculations from a small amount of reference data, is the standard enthalpy change in a reaction in which a compound is formed from its constituent elements. This is known as the *standard enthalpy of formation*, or more commonly as the *standard heat of formation* $\Delta H_f°$, and is defined in the following manner:

> **STANDARD HEAT OF FORMATION ($\Delta H_f°$): THE ENTHALPY CHANGE IN THE REACTION IN WHICH ONE MOLE OF A SUBSTANCE IS FORMED AT CONSTANT TEMPERATURE FROM ITS CONSTITUENT ELEMENTS, ALL SPECIES BEING IN THEIR STANDARD STATE.**

The units of $\Delta H_f°$ are calories per mole and, unless otherwise indicated, the temperature is 298 °K.

By convention we take $\Delta H_f°$ *of the elements as zero.* When we use standard heats of formation to calculate reaction enthalpy changes, the zero value for the elements does not affect the calculations since we will be dealing with *differences*. We take zero as the reference point simply for convenience in these calculations. Values for $\Delta H_f°$ of selected substances are given in Table 1-2.

In specifying a standard heat of formation we are stating an "ordinary ΔH" for a reaction occurring under defined conditions. Thus the information

TABLE 1-2.   STANDARD HEATS OF FORMATION AT $298\,°K^a$

Substance	ΔH_f°, kcal/mole	Substance	ΔH_f°, kcal/mole
$Al_2O_3(s)$	-400	$Mn_3O_4(s)$	-331
$CO(g)$	-26.4	$H_2O(l)$	-68.3
$CO_2(g)$	-94.1	$H_2O(g)$	-57.8
$CO(NH_2)_2(s)$	-79.6	$HCl(g)$	-22.1
$C_2H_4(g)$	$+12.5$	$NH_3(g)$	-11.0
$CH_3OH(l)$	-57.0	$NH_4Cl(s)$	-75.4
$C_2H_5OH(l)$	-66.4	cyclo-$C_6H_{12}(l)$	-37.3
$HCHO(g)$	-27.7	$C_6H_6(l)$	$+11.7$
$CH_3COOH(l)$	-116	$SO_2(g)$	-70.9
$HCOOH(l)$	-97.8	$CH_3SH(g)$	-2.97
$CH_3S_2CH_3(l)$	-5.71	$C_2H_5O_2N(s)$	-126
$CCl_4(l)$	-33.3	$C_2H_2O_4(s)$	-198

^{a} From *Selected Values of Chemical Thermodynamic Properties* (Circular 500, National Bureau of Standards, 1952) and *Selected Values of Properties of Hydrocarbons* (Circular C461, National Bureau of Standards, 1947).

that $\Delta H_{f,CH_3OH(l)}^\circ = -57,040$ cal/mole means that the reaction

$$2H_2(g, 1 \text{ atm}) + C(s) + \tfrac{1}{2}O_2(g, 1 \text{ atm}) \rightarrow CH_3OH(l)$$

has $\Delta H_{298} = -57,040$ cal.

An important advantage of standard heats of formation is that they allow direct calculation of ΔH° values for reaction. This eliminates the necessity of manipulating numbers of thermochemical equations. We calculate the total enthalpy change in going from constituent elements to products and subtract from this the total ΔH in going from the same elements to reactants. Since the reference point is the same for all elements, this difference in ΔH_f° values gives directly the ΔH° value for the overall reaction.

Let us apply this technique to the reaction

$$C_2H_4(g) + H_2O(l) \rightarrow C_2H_5OH(l)$$

using the data in Table 1-2. Beginning at the reference point on the enthalpy scale, the change in H in forming 1 mole of gaseous C_2H_4 from elemental carbon and hydrogen is $+12,496$ cal, and that for water from oxygen and hydrogen is $-68,317$ cal. Thus in the formation of 1 mole each of these reactants from their elements, the total ΔH is $12,496 - 68,317 = -56,021$ cal. For the product ethyl alcohol, we have directly from the ΔH_f° value that the enthalpy change in proceeding from C, H_2, and O_2 to C_2H_5OH is $-66,360$ cal. Finally, the enthalpy change in the reaction—the difference in H between product and reactants—is the difference between these two ΔH terms:

$$-66,360 - (-56,021) = -10,339 \text{ cal}$$

The example illustrates the general technique of proceeding, for purposes of calculation only, from the initial to final state via one or more hypothetical paths. The alteration in the value of the state function for the system is the same for all possible paths.

For any number of such examples it readily can be demonstrated that in any process the standard enthalpy change is the total standard enthalpy of formation of products less that of reactants. Since ΔH_f° values are in calories per mole, each must be multiplied by the number of moles of that substance that appears in the reaction equation. The calculation can be summarized in the equation

$$\Delta H^{\circ} = \sum_{\text{products}} n\,\Delta H_f^{\circ} - \sum_{\text{reactants}} n\,\Delta H_f^{\circ} \tag{1-33}$$

where each $\sum n\,\Delta H_f^{\circ}$ term represents the sum of the standard heats of formation in calories per mole times number of moles n.

EXAMPLE 1-3. Calculate the standard enthalpy change in the process $2\,CH_3SH(g) \rightarrow H_2(g) + CH_3S_2CH_3(l)$.

$$\Delta H^{\circ} = \Delta H_{f,CH_3S_2CH_3(l)}^{\circ} + \Delta H_{f,H_2(g)}^{\circ} - 2\,\Delta H_{f,CH_3SH(g)}^{\circ}$$
$$\Delta H^{\circ} = -5.71 + 0 - 2 \times (-2.97) = -5.71 + 5.94$$
$$\Delta H^{\circ} = +0.23 \text{ kcal}$$

Equation (1-33) is representative of a number of equations of identical form encountered throughout thermodynamics. The change in the value of a state function can be calculated by taking the difference between the sum of such functions for the products and reactants.

1-4-4 Enthalpy Changes in Physical and Chemical Processes

In the previous section we encountered the idea of defining an enthalpy change in some process of interest occurring under specified conditions. While ΔH_f° is perhaps the most important such designation, there are several other specific ΔH values.

It is also noteworthy that all reactions and processes discussed here are changes in the thermodynamic state of the system. Consequently, the definitions stated explicitly for enthalpy can be applied to any state function. Thus we could, for example, define a standard energy of formation ΔE_f° as the energy change in the reaction wherein 1 mole of a compound is formed from its constituent elements in their standard state.

Perhaps the most widely used ΔH in biological discussions is the enthalpy change associated with the complete combustion of a compound. Strictly speaking, the quantity is a ΔH°, but in general it is referred to simply as the *heat of combustion* and is defined as

> HEAT OF COMBUSTION ΔH_{comb}: THE ENTHALPY CHANGE IN THE REACTION IN WHICH ONE MOLE OF A COMPOUND IS COMPLETELY OXIDIZED TO CARBON DIOXIDE AND WATER BY OXYGEN AT ONE ATMOSPHERE PRESSURE AT A CONSTANT SPECIFIED TEMPERATURE.

Heats of combustion are the most frequently measured thermochemical data (see Sec. 1-4-6). A few representative values are given in Table 1-3.

All combustion reactions are evidently very exothermic. The data in the table also indicate the direct relation between molecular weight of the paraffinic and olefinic hydrocarbons and their heat of combustion.

The heat energy available from an oxidation process, whether "burning" in air or a metabolic process in the human body, may be released in one or several stages. The usefulness of the ΔH_{comb} value is that it indicates the maximum energy available when the process is carried to completion. Frequently we are not directly concerned with the number or type of steps involved, only with the total energy.

A comparison of the direct and stepwise oxidation of methane is shown in Fig. 1-11. In the stepwise procedure the energy is released in four separate processes as CH_4 is converted successively to ethanol, formaldehyde, acetic acid, and finally CO_2 and water. The total enthalpy change is identical with that obtained when the conversion

$$CH_4(g) + 2\ O_2(g) \rightarrow CO_2(g) + H_2O(l) \qquad \Delta H = -212 \text{ kcal}$$

is carried out directly. From another viewpoint the example represents a graphical representation of Hess' law as it applies to the system.

We are also concerned with the enthalpy change in physical processes. In Sec. 1-3-5 vaporization was discussed. Enthalpy changes similarly can be defined for other phase transitions, for example, ΔH_{fusion} and $\Delta H_{sublimation}$.

TABLE 1-3.   HEATS OF COMBUSTION AT 298 °K[a]

Substance	ΔH_{comb}, kcal/mole	Substance	ΔH_{comb}, kcal/mole
C(s)	−94.1	n-$C_4H_{10}(g)$	−688
$CH_4(g)$	−212	cyclo-$C_5H_{10}(l)$	−787
$C_2H_2(g)$	−310	n-$C_5H_{12}(g)$	−845
$C_2H_4(g)$	−337	cyclo-$C_6H_{12}(l)$	−937
$C_2H_6(g)$	−373	n-$C_6H_{14}(l)$	−998
$C_3H_6(g)$	−492	n-$C_7H_{16}(l)$	−1151
$C_3H_8(g)$	−531	n-$C_8H_{18}(l)$	−1308
i-$C_4H_8(g)$	−649	$C_6H_6(l)$	−781
i-$C_4H_{10}(g)$	−686	$C_6H_5CH_3(l)$	−935

[a] From *Selected Values of Properties of Hydrocarbons* (Circular C461, National Bureau of Standards, 1947).

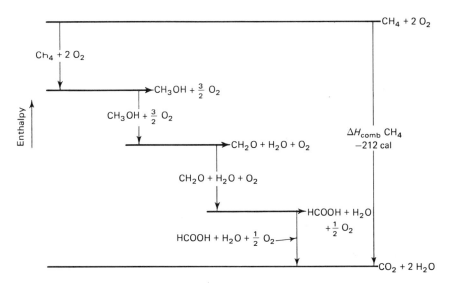

FIG. I-II Schematic representation of the enthalpy changes in the direct and stepwise oxidation of methane.

A general ΔH of this type may be defined as follows:

HEAT OF PHASE CHANGE: THE ENTHALPY CHANGE FOR THE PROCESS IN WHICH ONE MOLE OF A SUBSTANCE IS CONVERTED FROM ONE PHASE TO ANOTHER AT CONSTANT TEMPERATURE AND PRESSURE.

This quantity is not a standard enthalpy change, but the difference between, for example, ΔH°_{vap} (liquid and vapor both at 1 atm) and ΔH_{vap} as defined by the above statement is normally quite small and can be neglected in all but the most exact computations.

A phase change also can be represented in the same form as a chemical reaction. Thus $\Delta H_{vap,C_2H_5OH} = 9{,}220$ cal/mole* is an alternative statement to

$$C_2H_5OH(l) \rightarrow C_2H_5OH(g) \qquad \Delta H = 9{,}220 \text{ cal}$$

There is one phase change, that of the solution process, which is treated somewhat differently. The heat of solution ΔH_{soln} is defined as the enthalpy change when one mole of a substance goes into solution in a *specified number of moles of solvent*.

The necessity of specifying the state of the species involved in thermochemical equations should now be obvious. The ΔH for the formation of a

---

* Heats of vaporization are not strongly temperature dependent, and it is usual to state their values without specifying the temperature, although usually the temperature is that of the normal boiling point, the temperature at which the vapor pressure is 1 atm.

compound in the gas phase will be greater than that of the reaction where it is produced as a liquid, since in the former additional energy is required to effect the liquid-to-vapor transition. The difference between the two ΔH values would be the heat of vaporization.

1-4-5 Variation of ΔH with Temperature

When the temperature of a substance is changed its enthalpy is altered. If the heat capacity is constant and no phase change occurs over the ΔT of interest, ΔH is given by

$$\Delta H = C_p(T_2 - T_1) \tag{1-34}$$

obtained by combining Eqs. (1-6) and (1-15). Alternatively, if the heating is isometric we could obtain a similar expression in terms of C_v from Eqs. (1-6), (1-10), and (1-12).

Confining ourselves to isobaric heating, we consider a process that occurs at a certain temperature, where it has ΔH_T. If we now change the operating temperature, ΔH for the process will be altered since the enthalpy of the products and that of the reactants will change by different amounts, depending on their heat capacities. On raising the temperature the enthalpy of both products and reactants is increased, but to different extents since normally the total heat capacities will not be the same. The result is that at the higher temperature the initial and final states of the system will have, on the absolute enthalpy scale, a separation that is not the same as that at the lower temperature.

To develop a general expression for the temperature dependence of ΔH we make use of the known enthalpy change at one T, which indicates the enthalpy separation between products and reactants, and the heat capacities of products and reactants. The latter allows calculation of the enthalpy change associated with the temperature alteration for each species involved.

For purposes of calculation, three processes are visualized: (1) change the temperature of the reactants from T_b to T_a, at which ΔH is known; (2) carry out the reaction at T_a; and (3) change the temperature of the products from T_a to T_b. The total ΔH for the three processes will be the desired quantity, ΔH at T_b. Outlining this procedure in the form of thermochemical equations and applying Hess' law gives

I. Reactants at $T_b \rightarrow$ reactants at T_a, ΔH_I

II. Reactants at $T_a \rightarrow$ products at T_a, ΔH_{II}

III. Products at $T_a \rightarrow$ products at T_b, ΔH_{III}

Reactants at $T_b \rightarrow$ products at T_b, $\Delta H_{T_b} = \Delta H_I + \Delta H_{II} + \Delta H_{III}$

To develop a working equation for ΔH at the new temperature, the only requirement now is the substitution of the appropriate expression for each ΔH term in the summation.

ΔH_I can be calculated from Eq. (1-34):

$$\Delta H_I = C_{p,\text{reactants}}(T_2 - T_1) = \sum_{\text{reactants}} nC_p(T_a - T_b)$$

Note the subscripts in this equation. In the general expression, T_2 is the final temperature, which for process I is T_a. The summation is of the heat capacity at constant pressure of each reactant times the number of moles participating in the reaction times the temperature difference.

ΔH_{II} is the enthalpy change in the reaction at T_a, ΔH_{T_a}. For process III Eq. (1-34) is

$$\Delta H_{III} = C_{p,\text{products}}(T_2 - T_1) = \sum_{\text{products}} nC_p(T_b - T_a)$$

where for this process the final temperature is T_b.

To obtain the final expression the expressions for the three ΔH terms are added:

$$\Delta H_{T_b} = \sum_{\text{reactants}} nC_p(T_a - T_b) + \Delta H_{T_a} + \sum_{\text{products}} nC_p(T_b - T_a)$$

$$\Delta H_{T_b} = \Delta H_{T_a} + [\sum_{\text{products}} nC_p - \sum_{\text{reactants}} nC_p](T_b - T_a) \tag{1-35}$$

The term in brackets represents the difference in heat capacity of products and reactants, ΔC_p. For the general reaction $bB + dD \rightarrow mM + nN$, ΔC_p is given by

$$\Delta C_p = \sum_{\text{products}} nC_p - \sum_{\text{reactants}} nC_p \tag{1-36}$$

$$\Delta C_p = mC_{p,M} + nC_{p,N} - bC_{p,B} - dC_{p,D}$$

Substitution of Eq. (1-36) into Eq. (1-35) for ΔC_p and conversion to the conventional temperature designations of T_1 for T_a (the temperature at which ΔH is known) and T_2 for T_b (the temperature at which the enthalpy change is to be calculated) give the final working equation

$$\Delta H_{T_2} = \Delta H_{T_1} + \Delta C_p(T_2 - T_1) \tag{1-37}$$

This is an extremely useful equation, but like all expressions in thermodynamics we must be careful to note its limitations. These arise not only from the assumptions made in arriving at the final equation itself but also from those inherent in the supplementary relations used throughout the derivation. For Eq. (1-37) these considerations lead to the restrictions that the reaction must occur at constant temperature at T_2 and at T_1 and that the heat capacities of all species involved remain constant between the two temperatures.

Finally, note that the equation can be written for standard enthalpy changes through the use of ΔH° values and for calculating ΔC_p° from C_p° values (those normally tabulated, as in Table 1-1).

EXAMPLE 1-4. ΔH_{298}° for the reaction $HCl(g) + NH_3(g) \rightarrow NH_4Cl(s)$ is $-42,300$. Using the heat capacity values given in Table 1-1, evaluate ΔH_{323}° for this process.

$$\Delta C_p^\circ = C_{p,NH_4Cl(s)}^\circ - C_{p,HCl(g)}^\circ - C_{p,NH_3(g)}^\circ$$
$$\Delta C_p^\circ = 20.1 - 6.96 - 8.52 = +4.62 \text{ cal/deg mole}$$
$$\Delta H_{323}^\circ = \Delta H_{298}^\circ + \Delta C_p^\circ(323 - 298)$$
$$\Delta H_{323}^\circ = -42,300 + 4.62 \times 25 = -42,300 + 116$$
$$\Delta H_{323}^\circ = \underline{-42,184 \text{ cal}}$$

By a series of completely analogous arguments the expression for the temperature dependence of the energy change can be written as

$$\Delta E_{T_2} = \Delta E_{T_1} + \Delta C_v(T_2 - T_1) \tag{1-38}$$

I-4-6 Experimental Determination of Heat Effects

The measurement of the amount of heat absorbed or evolved in a physical or chemical process is known as *calorimetry*. Although the techniques involved can vary from those that are quite straightforward (as exemplified by the classical calorimetry experiments) to more complex measurements, all such experiments are characterized by an essential measurement of a temperature change in some system of known heat capacity.

The number of chemical reactions that can be studied conveniently by calorimetry is relatively small, again emphasizing the value of the calculation techniques based on Hess' law. Normally the reaction must proceed in a short time to allow measurement of the temperature change without inordinate difficulty. Furthermore unless difficult corrections are to be applied, the process must go to completion and be free from any complicating side reactions.

In its simplest form a *calorimeter* suitable for the study of a liquid phase reaction can be a container thermally insulated from the surroundings and open to the atmosphere. We carry out the process with a known quantity of material and measure the temperature change in a system defined as the calorimeter and its contents. A simple calorimeter of this type is shown in Fig. 1-12. To calculate the heat effect it is necessary to know the heat capacity of the substances involved in the process as well as that of the calorimeter and to prevent, or make corrections for, any significant thermal transfer to the surroundings. The heat capacity of the calorimeter is usually determined by noting the temperature rise produced by the input of a measured amount of electrical energy to a resistance heater immersed in the vessel.

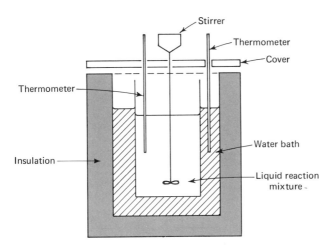

FIG. 1-12 A simple calorimeter for the study of heat effects in a liquid phase reaction.

For a process such as the neutralization of an acid by a base in aqueous solution, a calorimeter like the one described above could be used to determine the heat of reaction as follows:

The acid and base solutions at known concentrations and initial temperatures are mixed in the calorimeter, and the temperature change is noted. From the total heat capacity of the system and the temperature change, the heat evolved in the reaction can be determined. Temperature readings are taken at intervals before and after mixing. By graphical analysis the exact temperature of the solutions at the time of mixing and the true temperature rise can be determined. The result is a measurement of the heat involved in the reaction of some known amount of material, from which the heat of neutralization in calories per mole can be readily obtained.

In the *adiabatic calorimeter* the reaction vessel and its jacket are placed in a water bath that is manually or automatically maintained at the same temperature as the vessel. In this way heat transfer between the calorimeter and the surroundings is reduced to levels that are insignificant except for the most accurate determinations.

For measurements of heats of combustion and other reactions involving gases, a *bomb calorimeter* is employed in which the material undergoing reaction is sealed in a metal cylinder constructed to withstand extremely high pressures. For combustion reactions oxygen is introduced under pressure and the process is initiated by electrical ignition of the sample. The bomb calorimeter can be operated simply by using the bomb as the reaction vessel in a simple apparatus such as that in Fig. 1-12, or for more accurate measurements the apparatus can be set up to operate adiabatically. Where heats of combustion are used to calculate other thermochemical data, such as enthalpies of formation, it is important that values of high accuracy be used since usually the desired quantity is obtained from small differences between large ΔH_{comb} values.

In bomb calorimetry, in general, the observed heat effect is related, via Eq. (1-10), to the change in energy rather than the change in enthalpy since the process occurs at constant volume. Thus appropriate corrections must be applied to obtain the desired ΔH.

I-4-7 Biological Applications of Thermochemistry

When a chemical reaction occurs in the human body it exhibits exactly the same changes in thermodynamic state functions as when the process takes place in a laboratory reactor. It is important to realize, therefore, that once we have obtained (by the methods discussed previously) either by calculation or experimental observation a change in energy or enthalpy, we have information directly applicable to the process occurring in a biological environment.

The energy liberated in a chemical reaction results in the production of heat and the performance of work. Biologically the work can take the form of the movement of substances from regions of low to high solution concentration, muscle contractions, and the like. Later discussions show how a process can be examined from the point of view of determining what fraction of the total enthalpy change can be extracted in a form capable of performing useful work. For the moment, however, it is necessary to be concerned only with the concept of the total energy produced—the gross energy or enthalpy change—in terms of the biological implications of chemical processes.

The basic energy requirements of the body are supplied by the oxidation of foods. Thus the most important aspect of biological thermochemistry is the "combustion" reaction in which compounds are converted to carbon dioxide and water. When examined from the biological point of view, the ΔH_{comb} value defined in Sec. 1-4-4 is the *caloric value* of the substance. Thus, for example, caloric values of 9,500 and 3,750 cal/gm for glucose and fat, respectively, in fact are heats of combustion for these materials and represent the maximum energy obtainable from oxidation in the body. In actual metabolism the amount of energy obtainable per unit weight of substance may be less than the heat of combustion for some materials because of incomplete absorption or ingestion. For carbohydrate and fat oxidation does go to completion, but for protein compounds the process is by no means complete and significant quantities of oxidizable carbon and hydrogen remain unused in energy metabolism. Caloric values for various foods can be determined in a bomb calorimeter in the same manner as for any other compound.

Reactions such as oxidation also can be studied as they occur in the body. In *direct calorimetry* rather elaborate calorimeters are designed to measure heat effects associated with the metabolic process. There is provision for measurement of both the heat transfer from the subject as well as determination of the conversion of oxygen to carbon dioxide. These measurements are complex, but the basic calorimetry principles outlined in the previous section are equally applicable here. The measurement of the CO_2/O_2 ratios is an alternative method of obtaining a value for the caloric output of the subject. Very good agreement between the direct method and calculations based on analysis of input and exit gases (*indirect calorimetry*) has been obtained. The latter approach is evidently less difficult experimentally. Indirect calorimetry makes use of the so-called *respiratory quotient RQ*, defined as the volume of CO_2 produced divided by the volume of O_2 consumed per unit time. The RQ of a compound undergoing oxidation can be calculated directly from the combustion equation. For the carbohydrate maltose, for example, we have

$$C_{12}H_{22}O_{11} + 12\ O_2 \rightarrow 12\ CO_2 + 11\ H_2O$$

where $RQ = 12\ CO_2/12\ O_2 = 1.0$. From the heat of combustion the number of calories produced for each liter of O_2 consumed can be calculated directly from the ideal gas law and the stoichiometry of the oxidation process. The RQ of fat is about 0.7, and that of protein is about 0.8. From measurements of the unoxidized nitrogen eliminated, along with the actual RQ of the subject, the volume of oxygen consumed and hence the heat produced in the oxidation of each of the three types of material ingested can be calculated, and thus the total caloric output can be determined.

This very brief account illustrates how in a relatively uncomplicated manner the calculation of enthalpy changes and heat effect in biological systems can be made either by direct experimental calorimetry or by utilizing thermochemical equations and simple gas analysis.

PROBLEMS

1-1. Calculate the standard energy and enthalpy change at 298 °K in the following reactions:
  (a) $C_2H_5OH(l) \rightarrow CH_4(g) + HCHO(g)$
  (b) $2\ Cl_2(g) + CH_4(g) \rightarrow 2\ H_2(g) + CCl_4(l)$
  (c) $C_2H_4(g) + H_2O(l) \rightarrow C_2H_5OH(l)$
  (d) $3\ C_2H_4(g) \rightarrow C_6H_6(l) + 3\ H_2(g)$

1-2. Calculate the heat absorbed when 11.2 liters O_2 at 0.5 atm and 25°C are heated to 35 °C at constant pressure.

1-3. For each of the steps in the oxidation of methane, as shown in Fig. 1-11, evaluate ΔH_{298} and hence calculate ΔH_{comb} of methane:

$$CH_4(g) + \tfrac{1}{2} O_2(g) \quad \rightarrow CH_3OH(l)$$
$$CH_3OH(l) + \tfrac{1}{2} O_2(g) \rightarrow HCHO(g) + H_2O(l)$$
$$HCHO(g) + \tfrac{1}{2} O_2(g) \rightarrow HCOOH(l)$$
$$HCOOH(l) + \tfrac{1}{2} O_2(g) \rightarrow CO_2(g) + H_2O(l)$$

1-4. Determine the heat of combustion [to $CO_2(g)$ and $H_2O(l)$] of $C_2H_5OH(l)$ using the standard heats of formation in Table 1-2.

1-5. Calculate the work done and the heat absorbed when 1 liter of ideal gas at an initial pressure of 7.50 atm is expanded at constant temperature to a final volume of 7.50 liters (a) in a vacuum, (b) against a constant external pressure of 1 atm, and (c) reversibly.

1-6. Find ΔH_{317}° and ΔE_{317}° for the reaction $CO(g) + \tfrac{1}{2}O_2(g) \rightarrow CO_2(g)$. Assume that C_p for each substance does not vary significantly with temperature between 298 and 317 °K.

1-7. The heat of fusion of benzene is 30.3 cal/gm at its melting point, 5.48 °C. If ΔC_p for the process in which solid benzene is converted to liquid benzene is 0.10 cal/deg gm, calculate the molar heat of fusion of C_6H_6 at 5.00°C.

1-8. 1.00 gm of n-$C_5H_{12}(l)$ is burned in a bomb calorimeter. If a temperature increase of 0.100 °C is observed for the system, calculate the total heat capacity of the calorimeter system.

1-9. Calculate ΔH_{298}° for the reaction $CO_2(g) + 2\,NH_3(g) \rightarrow CO(NH_2)_2(s) + H_2O(l)$ using the standard heat of formation of carbon dioxide and the information given in the discussion of Fig. 1-10.

1-10. Calculate the heat evolved in the oxidation of 10.0 gm of benzene to $CO_2(g)$ and $H_2O(l)$ at 298 °K.

1-11. 2.5 moles of monatomic ideal gas is taken from its initial state at $P = 10$ atm and $T = 27$ °C to a final state through the following sequence of processes:

(a) isometric heating to 127 °C
(b) isothermal reversible expansion to twice its initial volume
(c) isometric cooling to 77 °C

Evaluate q, w, and ΔE for the overall process.

1-12. For the reaction $C(s) + H_2O(g) \rightarrow CO(g) + H_2(g)$, calculate the standard enthalpy change at 25 °C. Taking $C_{p,C(s)} = 2.07$ cal/deg mole and additional C_p data from Table 1-1, compute the necessary data and plot a graph of ΔH° vs. T over the range 25 to 50 °C. If this reaction were studied calorimetrically over this temperature range, how would you interpret an observed curvature in the ΔH vs. T plot?

1-13. Reconsider the derivation of Eq. (1-37) as it applies to a process $A(l) \rightarrow B(l)$. If B undergoes vaporization at T' (a temperature between T_1 and T_2), derive a general expression for the temperature dependence of ΔH for this system.

1-14. Manganese could be formed by the process

$$3 \; Mn_3O_4(s) + 8 \; Al(s) \rightarrow 9 \; Mn(s) + 4 \; Al_2O_3(s)$$

Determine the amount of heat absorbed or evolved when 100 gm of manganese are formed.

1-15. Calculate the work done in the expansion of 1 mole of ideal gas under reversible conditions at a constant temperature of 25 °C (a) by using Eq. (1-24) and (b) by plotting a graph of P (calculated from the ideal gas law) as a function of V for values of the volume of 1, 1.5, 2, 3, . . . , 10 and by measuring the area under the curve. (To measure the area, plot the graph and count the number of squares under the curve between the two volume limits and divide by the number of squares representing 1 liter atm of PV work.)

Thermodynamics:
Entropy and Free Energy

2

Why do we need a second law of thermodynamics? The fundamental reason is that it allows us to make predictions. We can answer the important question: "Can this reaction possibly occur?"

On the basis of everyday experience we could enumerate many processes that are spontaneous, that is, which occur without the application of an external force. There also are many processes for which this characteristic is not obvious. A mechanism by which we could decide if any particular process is spontaneous would be very valuable. The first law permits calculation of the enthalpy and energy change in a process but does not indicate whether it can actually happen.

We have seen, for example, how the change in energy for a gas compression can be evaluated. For a particular gaseous system we determine that there is a ΔE of some number of calories between initial and final states. On the basis of the first law alone, nothing more can be said. It cannot be stated that the compression will or will not occur. The law allows calculation of the change in a function but does not *prohibit* any state-to-state transition we might care to imagine. A gas will not compress of its own volition, although there is nothing in the first law to indicate that such an occurrence is not spontaneous. In this simple case, as well as in any of the systems examined previously, the necessity of some *additional* requirement or regulation for spontaneity is apparent.

2-1-1   The Concept of Entropy

As we did for the first law, we begin our arguments by making deductions from experience.

Consider the process illustrated in Fig. 2-1. Such a process will not happen unless we apply some external force to the system. Heat will not flow "uphill," even though the amount of thermal energy transferred could be calculated quite simply from Eq. (1-6) or (1-7). Experience indicates that if we begin with the initial state shown in the figure, and if the system is left to itself, the final state will be a uniform temperature of 75°C. This is the process that will occur *spontaneously*.

A solute, in amounts less than the maximum solubility, placed in a solvent will diffuse spontaneously throughout the system to yield eventually a homogeneous mixture. Subsequently, we do not observe a spontaneous coagulation of solute particles in a particular part of the solution.

Among chemical reactions there are numerous examples of systems in which a reaction occurs spontaneously, but when the products are placed together in a reaction vessel, the spontaneous reformation of the reactants to a detectable extent is not observed.

The second law has its origin in the attempt to find a characteristic of nature and processes occurring in nature that can be used to decide whether an event can occur spontaneously. What is needed is a state function and a way of relating calculated changes in its value to spontaneity. Suppose some suitable function X were obtained. It then should be possible to calculate ΔX for a number of spontaneous and nonspontaneous processes and to extract from these data a general property of ΔX in each of the two types of process. Anticipating the final result of the arguments, we might find that it is the *sign* of ΔX rather than its magnitude that is critical.

It is not necessary to examine in detail how such a function is obtained. However, we can, acquire some qualitative idea of its origin. Consider how

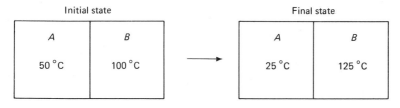

FIG. 2-1 *An impossible heat transfer process. The squares represent amounts of material A and B with the same total heat capacity.*

this approach could be applied to the physical process shown in Fig. 2-1. What we wish to find is a function of the system that changes in a character- istic manner as the final state of $T = 75\ °C$ in A and B. If subsequently for some other process the same function behaves identically, we might surmise tentatively that the second process also is spontaneous.

What kind of function might be used for a chemical reaction? By analogy with mechanical spontaneity, potential energy would appear to be the determining factor. This function invariably decreases in a spontaneous mechanical process. This suggests using ΔH for reactions, since it might be expected that if a reaction evolved heat through a decrease in enthalpy, an exothermic process, it would be spontaneous. If the enthalpy change is large, this rule appears to hold reasonably well in general. However, if ΔH is relatively small, the correlation is by no means general. There are many exothermic reactions that are not spontaneous and many endothermic reactions that are. There must be some additional factor outweighed when the enthalpy change is large. It is necessary, therefore, to examine in greater detail the characteristics of a spontaneous process.

One very useful way of examining a system before and after a spontane- ous process is to consider its *disorder* or *randomness*. If we isolate the system, *permit no energy or mass transfer across the system boundaries*, a spontaneous process appears to be characterized by an *increase in the randomness of the isolated system*. The process involves a change, sometimes obvious and sometimes more subtle, from a more ordered to a less ordered state.

The idea of an increase in randomness can be illustrated with two simple examples. In Fig. 2-2, as the gas expands to fill the available space, the system goes from a more ordered (all gas molecules in flask A) to a less ordered condition (gas molecules in both flasks). Thus the disorder or randomness has increased.

In a spontaneous solution process the system is more ordered in its initial state, where the solute is confined to the portion of the solution to which it was added, than in its final state, where the solute distribution is

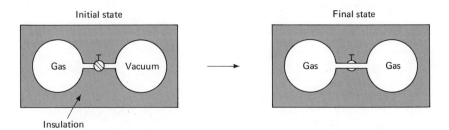

FIG. 2-2 *The spontaneous expansion of a gas into a vacuum. The system is isolated since there is no energy or mass transfer with the surroundings.*

homogeneous. Both this and the gas system discussed above are isolated. The process occurs in each case without energetic or thermal interactions with the surroundings. In physical processes of this kind the increase in randomness is obvious. In chemical reactions the change in order of the system may not be apparent. In addition, there is almost invariably energy transfer between system and surroundings, and it is more difficult to envisage an isolated system.

More detailed analyses, however, indicate that if *any* process occurs spontaneously and the system is isolated, there is an increase in randomness. If we accept this as a criterion of spontaneity, a function that measures the degree of disorder in the system will satisfy the conditions specified for the function we wish to use as the basis for the second law. The thermodynamic function that is a direct measure of randomness is the *entropy S*. Since thermodynamic laws are deduced, and not derived in the sense of a mathematical derivation of an equation, we can rationalize our approach in the following way. As a working hypothesis the existence of this function is assumed. It need be characterized initially in no more specific terms than to indicate that it is a measure of randomness. The second law can then be formulated in terms of entropy, determining as well how a change in entropy can be calculated. The results of this approach demonstrate that the treatment based on entropy does, in fact, provide a workable criterion of spontaneity.

2-1-2 Statement of the Second Law

The preceding discussion can now be formalized in the statement:

THE SECOND LAW OF THERMODYNAMICS: FOR ANY SPONTANEOUS PROCESS IN AN ISOLATED SYSTEM THERE IS AN INCREASE IN THE VALUE OF A THERMODYNAMIC STATE FUNCTION, ENTROPY S, WHICH IS A MEASURE OF THE RANDOMNESS OF THE SYSTEM.

Considering the whole universe as an isolated system, any actual change that occurs will be accompanied by an increase in randomness—an increase in entropy. In discussions of the first law it was noted that one expression of the law was the principle of conservation of energy. The idea of a constant amount of energy and increasing entropy is embodied in the classic thermodynamic statement that *the energy of the universe is constant, the entropy is increasing.*

For a closed but *nonisolated* system, one in which there may be energy transfer but no mass transfer with the surroundings, the restriction that ΔS be positive for a spontaneous process does not apply. Here the entropy change in the system can be positive, negative, or zero, but in any allowable process the entropy of the entire universe must increase. It is evident that this is not a particularly easy criterion to apply. Later discussions indicate how this difficulty leads to the development of a more practical criterion, but the entropy concept remains a fundamentally important one and is incorporated into the new function.

Within this limitation, if some means of calculating the entropy change can be developed, then the desired prediction of whether a process can occur could be made.

The approach to the defining equation for ΔS begins with a consideration of the *energy* of a system. It can be demonstrated, first, that every system possesses some energy that is *unavailable for work at constant temperature* and, second, that this is related to the entropy.

Consider a system comprised of ice in a suitable container. Sufficient thermal energy is added to convert 75 percent of the ice to water. Note initially that if the container is open to the atmosphere, the process will occur at constant pressure. What is more significant is that the temperature will remain constant at 0 °C and will not change until all of the ice has been converted to liquid. The amount of thermal energy put into the system is the product $0.75 \times$ mass (gm) $\times$ heat of fusion (cal/gm). This amount represents energy unavailable for work in any subsequent constant temperature process. Suppose we wish to utilize the energy added to the system to do the work of gas expansion. We could place in thermal contact with the ice water system a cylinder with a movable piston that confined some amount of gas. If the cylinder system is initially at 0 °C, the heat energy will not flow out of the ice water mixture into the cylinder and thereby cause the gas to expand. The only way in which this energy could be extracted by such a device is to introduce the cylinder system initially at some temperature less than 0 °C. The heat would then flow to the cylinder and cause the gas to expand, but this would not be an isothermal process. The energy of the system resulting from the melting of the ice is unavailable in any *constant temperature process*. This is, of course, just one example of a property generally characteristic of the energy of all systems.

How can we measure the amount of this particular kind of energy? Let us examine a perfectly general thermodynamically reversible process for which the first law can be written in the form

$$\Delta E = q_{\text{rev}} - w_{\text{max}} \qquad (2\text{-}1)$$

Just as for the isothermal expansion of an ideal gas, if *any* process is carried out under reversible conditions, the maximum amount of work is accomplished. No matter how the change of state occurs in practice, we will never obtain more work than the w_{max} calculated for the reversible isothermal path. In addition, since ΔE for a given initial and final state will be invariant, the first law in the form $q = \Delta E + w$ shows that when $w = w_{\text{max}}$, q is also at its maximum value. Thus q_{rev} in Eq. (2-1) represents the maximum amount of heat that can be absorbed. Because of this q_{rev} term, the maximum amount of work in any process is not equal to the energy change. In other words, only part of the energy alteration associated with a particular state-to-state transition can appear as work. The remainder q_{rev} therefore represents the energy unavailable unless there is a subsequent nonisothermal process.

The final step in the formulation of the entropy equation is to relate the energy represented by q_{rev} to the entropy. As indicated explicitly with heat ($q = \dot{C} \, \Delta T$) and work ($w = P \, \Delta V$), any form of energy can be expressed as the product of an intensity and capacity factor. For q_{rev} we proceed in the same manner:

Intensity factor $\times$ capacity factor = isothermally unavailable energy

Temperature $\times$ entropy change = q_{rev}

we write this expression as an equation for an infinitesimal amount of heat absorbed:

$$T \, dS = dq_{rev} \tag{2-2}$$

or rearrange to obtain a "definition of the entropy change":

$$dS = \frac{dq_{rev}}{T} \tag{2-3}$$

For a finite change in entropy we sum a series of infinitesimal changes, that is, we integrate Eq. (2-3):

$$\Delta S = S_2 - S_1 = \int_{T_1}^{T_2} \frac{dq_{rev}}{T} \tag{2-4}$$

For a constant temperature process we can proceed in a manner analogous to that employed for Eq. (1-3) at constant pressure and remove the invariant $1/T$ terms from the "summation" to obtain

$$\Delta S = \frac{1}{T} \int_1^2 dq_{rev} = q_{rev}/T \quad (T \text{ constant}) \tag{2-5}$$

The sum of the dq_{rev} terms from initial state to final state, represented by the integral between state 1 and state 2 in Eq. (2-4), is just the total heat absorbed reversibly, q_{rev}. The form of Eq. (2-5) gives the units of entropy as calories per degree.

Keeping in mind the restriction of constant temperature, Eq. (2-5) can be rearranged to the useful form

$$q_{rev} = T \Delta S \tag{2-6}$$

This relation is essentially Eq. (2-2) written for a finite change. It also is important because it will be utilized again in developing a more useful criterion of spontaneity.

Entropy is a state function. In a particular state a system will have a definable amount of isothermally unavailable energy—a certain degree of randomness—that is independent of the manner in which the state was attained. The change in this amount in a given process can be calculated from Eq. (2-4).

In the equation $\Delta S = q_{rev}/T$ the situation is analogous to the one encountered with the equations $\Delta H = q_p$ and $\Delta E = q_v$. The change in the value of the state function is independent of path but can be equated to the change in a path-dependent function *along a specific path*. The entropy change in a system undergoing a process along *any* path is numerically equal to the heat absorbed reversibly at constant temperature divided by the temperature at which the process occurs. If the temperature varies, ΔS can be evaluated by using Eq. (2-4).

Consider now an actual and therefore thermodynamically irreversible process in which an amount of heat q_{irr} is absorbed. Since q_{rev} is the maximum amount of heat that can be absorbed, it follows that $q_{rev} > q_{irr}$ and therefore

$$\Delta S = \frac{q_{rev}}{T} > \frac{q_{irr}}{T} \quad (T \text{ constant}) \tag{2-7}$$

Since we saw in Sec. 1-3-3 that no reaction is truly reversible, it might appear that entropy is not a useful function because we could not calculate the change in its value in a real situation. The value of ΔS, however, is path independent. All that the inequality in Eq. (2-7) indicates is that we cannot calculate ΔS from q_{irr}/T and that some reversible means of going from initial state to final state must be imagined for the purposes of making the calculation of ΔS. The real transition may proceed by any path and still have the same ΔS.

There are a number of physical processes that are either by nature reversible or for which reversible paths for calculation purposes can easily be envisaged. Before examining methods for determining ΔS in chemical reactions, we will deal with a number of such processes. This approach will illustrate a number of techniques and at the same time derive a number of quite useful equations.

2-1-3   Calculation of Entropy Changes in Physical Processes

Phase changes. Such processes occur at constant temperature and pressure. If equilibrium conditions are maintained (for example, in the vaporization of a liquid the external pressure and vapor pressure must be equal) the process is reversible and $q_{rev} \equiv q_p \equiv \Delta H_{phase\ change}$. From Eq. (2-6) we can write

$$\Delta S = \frac{\Delta H_{phase\ change}}{T} \tag{2-8}$$

$$\Delta S = \frac{\Delta H}{T} \times m \quad \frac{cal}{deg\ mole}$$

The applicability of this relation is limited strictly to phase changes under reversible conditions. It cannot be employed for an ordinary chemical reaction.

EXAMPLE 2-1.   The heat of vaporization of cyclohexane at 60.8 °C is found to be 7,170 cal/mole. Calculate the molar entropy of vaporization ΔS_{vap} at that temperature.

$$T = 61 + 273 = 334 \ ^\circ K$$

$$\Delta S = \frac{\Delta H_{vap}}{T} = \frac{7,170}{334} = \underline{21.4 \ cal/deg \ mole}$$

Isothermal ideal gas volume change. We have already determined that the heat absorbed in a reversible isothermal expansion or compression of an ideal gas is given by

$$q_{rev} = 2.303 \ nRT \ \log \frac{V_2}{V_1} \tag{1-24}$$

To obtain the entropy change, Eq. (2-6) indicates that we need only to divide this relation by T to obtain

$$\Delta S = 2.303 \ nR \ \log \frac{V_2}{V_1} \tag{2-9}$$

This equation can be used for *any* isothermal volume change for an ideal gas, reversible or irreversible. Although the equation yields the entropy change along the reversible path, ΔS via any path must be the same.

EXAMPLE 2-2.   2.50 moles of an ideal gas is expanded isothermally to 5.00 times its initial volume. Calculate the change in entropy.

$$V_2/V_1 = 5.00 \qquad \log V_2/V_1 = 0.699 \qquad n = 2.50 \ moles$$

$$R = 1.987 \ cal/deg \ mole \qquad \Delta S = 2.303 \ nR \ \log \frac{V_2}{V_1}$$

$$\Delta S = 2.303 \times 2.50 \times 1.987 \times 0.699$$

$$\Delta S = \underline{7.99 \ cal/deg}$$

Note that it was not necessary to specify the temperature, if the expansion is reversible, or the absolute values of V_1 and V_2.

Temperature alterations. Changing the temperature of a system results in a change in entropy. Calculation of ΔS requires the use of Eq. (2-4). To obtain the total ΔS we have to sum an infinite number of processes in which an amount of heat dq_{rev} is absorbed. As in earlier encounters with integrations of this type, we must be able to alter the equation to get a single variable so that we can quote and use the applicable integration rule. Recall that for a constant temperature process we have from Eqs. (1-15) and (1-34)

$q_p = nC_p \Delta T$, which for each infinitesimal step of the process can be written

$$dq_p \equiv nC_p \, dT \equiv dq_{\text{rev}} \tag{2-10}$$

We are now taking advantage of the fact that if the process is taking place via an infinitely small increment, it is by definition reversible. An infinitesimal amount of heat can be absorbed only reversibly. Substitution for dq_{rev} in Eq. (2-4) gives

$$\Delta S = \int_{T_1}^{T_2} \frac{nC_p}{T} \, dT \tag{2-11}$$

This is a valuable equation because it can be used to calculate the absolute entropy of a substance. This important topic is considered in the next section. For the moment we confine ourselves to a temperature range where C_p can be assumed to remain constant within experimental error and where there is no phase change. We can remove the constant terms, n and C_p, from the integral and proceed as in Sec. 1-3-4 to obtain the desired equation, again making use of the integration rule for the general function dx/x.

$$\Delta S = nC_p \int_{T_1}^{T_2} \frac{dT}{T} = nC_p(\ln T_2 - \ln T_1)$$

$$\Delta S = nC_p \, 2.303 \log \frac{T_2}{T_1} \tag{2-12}$$

Again the equation is derived for the reversible path, but the same entropy change will result from the transition along any path.

> EXAMPLE 2-3.   1.00 mole of water is heated at constant pressure from 25 to 50 °C. Calculate the change in entropy.
>
> $n = 1.00$ $C_p = 18.0$ cal/deg mole   (Table 1-1) $T_1 = 298°$
>
> $T_2 = 323$ °K $\Delta S = nC_p \, 2.303 \log \dfrac{T_2}{T_1} = 1.00 \times 18.0 \times 2.303 \times 0.0350$
>
> $\Delta S = 1.45$ cal/deg

In all three examples discussed in this section the change in system entropy is positive. This is consistent with the physical interpretation of entropy as a measure of randomness. As a substance is vaporized, or simply undergoes an increase in temperature, it moves from a more ordered to a less ordered state and therefore exhibits a positive ΔS. Similarly, when a gas expands to a larger volume, there is a transition to a state of greater disorder and therefore a state characterized by increased entropy.

2-2 THE THIRD LAW—ABSOLUTE ENTROPIES

Equation (2-12) indicates that as the temperature of a substance rises there is an increase in its entropy. Viewing entropy as a measure of disorder consider, for example, a system that is a pure gas at room temperature, and examine qualitatively how the degree of randomness is altered as the temperature is lowered. As the system is cooled through the condensation point and then through the freezing point, it is apparent that the order of the system is increasing. The molecules are first constrained in the liquid phase and then in the solid crystal lattice. As the temperature is further decreased, the extent of molecular vibration in the crystal is reduced until, at absolute zero, vibration presumably ceases. If the system forms a crystal in which there are no imperfections, it can be regarded as being perfectly ordered at 0 °K. Consequently, the entropy, a measure of *dis*order, should be zero. This idea is formalized in the *Third Law of Thermodynamics*:

> THIRD LAW: THE ENTROPY OF A PURE, PERFECTLY CRYSTAL-LINE SUBSTANCE IS ZERO AT ABSOLUTE ZERO.

For other thermodynamic functions we either ignore the origin or establish an arbitrary zero point for purposes of calculation. Entropy, however, has a definite reference point in temperature and hence we can calculate its *absolute value*.

The absolute entropy of a substance evidently depends on its physical state and temperature. To calculate $S_{absolute}$ at a specific temperature, we must compute the change in entropy resulting from the increase in temperature from 0 °K to T and add to this the ΔS due to any phase change that occurs in this temperature range.

Let us first examine the simple case of the absolute entropy of a substance that is a single-phased* solid up to some temperature of interest, for example, 298 °K. Here only the ΔT need be taken into account as a source of entropy alteration, and hence Eq. (2-11) can be used to obtain the value of S. The relation is written specifically for the temperature interval of 0 to 298 °K:

$$\Delta S = S_{298} - S_0 = \int_0^{298} \frac{C_p}{T} \, dT$$

Since by the third law $S_0 = 0$ (the entropy at absolute zero), this expression can be written as

$$S_{298} = \int_0^{298} \frac{C_p}{T} \, dT$$

---

* Although a substance may remain in the solid phase between two specified temperatures, it is still possible that it may undergo a solid → solid phase change, for example, sulfur (monoclinic) → sulfur (rhombic), and this would have to be included in the calculation of absolute entropy. However, we do not encounter such situations here.

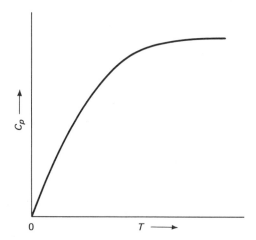

FIG. 2-3 Typical variation of the heat capacity at constant pressure of a substance between 0 °K and room temperature. The substance exists as a single phase solid within this temperature range.

If the state of the compound at 298 °K is the most stable form and the pressure is 1 atm, the value obtained is the *standard absolute entropy*, designated in general as S_T°, or for this case specifically S_{298}°. Since the temperature range covered is large, C_p cannot be assumed to be constant, nor can it easily be expressed as a function of temperature to allow direct integration to be performed as in the derivation of Eq. (2-12). A typical variation of C_p with temperature is shown in Fig. 2-3. Note that the heat capacity is zero at 0 °K.*

For a substance that undergoes one or more phase changes between 0 °K and T, the absolute entropy at T includes a number of terms. The S value is the sum of the ΔS for the phase change (given by $\Delta H/T$), the entropy change owing to temperature in the temperature interval between each transition [calculated from Eq. (2-11)], and the absolute entropy at the initial transition temperature. At the higher temperatures the C_p value usually remains sufficiently constant for Eq. (2-12) to be used.

EXAMPLE 2-4. The following data have been obtained† for fluorobenzene, C_6F_6:

$$\Delta H_{\text{fusion}} = 2{,}770 \text{ cal/mole} \qquad mp = 278.3 \text{ °K}$$

$$C_{p,C_6F_6,279-298 \text{ °K}}^\circ = 52.5 \text{ cal/deg mole.}$$

* Experimental determinations of heat capacity are not usually made below about 10 °K. Several methods are available to extrapolate experimental curves to 0 °K, but for calculations where great accuracy is not required, simple extrapolation "by eye" is adequate.

† J. F. Counsell, J. H. S. Green, J. L. Hales, and J. F. Martin, *Trans. Faraday Soc.* **61**, 212 (1965).

The table gives the C_p° values for the solid phase:

T (°K)	10	15	20	25	45	70	130	190	230	278.3
C_p° (cal/deg mole)	1.06	2.75	4.37	5.82	10.2	15.1	26.3	36.1	41.1	45.6

Calculate the standard absolute entropy of hexafluorobenzene at 298 °K.

Absolute entropy of solid.
Calculate C_p°/T for each datum given and plot C_p°/T vs. T (°K) to obtain the graph shown in Fig. 2-4. Mechanical determination of the area under the curve between 0 and 278.3 °K gives $S_{278.3}^\circ = 53.3$ cal/deg mole.

Entropy change in fusion

$$\Delta S_{\text{fusion}} = \Delta H_{\text{fusion}}/T_{mp}$$

$$\Delta S_{\text{fusion}} = \frac{2,770}{278.3} = 9.95 \text{ cal/deg mole}$$

Entropy change in heating of liquid

$$\Delta S = nC_p \, 2.303 \log \frac{T_2}{T_1}$$

$$\Delta S = 1.0 \times 52.5 \times 2.303 \times 0.0294 = 3.55 \text{ cal/deg mole}$$

$$S_{298,\text{C}_6\text{F}_6}^\circ = S_{\text{solid}}^\circ + \Delta S_{\text{fusion}} + \Delta S_{278 \to 298}$$

$$S_{298,\text{C}_6\text{F}_6}^\circ = 53.3 + 9.95 + 3.55 = \underline{66.8 \text{ cal/deg mole}}$$

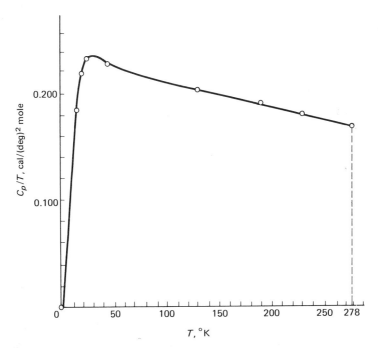

FIG. 2-4 *Variation of C_p/T with T(°K) for hexafluorobenzene (see Example 2-4).*

TABLE 2-1. STANDARD ABSOLUTE ENTROPIES AT $298\,^{\circ}K^a$

Substance	S°_{298}, cal/deg mole	Substance	S°_{298}, cal/deg mole
$Al_2O_3(s)$	12.14	$H_2O(g)$	45.14
$CO(g)$	47.30	$H_2O(l)$	16.72
$CO_2(g)$	51.06	$HCl(g)$	44.62
$Cl_2(g)$	53.29	$N_2(g)$	45.78
$CO(NH_2)_2(s)$	25.00	$NO(g)$	50.3
$CH_4(g)$	44.50	$NO_2(g)$	57.5
$C_2H_4(g)$	52.45	$N_2O_4(g)$	72.3
$CH_3OH(l)$	30.3	$NH_3(g)$	46.0
$C_2H_5OH(l)$	38.4	$NH_4Cl(s)$	22.6
$H_2(g)$	31.21	$O_2(g)$	49.00
$HCHO(g)$	52.26	cyclo-$C_6H_{12}(l)$	71.28
$HCOOH(l)$	30.82	$C_6H_6(g)$	64.34
$CH_3COOH(l)$	38.2	$SO_2(g)$	59.40
$CCl_4(l)$	51.25	$SiF_4(g)$	68.0

[a] From *Selected Values of Chemical Thermodynamic Properties* (Circular 500, National Bureau of Standards, 1952) and *Selected Values of Properties of Hydrocarbons* (Circular C461, National Bureau of Standards, 1947).

Through use of experimental data of this type, standard absolute entropy values at 25°C have been determined for numerous substances, including elements. Some values are given in Table 2-1. Note that unlike ΔH°_f, where ΔH°_f (elements) = 0 was taken as an arbitrary zero point, $S^{\circ}_{\text{elements}}$ has a finite value.

To emphasize the physical significance of entropy as a measure of randomness it is interesting to compare some of the values in the table. We saw in the preceding section that alterations in the state of the same substance leading to greater disorder showed a positive ΔS. Here we compare the absolute entropy of different substances under the same conditions and rationalize the differences in the values in terms of randomness. For example, the S°_{298} values for N_2, NO, NO_2, and N_2O_4 are 45.6, 50.3, 57.5, and 72.3 cal/deg mole, respectively. There evidently is a parallel between increasing entropy and increasing molecular complexity. As the number and type of atoms in the molecular system increases so does the number of possible modes of intramolecular vibration and rotation. This increase in possibilities for disorder is reflected in an increase in absolute entropy.

2-3 ENTROPY CHANGES IN CHEMICAL REACTIONS

With absolute entropies of elements and compounds available, it is a relatively simple matter to calculate the change in entropy of a system undergoing a chemical reaction. The ΔS for the process will be the difference between the total absolute entropy of the products and that of the reactants. If we have a reaction at 298 °K in which all species are present in their standard states, the appropriate equation is

$$\Delta S^\circ_{298} = \sum_{\text{products}} n S^\circ_{298} - \sum_{\text{reactants}} n S^\circ_{298} \qquad (2\text{-}13)$$

The terms in each summation represent the standard absolute entropy per mole times the number of moles of that substance in the reaction as written. The units of ΔS° are cal/deg. Note that the form of this equation is of the type encountered previously, for example, Eq. (1-33), in which the change in a thermodynamic function in a reaction is calculated by taking the difference between the function for products and reactants.

EXAMPLE 2-5.   Calculate the standard entropy change at 298 °K in the reaction $C_2H_5OH(l) + O_2(g) \rightarrow CH_3COOH(l) + H_2O(l)$ using S° values from Table 2-1.

$$\Delta S^\circ_{298} = S^\circ_{298,\,CH_3COOH(l)} + S^\circ_{298,\,H_2O(l)} - S^\circ_{298,\,C_2H_5OH(l)} - S^\circ_{298,\,O_2(g)}$$

$$\Delta S^\circ_{298} = 38.2 + 16.7 - 38.4 - 49.0$$

$$\Delta S^\circ_{298} = \underline{-32.5 \text{ cal/deg}}$$

An equation such as (2-13) also can be used to obtain the change in entropy in some process wherein the species are not in their standard state simply by use of the absolute entropies in that state. In that case the equation becomes

$$\Delta S = \sum_{\text{products}} n S - \sum_{\text{reactants}} n S \qquad (2\text{-}14)$$

Normally we would have available data such as that in Table 2-1, and therefore it would be necessary to convert a given S° value to that for the nonstandard state and for some temperature other than 25 °C, if this also were required. Although the general equations for this calculation can be worked out, the computation is seldom encountered. The several physical change entropy relations developed for the species individually can be used when this kind of problem must be solved.

    For a temperature other than 25°C we can make use of Eq (2-11) or (2-12) to calculate for each reactant and product the ΔS resulting from the temperature change and add this to the S° value in the table.

    If the pressure is not 1 atm, the effect generally can be neglected for liquids and solids. The required additional entropy term for gases is calculated from the equation

$$\Delta S = 2.303 \, nR \log \frac{P_1}{P_2} \qquad (2\text{-}15)$$

This relation is obtained from a combination of Eq. (2-5), $\Delta S = q_{\text{rev}}/T$, and Eq. (1-25), $q_{\text{rev}} = 2.303 \, nRT \log (P_1/P_2)$.

Having obtained a value of ΔS, we can readily calculate $T \Delta S$, the heat that would be absorbed by the system were the process to occur reversibly. Frequent use of q_{rev} is made in subsequent discussions.

2-4 THE FREE ENERGY

Although the discussion has dealt in detail with the calculation of entropy changes, recall that we are searching for an easily applied criterion of spontaneity. We saw that according to the second law a spontaneous process is accompanied by an increase in entropy of an isolated system. For chemical reactions it is difficult if not impossible to isolate a system for purposes of calculation. Hence the criterion $\Delta S_{\text{isolated system}} > 0$ would normally take the form

$$\Delta S_{\text{nonisolated system}} + \Delta S_{\text{surroundings}} \equiv \Delta S_{\text{tot}} > 0$$

The implication is that if the system is not isolated, ΔS_{system} may be positive, negative, or zero. It is the total entropy that must increase in a spontaneous process. For chemical reactions, therefore, as well as for other situations where it is not convenient to deal with an isolated system, we require a criterion that can be more easily applied. To fill this need an additional function has been developed whose change for the *nonisolated system alone* determines if the process is spontaneous.

One function that has been found to serve this important purpose is the *free energy G*. The function is also known as the *Gibbs free energy* and often is designated as F, but there is no general agreement on nomenclature. The term free energy and the letter G are used in this book.

2-4-1 Definition of Free Energy

The free energy is essentially an invented function, obtained by a more or less pragmatic approach to the task of acquiring a workable criterion of spontaneity. The defining equation is

$$G = H - TS \tag{2-16}$$

Since H, T, and S are all state functions, the free energy as defined by this relation is also a state function. It is the *change* in free energy that is of primary interest, and therefore a more useful working relation is

$$\Delta G = \Delta H - \Delta(TS) \tag{2-17}$$

Both ΔH and $\Delta(TS)$, and therefore ΔG, have units of calories. For a constant temperature process—the type with which we are normally concerned—Eq. (2-17) becomes

$$\Delta G_T = \Delta H - T \Delta S \tag{2-18}$$

The free energy is not a "new function" in our development but rather a new functional relationship between state functions already defined. For a constant temperature process Eq. (2-18) indicates immediately that ΔG can be calculated directly from the enthalpy change—obtained from Hess' law— and the $T \Delta S$ term whose evaluation was discussed in the preceding section. Calculation of free energy changes in general are dealt with is subsequent sections.

2-4-2 Significance of ΔG

We will first look for a physical interpretation of the free energy change and then see how ΔG values can be utilized in thermodynamics. We proceed by making a number of successive alterations in the defining equation, Eq. (2-17), to relate ΔG to quantities that are more readily visualized.

If the defining equation for the enthalpy change is substituted into Eq. (2-17), the result is

$$\Delta G = \Delta E + \Delta(PV) - \Delta(TS)$$

As it stands this equation is perfectly general. If we confine ourselves to the normal situation where the temperature and pressure are constant, then

$$\Delta G_{TP} = \Delta E + P \Delta V - T \Delta S \tag{2-19}$$

On the basis of Eq. (2-1), $\Delta E = q_{rev} - w_{max}$, and Eq. (2-6), $q_{rev} = T \Delta S$, we have

$$\Delta E = T \Delta S - w_{max} \tag{2-20}$$

Substitution of ΔE as given by this equation into Eq. (2-19) yields

$$\Delta G_{TP} = T \Delta S - w_{max} + P \Delta V - T \Delta S$$

or

$$-\Delta G_{TP} = w_{max} - P \Delta V \tag{2-21}$$

This important relation indicates a fundamental significance of the free energy change. The w_{max} term represents the work that would be obtained if the process were to occur reversibly. The second term, $P \Delta V$, is the work done in expansion against the applied pressure. In chemical systems the PV work done seldom can be utilized in a practical way. It therefore is useful to envisage the difference between w_{max} and $P \Delta V$ as the *useful* or *net work* that could be performed by a chemical reaction.

If we were concerned with an electrochemical reaction in which a gas is being formed, the maximum electrical work (net work) we could obtain would be w_{max} (the total work accomplished under reversible conditions) less the $P \Delta V$ term (the normally nonrecoverable work of expansion against the surroundings). Similarly, in biological systems we might be interested in the work of muscle contraction. It is $w_{max} - P \Delta V$, rather than w_{max} itself, that indicates the maximum number of calories of *useful* work obtainable since an amount $P \Delta V$ of the total is "lost" in expansion.

Equation (2-21) indicates, then, that the

> *FREE ENERGY DECREASE IS EQUIVALENT TO THE MAXIMUM AMOUNT OF USEFUL WORK OBTAINABLE FROM A GIVEN PROCESS AT CONSTANT TEMPERATURE AND PRESSURE.*

When the process is carried out irreversibly, the actual work obtained will be less than w_{max}, but $-\Delta G_{TP}$ is a change in a state function. It always indicates the amount of work obtainable if the system were taken from initial to final state via a reversible path.

We can now establish the important criteria for spontaneity by considering how Eq. (2-18) would apply to a process in which both pressure and temperature remain constant. The ΔH for such a process will be equal to q, the actual amount of heat absorbed. The second term in Eq. (2-18), $T \Delta S$, can be replaced by q_{rev}. With these substitutions we have

$$\Delta G_{TP} = q - q_{rev} \tag{2-22}$$

From this equation and the previously established characteristics of q_{rev}, we can arrive at a quantitative distinction between three types of thermodynamic process.

Spontaneous process. An actual spontaneous process that occurs in nature must do so irreversibly. Thus the q term can be written explicitly as q_{irr}. Furthermore, we have seen that the amount of heat absorbed in an irreversible process is *less* than that absorbed in the same state-to-state transition reversibly conducted. Stated formally,

$$\Delta G_{TP} = q_{irr} - q_{rev} \qquad q_{irr} < q_{rev} \qquad \Delta G_{TP} < 0$$

We therefore can state the criterion for a spontaneous process:

> *A PROCESS FOR WHICH ΔG_{TP} IS NEGATIVE IS SPONTANEOUS.*

What distinguishes this principle from that based on entropy change and makes it considerably more useful is that the determining factor is the free energy change *in the system alone*. We need not consider the surroundings as required in an application of the $\Delta S_{tot} > 0$ rule.

A great deal of the utility of the thermodynamics is distilled into this single statement that a spontaneous process at constant temperature and pressure is accompanied by a decrease in the free energy of the system.

Since free energy is defined so that its value decreases in a spontaneous process, the function can be profitably regarded as a measure of the *chemical potential* of the system. Strictly, the chemical potential μ is defined in a rigorous fashion, but for our purposes it is useful nevertheless to think of G

as a potential. As a reaction proceeds, the system can be envisaged as moving from a state of higher to one of lower chemical potential. This is the fundamental driving force for any process—an unalterable tendency to move to a state of lowest free energy.

Nonspontaneous process. If a particular process exhibits a positive free energy change, Eq. (2-18) again can be utilized to demonstrate that such a process is thermodynamically impossible. If follows from that equation that if ΔG_{TP} is positive, the heat actually absorbed would be greater than the amount associated with reversibility. Since $q_{rev} \equiv q_{max}$, the condition that $q_{actual} > q_{rev}$ is clearly impossible. In terms of the chemical potential, it is not possible to move spontaneously to a state of higher chemical potential. This establishes the second criterion:

> *A PROCESS IN WHICH ΔG_{TP} IS POSITIVE IS THERMODYNAMICALLY IMPOSSIBLE.*

Reversible process. Here the heat actually absorbed is identical with q_{rev} and, from Eq. (2-18), $\Delta G_{TP} = 0$. The third and final criterion therefore can be stated:

> *A PROCESS IN WHICH ΔG_{TP} IS ZERO IS THERMODYNAMICALLY REVERSIBLE.*

In Sec. 1-1-3, it was seen that a system undergoing a reversible process is in continuous equilibrium. We therefore can restate the above conclusion in more widely applicable terms;

> *A SYSTEM WHERE THERE IS ZERO FREE ENERGY DIFFERENCE BETWEEN INITIAL AND FINAL STATES IS IN EQUILIBRIUM.*

This important statement is utilized frequently in subsequent discussions of chemical equilibrium and is treated in detail in Chapter 3. It is valuable to appreciate, however, how this criterion is related to the general principles of free energy of the system.

One immediate application of the these statements that can be made directly is to the free energy characteristics of a phase change. If conditions are such that Eq. (2-8), $\Delta S = \Delta H/T$, applies (that is, if the process is reversible), then $\Delta H = T\Delta S$, and therefore from Eq. (2-18) ΔG_{TP} must be zero. Alternatively, since it was stated in the earlier discusssion that the system was in equilibrium, we could now base the argument directly on the free energy change criterion. On this basis it follows that since ΔG_{TP} is zero, ΔH and $T\Delta S$ are equal. These two approaches are equivalent but illustrate that the previous identification of ΔH in the phase change with q_{rev} is consistent with the free energy criteria developed here.

Since there is no free energy difference, we can state the important principle:

THE FREE ENERGY PER MOLE OF ANY SUBSTANCE IN ONE PHASE IS THE SAME AS THAT OF THE SAME SUBSTANCE IN ANY OTHER PHASE IN EQUILIBRIUM WITH IT.

2-4-3 Standard Free Energy Changes

From the discussion to this point it is evident that the knowledge of ΔG in a process is information of considerable utility. However, it is, ΔG *under the conditions of the reaction* that we must obtain if the criteria are to be applied. In Sec. 2-4-1 it was indicated that ΔG could be calculated directly from the defining equation $\Delta G = \Delta H - T \Delta S$ if the enthalpy and entropy change values were available for the given conditions. This situation unfortunately is not encountered very often. Consequently, other methods must be developed for the calculation of ΔG. These procedures are usually based on a computation of the *standard* free energy change $\Delta G°$—the difference in the free energy of products and reactants in their *standard* state. Like other "standard state reactions," such a process may be hypothetical, but this does not prevent its being used to obtain the free energy alteration in a real process. It is worthwhile to emphasize that the criterion for spontaneity makes use of the free energy change under the conditions of the reaction and *not* the standard free energy change.

If standard enthalpy and entropy values are available, $\Delta G°$ can be evaluated via Eq. (2-18) written in the form

$$\Delta G° = \Delta H° - T \Delta S° \qquad (2\text{-}23)$$

where for this and subsequent equations the subscripts on ΔG and $\Delta G°$ usually are omitted, assuming constant temperature and pressure conditions unless indicated otherwise.

EXAMPLE 2-6.   Calculate $\Delta G°_{298}$ for the reaction $HCl(g) + NH_3(g) \rightarrow NH_4Cl(s)$, using a $\Delta H°_{298}$ value of $-42,300$ cal and $S°_{298}$ data from Table 2-1.

$$\Delta S°_{298} = S°_{298,NH_4Cl(s)} - S°_{298,NH_3(g)} - S_{298,HCl(g)}$$
$$\Delta S°_{298} = 22.6 - 46.0 - 44.6 = -68.0 \text{ cal/deg}$$
$$\Delta G°_{298} = \Delta H°_{298} - T \Delta S°_{298}$$
$$\Delta G°_{298} = -42,300 - 298 \times (-68.0) = -42,300 + 2,026$$
$$\underline{\Delta G°_{298} = -40,274 \text{ cal}}$$

A more direct method of obtaining $\Delta G°$ is through the use of a Hess' law-type manipulation of the *standard free energy of formation* of the species involved in the reaction. Proceeding in a manner identical to that employed previously for ΔH_f°, we first define this specific free energy change as follows:

STANDARD FREE ENERGY OF FORMATION ΔG_f° IS THE FREE ENERGY CHANGE IN THE REACTION IN WHICH ONE MOLE OF A COMPOUND IS FORMED AT CONSTANT TEMPERATURE, ALL SPECIES BEING IN THEIR STANDARD STATE.

Since all standard states are characterized by a pressure of 1 atm, it is not necessary to specify isobaric conditions in the definition. The units of ΔG_f° are calories per mole, and by convention we take $\Delta G_{f,298,\text{elements}}^\circ$ as zero. Values for the standard free energy of formation of a number of compounds are given in Table 2-2.

By analogy with the use of ΔH_f° data, the values in this table can be employed to calculate ΔG° for a reaction using the equation

$$\Delta G^\circ = \sum_{\text{products}} n\,\Delta G_f^\circ - \sum_{\text{reactants}} n\,\Delta G_f^\circ \qquad (2\text{-}24)$$

EXAMPLE 2-7. Calculate ΔG° for the reaction $C_2H_5OH(l) + O_2(g) \rightarrow CH_3COOH(l) + H_2O(l)$, using standard free of formation values from Table 2-2.

$$\Delta G_{298}^\circ = \Delta G_{f,298,CH_3COOH(l)}^\circ + \Delta G_{f,298,H_2O(l)}^\circ - \Delta G_{f,298,C_2H_5OH(l)}^\circ$$

$$\Delta G_{298}^\circ = -93{,}800 - 56{,}690 + 41{,}770$$

$$\Delta G_{298}^\circ = \underline{-108{,}720 \text{ cal}}$$

There is an additional and very useful method of obtaining ΔG° for reactions that are chemically reversible to a measureable extent, based on the equilibrium constant. This is discussed in Chapter 3.

TABLE 2-2. STANDARD FREE ENERGIES OF FORMATION AT $298\ ^\circ K^a$

Substance	ΔG_f° kcal/mole	Substance	ΔG_f° kcal/mole
$Al_2O_3(s)$	−376.8	$H_2O(g)$	−54.64
$CO(g)$	−32.81	$H_2O(l)$	−56.69
$CO_2(g)$	−94.26	$HCl(g)$	−27.77
$CO(NH_2)_2(s)$	−47.12	$NH_3(g)$	−3.98
$CH_4(g)$	−12.14	$NO(g)$	+20.72
$C_2H_4(g)$	+16.28	$NO_2(g)$	+12.39
$CH_3OH(l)$	−39.73	$N_2O_4(g)$	+23.49
$C_2H_5OH(l)$	−41.77	$NH_4Cl(s)$	−48.73
$HCHO(g)$	−26.2	cyclo-$C_6H_{12}(l)$	+6.37
$CH_3COOH(l)$	−93.8	$C_6H_6(g)$	+30.99
$HCOOH(l)$	−82.7	$SO_2(g)$	−71.79
$CCl_4(l)$	−16.4	$SiF_4(g)$	−360

^{a} From *Selected Values of Chemical Thermodynamic Properties* (Circular 500, National Bureau of Standards, 1952) and *Selected Values of Properties of Hydrocarbons* (Circular C461, National Bureau of Standards, 1947).

Having evaluated the standard free energy change, it is then necessary to make the additional calculations required to obtain the value of ΔG under the conditions of the experiment. Essentially we must calculate the additional free energy terms arising from the difference in G between the standard state and nonstandard state for each species.

If a particular reaction involves only species in their standard state at 298 °K, then ΔG and the ΔG°_{298} calculated directly from tabulated data are identical. The criterion for spontaneity then can be applied directly. In general, however, such is not the case, and the free energy change under the conditions of interest must be evaluated before attempting to utilize the ΔG value for predictions.

2-4-4 Free Energy Change in Physical Processes

Reaction conditions may require that reactants and products be in a nonstandard state at 298 °K or some other temperature. Alternatively, reactants and products may be in standard states, but at a temperature other than 298 °K. The transition from the standard state at 298 °K to the state of interest will involve one or more physical processes. Equations therefore are required to allow computation of the ΔG for these changes of state.

In Sec. 1-4-3 the standard state conditions were tabulated for ideal gases, pure liquids and solids, and solution components. The effect of pressure on G of liquids and solids is generally negligible, but equations giving ΔG for a change in gas pressure or solution composition are required.

The effect of temperature also should be considered in this context. It is usual, however, to deal with the temperature dependence of the free energy in terms of ΔG for the reaction rather than for the individual species.

The equations derived here will be useful both in the development of a general equation for ΔG and also in later discussion of physical processes whose occurrences can be rationalized easily in terms of free energy.

Pressure dependence. For an isothermal process involving only an ideal gas, the enthalpy change is zero (Sec. 1-3-4). Therefore, according to Eq. (2-18), the free energy change will be given by

$$\Delta G = -T \Delta S \quad \textit{(ideal gas at constant T)} \qquad (2\text{-}25)$$

Equation (2-15) relates ΔS and pressure for such a process. Substituting that expression for the entropy change in Eq. (2-25) gives

$$\Delta G = -T \left(2.303 \, nR \log \frac{P_1}{P_2} \right)$$

or

$$\Delta G = 2.303 \, nRT \log \frac{P_2}{P_1} \qquad (2\text{-}26)$$

For the ΔG resulting from a change in pressure from the standard state at $P_1 = 1$ atm to another state at $P_2 = P$, Eq. (2-26) becomes

$$\Delta G = G - G^\circ = 2.303 \, RT \log P$$
$$G = G^\circ + 2.303 \, RT \log P \qquad (2\text{-}27)$$

In this equation G is the free energy per mole in the nonstandard state where the pressure is P atm. Note, however, that P in Eq. (2-27) in effect is a dimensionless quantity since it is really the value of the ratio x atm/1 atm. The free energy in the state of interest, then, is a sum of G in the standard state and the free energy alteration resulting from the change in pressure. Most often we will be interested in the free energy of formation under nonstandard conditions. For this calculation we evidently can use

$$\Delta G_f = \Delta G_f^\circ + 2.303 \, RT \log P \qquad (2\text{-}28)$$

In this equation $n = 1$ by definition since we are dealing with a free energy change *per mole*. The use of this equation for compounds is quite straightforward. We add to the ΔG_f° value from Table 2-2 the correction term resulting from the pressure effect. What is being done here is to divide the process of formation into two steps: (1) the formation of the compound in its standard state and (2) a change from the standard to the nonstandard state—adding the ΔG terms to obtain the overall free energy change. The validity of such a procedure stems from G being a state function. While the value of ΔG_f° will be unique for the substance in question, the correction term is independent of the nature of the gas.

For elements it is the *standard* free energy of formation at 298 °K that is set equal to zero by convention. ΔG_f(elements) in nonstandard states therefore also must be calculated using Eq. (2-28), which in this application simplifies to $\Delta G_f = 2.303 \, RT \log P$, since $\Delta G_f^\circ = 0$.

EXAMPLE 2-8. Calculate the free energy of formation of $N_2(g)$ at 298 °K and 10 atm pressure.

$$\Delta G_{f,10\text{atm},298} = 2.303 \, RT \log P = 2.303 \, RT \log 10$$
$$\Delta G_f = 2.303 \times 1.987 \times 298 \times 1.0$$
$$\underline{\Delta G_f = 1{,}364 \text{ cal/mole}}$$

Concentration dependence. The free energy of components of a solution depends on their concentration expressed either as mole fraction X or molarity C. The standard state is defined in terms of a specific concentration, X or $C = 1.0$, and therefore we have to determine the ΔG term that arises when the concentration is changed to some other value.

We have not previously encountered any equations applicable specifically to solution components. The derivation of a whole new series of equations for solutions, however, can be neatly avoided by taking advantage of the principle discussed in Sec. 2-4-2 that the free energy of a substance is the same in any two phases in equilibrium. Every substance has a finite vapor pressure, even though it may be extremely small. Any change in the free energy of a solution component in the condensed phase will be accompanied by an identical change in the free energy of that substance in the vapor phase in equilibrium with the solution. If ideal gas behavior of the vapor is assumed and the change in vapor pressure owing to the change in concentration can be calculated, we can utilize Eq. (2-26) to calculate the ΔG term for the vapor and obtain thereby the ΔG for the component in solution.

Raoult's law as it applies to the vapor pressure of a solvent is discussed in detail in Chapter 9. For the moment we can write it in its simplest form, $P = kX$, which indicates the direct proportionality between the vapor pressure of the solvent and its mole fraction. We therefore can substitute in Eq. (2-26) the term kX for pressure to obtain

$$\Delta G = 2.303 \, nRT \, \log \frac{kX_2}{kX_1}$$

$$\Delta G = 2.303 \, nRT \, \log \frac{X_2}{X_1} \tag{2-29}$$

The ΔG is interpreted as that of the solution component, since $G_{\text{solvent }(l)} \equiv G_{\text{solvent }(g)}$. Taking X_1 as unity, the standard state, and the nonstandard state as $X_2 = X$, Eq. (2-29) written for molar quantities becomes

$$\Delta G = G - G^\circ = 2.303 \, RT \log X$$

or

$$G = G^\circ + 2.303 \, RT \log X \tag{2-30}$$

The free energy per mole in the standard state is G°, and G is the value when the mole fraction has some nonunity value X. For free energies of formation, Eq. (2-30) takes the form

$$\Delta G_f = \Delta G_f^\circ + 2.303 \, RT \log X \tag{2-31}$$

This relation also can be used for the solute if we wish to consider a solution as simply a mixture of components and to take the pure substances themselves as standard states.

For the same type of calculation for a solute, molarity can be used as the measure of concentration. By a series of similar arguments, beginning with Raoult's law* written in the form $P = k'C$, we can obtain the relations analogous to Eqs. (2-30) and (2-31);

$$G = G^\circ + 2.303 \, RT \log C \tag{2-32}$$

$$\Delta G_f = \Delta G_f^\circ + 2.303 \, RT \log C \tag{2-33}$$

---

* Written for a solute in this way, the relation is usually designated as Henry's law. Details of the distinction are discussed in Chapter 9.

where C is the concentration in moles per liter but in fact is dimensionless in this equation, having been divided by 1 mole/liter.

EXAMPLE 2-9.   Calculate the change in free energy of a pure liquid ($X = 1$) to which sufficient solute is added at 298 °K such that $X_{solvent}$ is reduced to 0.80.

$$\Delta G = 2.3\ RT \log \frac{X_2}{X_1} = 2.303\ RT \log \frac{0.80}{1}$$

$$\Delta G = 2.303 \times 1.987 \times 298 \times (-0.0969) = \underline{-132\ \text{cal/mole}}$$

It is apparent from Eqs. (2-27), (2-30), and (2-32) that the term that corrects for the free energy difference per mole between standard and non-standard state is of the same form in each case:

$$2.303\ RT \log\ (measure\ of\ concentration)$$

where the measure of concentration is pressure, mole fraction, or molarity, depending on the substance involved. It is convenient, therefore, to write a single general expression that could be used for all cases simply by sub-stitution of the appropriate concentration term.

There is a quantity called the *activity a*, that does measure concentration in a general sense. In its strict definition the activity represents exactly the *effective concentration* of a substance in a chemical reaction and is obtained by correcting the nominal concentration, for example, a pressure calculated assuming ideal gas behavior, for deviations from ideality, We are concerned later with such corrections and the general use of activities. We can make use of the concept immediately by neglecting the correction terms and equating activity with the nominal concentration—the pressure of a gas in atmospheres, the mole fraction of solvent, or the molarity of a solute. Thus, although activity is a fundamentally important concept, it is being used for the present as nothing more than a general representation of concentration. If it were not for the extensive use to be made later of the formal activity concept, we could replace a here by any symbol.

The use of activities allows us to use a single definition of the standard state of all three types of system as the *stable state at unit activity*. For example, then, unit activity for a gas would be interpreted as simply $P = 1$, since $a = P$. Since for a substance that can act as a solvent we take $a = X = 1$ (that is, the pure substance as the standard state), it follows that any *pure* liquid or solid participating in a process will invariably be at unit activity.

With the substitution $a = P$, X, or C, the three free energy equations developed previously can be combined into the general form

$$G = G° + 2.303\ RT \log a \tag{2-34}$$

and the equation for the free energy of formation will be

$$\Delta G_f = \Delta G_f° + 2.303\ RT \log a \tag{2-35}$$

In practice these two equations assume the form of those obtained for gases and solution components individually, immediately on substitution of the relevant term for *a*. The subsequent calculations are carried out as illustrated in the examples. It is advantageous, however, to be able to employ the general expression in the derivation of the equation for the change in free energy in the overall process.

2-4-5 Calculation of ΔG for a Chemical Reaction

We now have seen how the change in free energy in a reaction under standard state conditions, as well as the ΔG for the transition from this to a nonstandard state by individual species, can be calculated. The combination of these techniques is the final step in the development of the relation that can be used to determine the free energy change under conditions of interest and hence allow us to predict from thermodynamics whether a reaction can possibly occur.

The approach that is taken to this problem is illustrated schematically in Fig. 2-5. The overall calculation determines the separation of products and the reaction on the absolute free energy scale by adding to the $\Delta G°$ for the reaction the difference between the standard-to-nonstandard state ΔG terms for the components.

The general equation is most easily obtained by considering a general reaction

$$bB + dD \rightarrow mM + nN$$

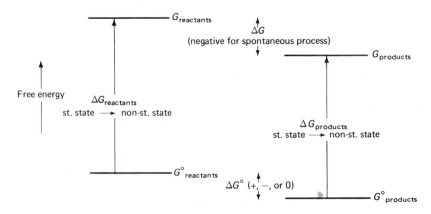

FIG. 2-5 Schematic representation of the calculation of ΔG for a chemical reaction, from $\Delta G°$ and the ΔG terms due to change from standard to non-standard states of products and reactants.

in which the free energy change is given by

$$\Delta G = mG_M + nG_N - bG_B - dG_D$$

Substitution for each free energy term from Eq. (2-34) gives

$$\Delta G = m(G_M^\circ + 2.303\ RT \log a_M) + n(G_N^\circ + 2.303\ RT \log a_N)$$
$$- b(G_B^\circ + 2.303\ RT \log a_B) - d(G_D^\circ + 2.303\ RT \log a_D)$$

where a_x is the activity of component x. This expression can be rearranged to a form that demonstrates the basic approach to the calculation, as illustrated in Fig. 2-5.

$$\Delta G = mG_M^\circ + nG_N^\circ - bG_B^\circ - dG_D^\circ + 2.303\ RT(m \log a_M + n \log a_N)$$
$$- 2.303\ RT(b \log a_B + d \log a_D)$$

The first four terms together evidently are just the free energy change in the process when all species are present in their standard states. The last two terms represent the ΔG for the standard-to-nonstandard state transition for products and reactants. Putting in explicitly the standard free energy change and rearranging the two correction terms to a more convenient form, we get

$$\Delta G = \Delta G^\circ + 2.303\ RT[\log(a_M)^m(a_N)^n - \log(a_B)^b(a_D)^d]$$

$$\Delta G = \Delta G^\circ + 2.303\ RT \log \frac{(a_M)^m(a_N)^n}{(a_B)^b(a_D)^d} \tag{2-36}$$

The ratio of activities, each raised to a power equal to the number of moles participating in the process as written, is often referred to as the *activity quotient* and is assigned the letter Q. Note very carefully that although Q has the same *form* as the general mass action law equilibrium constant the two are *not* equivalent. Q represents the activities of products and reactants in a process for which we desire a ΔG value. As noted in Chapter 3, Q and the equilibrium constant are closely related. Here we are concerned not with equilibrium but rather with viewing a given process as a transition between an initial state, the reactants, and a final state, the products. We can then calculate the difference in free energy between these states to apply the criterion for spontaneity.

Substituting Q for the ratio of activities in Eq. (2-36), we get the basic working equation for the calculation of free energy:

$$\Delta G = \Delta G^\circ + 2.303\ RT \log Q \tag{2-37}$$

It is apparent that this has the same form as the relations developed for the individual components, where the free energy change is obtained by the addition of a correction term to the value applicable to the system in the standard state.

In the application of this equation, note that the temperature substituted in the second term must be the same as that at which $\Delta G°$ is calculated; particularly in a lengthy calculation, it is sometimes possible to use inadvertently different temperature values in the two terms.

EXAMPLE 2-10. Calculate the free energy change at 298 °K in the process $2\,CO(g, 5\,atm) + O_2(g, 5\,atm) \rightarrow 2\,CO_2\,(g, 5\,atm)$.
The procedure here is to apply Eq. (2-24) to obtain $\Delta G°$ and then Eq. (2-37) to obtain ΔG.

$$\Delta G° = 2\,\Delta G°_{f,CO_2(g)} - 2\,\Delta G°_{f,CO(g)} = 2 \times (-94,300) - 2 \times (-32,800)$$
$$\Delta G° = -123,000\ cal$$
$$Q = \frac{(a_{CO_2})^2}{(a_{CO})^2(a_{O_2})} = \frac{P^2_{CO_2}}{P^2_{CO}P_{O_2}} = \frac{(5)^2}{(5)^2(5)} = 0.20$$

$$\Delta G = \Delta G° + 2.303\,RT\log Q = \Delta G° \times 2.303\,RT\log 0.20$$
$$\Delta G = -123,000 + 2.303 \times 1.987 + 298 \times (-0.699)$$
$$\Delta G = -123,000 - 953 = -123,953\ cal$$

EXAMPLE 2-11. Calculate the free energy change at 298 °K in the reaction $C_2H_5OH(l) + O_2(g, 40\,atm) \rightarrow CH_3COOH(l) + H_2O(l)$ using the $\Delta G°$ value of $-108,700$ cal obtained in Example 2-7.

$$\Delta G = \Delta G° + 2.303\,RT\log \frac{(a_{H_2O})(a_{CH_3COOH})}{(a_{C_2H_5OH})(a_{O_2})}$$

Since $a = X$ for the liquid components and they are present as pure materials, a_{H_2O}, a_{CH_3COOH}, and $a_{C_2H_5OH}$ are all unity. For oxygen, $a = P = 40$ atm. Q therefore becomes $1/P_{O_2} = 1/40$.

$$\Delta G = \Delta G° + 2.303\,RT\log 0.025$$
$$\Delta G = -108,700 - 2,182 = -110,882\ cal$$

2-4-6   Temperature Dependence of ΔG

Thus far systems at 298 °K have been dealt with exclusively. Since the free energy of individual components of the reaction mixture will vary with temperature, the overall ΔG for the process will be temperature dependent. Consequently, we must be able to evaluate the free energy change at one temperature from its value at some other temperature where data are available. Note that although the base temperature for $\Delta G°$ is normally 298 °K, we can legitimately have a reaction with all species in their standard state at any temperature. Thus the determination of the variation of either ΔG or $\Delta G°$ with temperature may be required. For example, if, we wish to calculate ΔG at 500 °K from a $\Delta G°$ value at 298 °K, we could proceed by either of two methods: (1) calculate $\Delta G°_{500}$ and then ΔG at this temperature using Eq.

(2-37) or (2) begin by evaluating ΔG_{298} through that equation and, subsequently, the temperature dependence. Since the enthalpy change is required to calculate ΔG as a function of T, the choice between the two methods usually depends on the type of ΔH information available.

The required relation is normally based on a differential equation giving the rate of change of ΔG with temperature at constant pressure. We can obtain the same result here, although in a somewhat less rigorous manner, directly from the defining equation for free energy, $\Delta G = \Delta H - T\,\Delta S$. Although this equation is applicable only to processes at constant temperature, we nevertheless can use it to obtain the temperature dependence of ΔG since the restriction is that *at each temperature considered* the process is to occur isothermally.

We designate two different temperatures by the subscripts 1 and 2 and find the relationship between ΔG_1 and ΔG_2. Equation (2-18) can be written explicitly for the process at each temperature as

$$\Delta G_1 = \Delta H_1 - T_1\,\Delta S_1 \qquad \Delta G_2 = \Delta H_2 - T_2\,\Delta S_2$$

or

$$\frac{\Delta G_1}{T_1} = \frac{\Delta H_1}{T_1} - \Delta S_1 \qquad \frac{\Delta G_2}{T_2} = \frac{\Delta H_2}{T_2} - \Delta S_2$$

Subtraction of these two equations gives

$$\frac{\Delta G_2}{T_2} - \frac{\Delta G_1}{T_1} = \frac{\Delta H_2}{T_2} - \frac{\Delta H_1}{T_1} - \Delta S_2 + \Delta S_1$$

We next assume, as is done in all derivations of this equation, that if the temperature difference is moderate the change in ΔH and ΔS can be neglected* since such variations will have such a small effect on the change in ΔG. This assumption is substantiated by actual calculations using exact values for ΔH and ΔS. If a very accurate ΔG value were required, the above equation could be used in full, providing the necessary enthalpy and entropy data were available.

The assumption of negligible variation in ΔH and ΔS values for the purposes of this particular calculation, however, must not be taken to indicate that accurate values for these functions are not required in general. Recall that the fundamental $\Delta G°$ value is calculated in the first instance from enthalpy and entropy values.

On the basis of this assumption the equation is simplified, since the entropy terms vanish and the two enthalpy terms can be replaced by a single

---

* We saw, for example, in Example 1-4 that for a ΔT of 25 °C, the $\Delta H°$ for the formation of $NH_4Cl(s)$ varied only slightly from $-42,300$ cal to $-42,184$. ΔS values behave similarly.

ΔH. The result is

$$\frac{\Delta G_2}{T_2} - \frac{\Delta G_1}{T_1} = \frac{\Delta H}{T_2} - \frac{\Delta H}{T_1}$$

which can be rearranged by simple algebra to a more convenient form:

$$\Delta G_2 = \Delta H + \frac{T_2}{T_1}(\Delta G_1 - \Delta H) \qquad (2\text{-}38)$$

Thus the only data required to calculate ΔG at any temperature is the free energy change at one other temperature and the enthalpy change in the reaction.

> EXAMPLE 2-12. Calculate the free energy change at 400 °K in the reaction $C_2H_5OH(l) + O_2(40\text{ atm}) \rightarrow CH_3COOH(l) + H_2O(l)$ using $\Delta H = -118{,}300$ cal and $\Delta G_{298}^\circ = -110{,}900$ cal.
>
> $\Delta G_{400}^\circ = \Delta H^\circ + (400/298)(\Delta G_{298}^\circ - \Delta H)$
>
> $\Delta G_{400}^\circ = -118{,}300 + 1.342(-110{,}900 + 118{,}300) = \underline{-117{,}318\text{ cal}}$

The task of developing the necessary formalism to accomplish one of the central aims of thermodynamics, the calculation of free energy change and the subsequent prediction of spontaneity or its absence on the basis of the ΔG value, is now complete. We will see in the next chapter how, for a system where we can take advantage of knowledge of equilibrium conditions, some of the calculations are simplified and some data are more easily obtained, but these applications are firmly based on the relations developed here.

Throughout the remainder of the book frequent use is made of the free energy concept in a variety of applications that interest the physical chemist in general and that have important biological implications, either direct or indirect. We have progressed sufficiently far, however, to examine here how the concepts developed can be used in connection with chemical reactions that occur in a biological environment.

2-5 APPLICATION OF THE FREE ENERGY CONCEPT TO BIOLOGICAL SYSTEMS

Nowhere perhaps is it more important to appreciate the distinction between ΔG and ΔG° than in a discussion of free energy changes in chemical reactions of biological interest. Confusion sometimes arises from the attempting to interpret the spontaneity of such processes in terms of the standard free energy change rather than ΔG under the conditions of the reaction in the actual system. As we have seen here, the latter quantity is the determining factor.

It is useful to begin the discussion by examining a number of general principles concerning the free energy that are particularly relevant to the type of process of interest in living systems.

According to Eq. (2-21) the decrease in free energy is a measure of the useful work that can be accomplished. Thus a process whose ΔG is negative has really two facets—it can proceed spontaneously and at the same time can accomplish, in theory at least, useful work. In biological systems the latter can take the form of the work of muscle contraction or the movement of substances against a concentration gradient (see Sec. 9-4-4).

If we consider a schematic reaction $A \rightarrow B$ occurring in aqueous solution, and which has a positive $\Delta G°$, there is an increase in free energy when the system goes from A in its *standard* state to B in its *standard* state. In a system containing these two species in some nonstandard state (that is, at concentrations other than 1 mole/liter) we must use Eq. (2-36) to evaluate ΔG and hence determine if the process is spontaneous.

For the example reaction we would write

$$\Delta G = \Delta G° + 2.303 \, RT \log \frac{[B]}{[A]} \qquad (2\text{-}39)$$

Let us assume that the value of $\Delta G°_{298}$ for this system is $+1000$ cal. This represents the free energy change when products and reactants, in a general process, are at concentrations of 1 mole/liter, in which case the logarithm term vanishes and $\Delta G \equiv \Delta G°$. For the particularly simple reaction being utilized in this discussion, ΔG and $\Delta G°$ will be identical in any case where $[A] = [B]$. If in a given set of circumstances B is present in a solution in relatively small amounts, for example, $[B] = 0.1$ mole/liter, while $[A]$ remains at 1.0 mole/liter, $\log [B]/[A]$ is equal to -1.0. Substitution of this value in Eq. (2-39) gives a value of -364 cal for ΔG. Thus we can see that this particular chemical conversion, exhibiting a positive $\Delta G°$ value, can be made to occur spontaneously by adjustment of the concentration. Furthermore, it is apparent that such an adjustment is accomplished quite easily.

An important implication of this kind of effect becomes apparent if we consider the nonthermodynamic parameter time. In the system $A \rightarrow B$, beginning with 0.1 mole/liter of B and 1 mole of A, the negative free energy change determined above will cause the spontaneous conversion of some A to B. After some time (determined by the rate of the reaction discussed in Chapter 6) the reaction will have increased the $[B]/[A]$ ratio to an extent such that ΔG is no longer negative and no further reaction will occur. Thus, if the conversion is to continue, there must be some mechanism by which we could reduce $[B]$ by removing this species from the system. This technique then would have the effect of constraining the ratio $[B]/[A]$ to small values and, hence, ΔG to a negative value. The maintenance of $[B]$ at low levels could be accomplished by its physical removal from the system or by adding some other species that would react spontaneously with B to yield some other product.

Despite the general precautions expressed above, it is quite possible to use $\Delta G°$ values to decide on the spontaneity of a process. As will be seen in the next chapter, $\Delta G°$ determines the distribution between reactants and products when equilibrium is attained in the system. By comparing the reactant–product distribution in a process of interest to that characteristic of equilibrium between these species, the direction in which the system must move to attain equilibrium will be apparent. Again, however, the sign of $\Delta G°$ does not indicate *directly* whether the process is spontaneous. $\Delta G°$ gives information about the equilibrium state only—the composition the system will attempt to attain. The use of $\Delta G°$ in this way is discussed in greater detail in Chapter 3.

In biological systems a number of reactions can occur concomitantly in a given "thermodynamic system." In so-called *coupled* or *tandem* reactions, the free energy change in one of the processes may be positive but the overall ΔG for the two reactions together can be negative. Consequently, the two processes may be spontaneous. Thermodynamics offers no indication of the mechanism of this coupling. The calculated changes in the free energy indicate only that the system can move from a state of higher to one of lower chemical potential when both reactions occur. One way in which this phenomenon can be rationalized is in terms of the free energy obtained from one reaction being used to do the "work" required to make the other, nonspontaneous reaction proceed.

Many important reactions in living systems involve the hydronium* ion H_3O^+ as either a reactant or a product. Since the concentration of this species is invariably restricted to a constant value by the various ionic equilibrium requirements, and since the $[H_3O^+]$ term appears in the Q ratio [as defined by Eq. (2-37)], it is common practice to quote a $\Delta G'$ rather than a $\Delta G°$ for a process. The former quantity refers to the situation where all species, except H_3O^+, are in their standard state, with the hydronium ion being at some specified low concentration. $\Delta G'$ values are quoted for the situation where $[H_3O^+] = 1 \times 10^{-7}$ (the value for pure water) at 25°C or, for physiological conditions, at 38 °C, where the concentration of the hydronium ion is slightly smaller. Care must be taken to clarify the conditions to which a particular quoted $\Delta G'$ value applies.

One other source of confusion in this general area of free energy data presentation is the use of the term "high-energy bond." From the physical chemistry point of view, characterization of chemical bonds as being high energy or low energy is usually made on the basis of the magnitude of the bond dissociation energy. This is the energy required for the homolytic

---

* The physical chemistry of the H_3O^+ ion is discussed in detail in Chapter 4.

scission of a bond, for example,

$$X:Y \rightarrow X\cdot + Y\cdot$$

that is, for the fragmentation of the molecule into two species, each having an unpaired electron. Certain biological compounds, which will be discussed presently, are said to possess high-energy bonds. This designation is made not because the bond dissociation energy as defined above is large but rather because of their participation in reactions that involve rupture of this bond and because they are characterized by a large decrease in free energy. Thus the nomenclature is based not on the energy *required* to break the bond but rather on the energy *released* when the bond is involved in a particular reaction.

In Chapter 1 the production of energy in living systems through the oxidation process was discussed. This energy is stored largely in organic phosphates, derivatives of the inorganic acid H_3PO_4. The most important of this class of compounds is adenosine-5′-triphosphate, ATP. The ionized form of this molecule is:

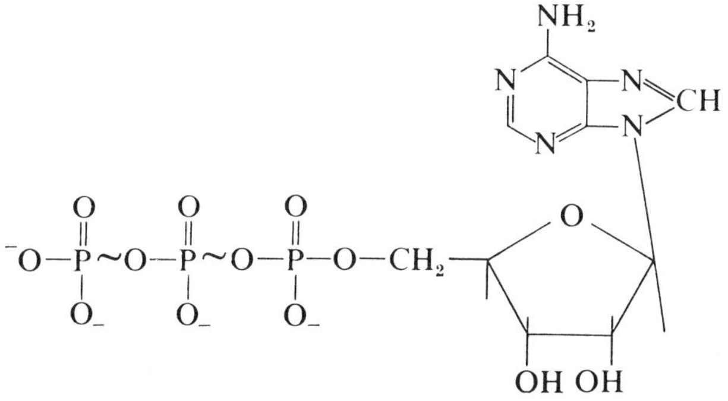

The symbol $\sim$ is used to designate the high-energy bonds in the molecule. In aqueous solution the molecule exists primarily in the ionized form indicated. Two other important phosphates are adenosine-5′-diphosphate, ADP, the molecule formed from ATP through loss of one phosphate group, and creatine phosphate;

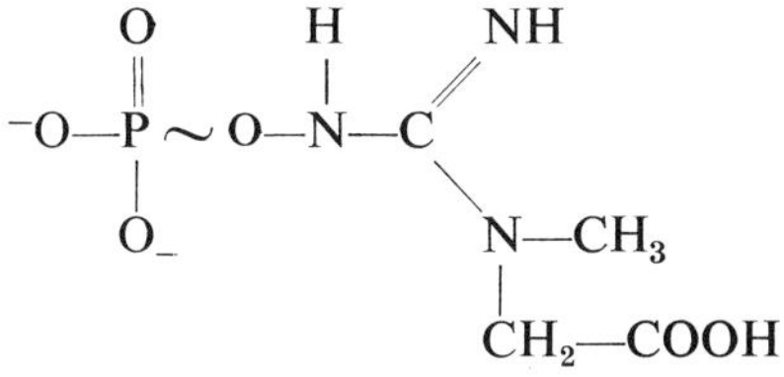

Even under standard conditions the free energy change in the hydrolysis of these compounds is quite negative. $\Delta G'$ (physiological conditions) for the hydrolysis of ATP,

$$ATP^{4-} + H_2O \rightarrow ADP^{3-} + P^{2-} + H^+$$

is $-7,400$ cal. This process, in conjunction with

Creatine phosphate^{2-} + ADP^{3-} + H$^+$ $\rightleftharpoons$ ATP^{4-} + creatine

which has $\Delta G' = -2,800$ cal, has been demonstrated to play a vital role in muscle contraction. In a muscle at rest, accumulation of ATP leads to a negative ΔG for the reverse of the reaction involving creatine. When contraction occurs the useful work is obtained by utilizing the free energy decrease in the hydrolysis of ATP to ADP. The decreased concentration of ATP, however, is immediately counteracted by its formation in the spontaneous reaction between creatine phosphate and ADP. In this way the energy required for the performance of work in effect is stored in the creatine phosphate.

A number of factors are involved in the large negative free energy decrease associated with some reactions of molecules containing one or more phosphate groups. The most important of these involves the formation of a particularly stable product. If there can be resonance stabilization of the product, as for example in the formation of the CH$_3$COO$^-$ ion through hydrolysis of acetyl phosphate, the free energy of the product is lowered and therefore the negative ΔG for the reaction is enhanced.

Phosphates also are involved in tandem reactions. For example, in the sequence

Glucose-6-phosphate $\xrightarrow{\text{I}}$ fructose-6-phosphate + ATP

$$\Bigg\updownarrow \text{II}$$

fructose-1,6-diphosphate + ADP

process I has a positive standard free energy change whereas that of II is negative. Thus, by utilization of the "high-energy bonds" in ATP, I and II are coupled with the net effect that reaction of glucose-6-phosphate is spontaneous.

PROBLEMS

2-1. Calculate the entropy change in the overall process described in Problem 1-11.

2-2. The ΔS° value for the reaction

$$C_6H_6(l) + 7.5\ O_2(g) \rightarrow 6\ CO_2(g) + 3\ H_2O(l)$$

is -51.96 cal/deg. Using *only* this information and the following absolute entropy values at 298 °K, in cal/deg mole, determine S° for benzene at 298 °K: $O_2(g)$, 49.00; $CO_2(g)$, 51.06; and $H_2O(l)$, 16.72.

2-3. Calculate the total change in entropy when 8.00 gm of O_2 is heated at constant pressure in the solid state from -229 to $-218.6\,°C$, the melting point, where it is converted completely to liquid. ($C_{p,O_2(s)} = 11.2\ \text{cal/deg mole}$; $\Delta H_{\text{fusion}} = 106.3\ \text{cal/mole.}$)

2-4. 4.00 gm of oxygen, initially at $100\,°C$, is expanded isothermally from 1 to 10 liters. Assuming ideal gas behavior, calculate the maximum work obtainable, the heat absorbed, and the value of ΔH, ΔS, and ΔG if the process is carried out reversibly.

2-5. Use the data given in Table 2-1 to calculate the entropy of vaporization of water at $298\,°K$.

2-6. Evaluate ΔG°_{298} in the following reactions from standard free energies of formation:
(a) $CH_4(g) + \frac{1}{2}\,O_2(g) \rightarrow CH_3OH(l)$
(b) $C_2H_5OH(l) \rightarrow CH_4(g) + HCHO(g)$
(c) $2\,Cl_2(g) + CH_4(g) \rightarrow 2\,H_2(g) + CCl_4(l)$

2-7. Calculate ΔG°_{298} for the reaction

$$CO_2(g) + 2\,NH_3(g) \rightarrow CO(NH_2)_2(s) + H_2O(l)$$

using the value of ΔH°_{298} determined in Problem 1-9 and ΔS°_{298} calculated from absolute entropies.

2-8. Calculate the free energy of formation at $298\,°K$ of $C_2H_5OH(l)$ in a solution in which there is a molecular solute of molecular weight 345 at a concentration of 100 gm/100 gm ethanol.

2-9. Calculate the free energy of formation of gaseous oxygen at $298\,°K$ at a pressure of (a) 10^{-3} atm and (b) 1.00 torr.

2-10. Calculate the change in molar free energy of glucose when 500 ml of water is added to 2 liters of a 1-molar aqueous glucose solution at $300\,°K$.

2-11. Write algebraic expressions for the activity quotient for each of the following reactions and then rewrite the expressions making the appropriate substitutions for the activity of each substance:
(a) $N_2(g) + 3\,H_2(g) \rightarrow 2\,NH_3(g)$
(b) $2\,NOBr(g) \rightarrow 2\,NO(g) + Br_2(g)$
(c) $S(s) + 2\,CO(g) \rightarrow SO_2(g) + 2\,C(s)$
(d) $NH_4Cl(s) \rightarrow NH_3(g) + HCl(g)$

2-12. Calculate ΔG°_{298} for the process

$$CO_2(g) + 2\,NH_3(g) \rightarrow CO(NH_2)_2(s) + H_2O(l)$$

using the data obtained in Problem 2-7.

2-13. Find ΔG for the process at 298,

$$CH_4(g) + \frac{1}{2}\,O_2(g) \rightarrow HCHO(g) + H_2(g)$$

when the pressure of each species is 1.00 torr.

2-14. ΔG_{298}° for $HCl(g, y \text{ atm}) + NH_3(g, y \text{ atm}) \rightarrow NH_4Cl(s)$ is $-22,980$ cal. Calculate the value of y (the pressure of HCl and NH_3) that would be associated with a free energy change of $-24,000$ cal for this process.

2-15. There is a thermodynamic state function, not discussed in this book, known as the *Helmholtz free energy* or *work content* and defined by the equation $A = E - TS$. Derive an expression relating the change in A to the change in the (Gibbs) free energy G. Use this relation to calculate ΔA_{298}° for the reaction $2 CO_2(g) \rightarrow 2 CO(g) + O_2(g)$.

2-16. Using only equations discussed in this book, derive an expression through which the entropy change in a reaction at T_2 can be calculated if ΔS at one temperature T_1 is known along with the heat capacities of all products and reactants. Take as the reaction $cC \rightarrow dD$, and begin by drawing a schematic diagram illustrating the entropy changes involved.

Chemical Equilibrium
in Molecular Systems

3

3-1 INTRODUCTION

Many chemical reactions are known to occur to a detectable extent in both directions. It is possible to observe experimentally in this kind of system both the "normal" conversion of reactants to products and the reverse process. Such systems attain a state of *dynamic equilibrium*. This is a dynamic state in which there is no cessation of reaction, but rather the composition of the system undergoes no further alteration unless there is a change in conditions. When a particular system, for example, $A + B \rightleftharpoons C + D$, comes to equilibrium at some temperature, the ratio of the concentrations of products to those of reactants has a unique value. This number is the value of the *equilibrium constant K*. Regardless of the proportions of species present initially, the same numerical value of K will characterize the system at that temperature.

If a chemical reaction reaches what appears to be an equilibrium state, we can use this property to determine if true equilibrium is attained. The reaction can be reinvestigated by beginning with several different sets of initial concentrations at the same temperature. When no further changes in composition can be observed, the ratio of product to reactant concentrations should be the same in every case. The concentration of individual reactants and products may vary—it is the value of their *ratio* that is fixed at equilibrium.

From the viewpoint of the classical *Law of Mass Action*, the equilibrium state can be envisaged as the situation where the rates of the overall forward and reverse reactions are equal. Both processes continue to occur, but without further alteration in the gross composition of the system.

Subsequent discussions in this chapter one confined to molecular systems in the illustrations used. In such cases very satisfactory results can be obtained by using nominal concentrations—pressure, mole fraction, or molarity. For ionic systems, on the other hand, equating activity with nominal concentration is frequently invalid, and the correction terms for nonideal behavior become significant. The relation between thermodynamic properties and the equilibrium constant to be developed here is perfectly general. Its application to ionic systems is dealt with in Chapter 4.

The general nature of the equilibrium constant can be described with reference to two simple examples. Consider first the reaction

$$C_2H_5OH(l) + CH_3COOH(l) \rightleftharpoons CH_3COOC_2H_5(l) + H_2O(l)$$

which can proceed to a detectable extent in both forward and reverse directions, as indicated by the notation $\rightleftharpoons$. When equilibrium is established at any particular temperature, the ratio of product to reactant concentrations assumes a value given by

$$K_c = \frac{[CH_3COOC_2H_5][H_2O]}{[C_2H_5OH][CH_3COOH]}$$

The brackets are used conventionally to represent concentrations in moles per liter, and the subscript c indicates that the numerical value of K_c quoted is for the case of molar concentrations. The actual value of K_c for this system at room temperature is about 4.0. The experimental facts are that regardless of the initial concentrations of these four species introduced into the reaction vessel, in an inert solvent, for example, at equilibrium the concentration ratio in the K_c defining equation will have a value of 4.0. The concentrations of the *individual* species may be different in every case. Consequently, all the situations listed in Table 3-1 represent possible equilibrium compositions since they all satisfy the condition that $[CH_3COOC_2H_5][H_2O]/[C_2H_5OH]$-$[CH_3COOH] = 4.0$.

TABLE 3-1. SOME POSSIBLE EQUILIBRIUM COMPOSITIONS IN THE ESTERIFICATION OF ACETIC ACID

Equilibrium concentrations, moles/liter				Calculated
$[C_2H_5OH]$	$[CH_3COOH]$	$[CH_3COOC_2H_5]$	$[H_2O]$	Value of K_c
0.1	0.1	0.1	0.4	4.0
0.05	0.02	0.8	0.05	4.0
0.375	0.173	0.346	0.742	4.0
3.50×10^{-3}	2.67×10^{-2}	6.11×10^{-3}	6.11×10^{-3}	4.0

For the gas phase reaction

$$2\,SO_2 + O_2 \rightleftharpoons 2\,SO_3$$

at equilibrium, $K_p = (P_{SO_3})^2/(P_{SO_2})^2(P_{O_2})$. The equilibrium constant is specified as being applicable to concentrations expressed as gas pressure in atmospheres through the subscript p. At equilibrium the partial pressure of each gas will be such that when the values are substituted into the equation, the ratio will have the value of K_p for this system at the applicable temperature.

3-2 CALCULATIONS INVOLVING THE EQUILIBRIUM CONSTANT

It will be seen later how equilibrium data can be used and how the characteristics of a system at equilibrium can be related to thermodynamic functions. As a preliminary, however, it is useful to examine the techniques involved in treating experimental equilibrium data. We will see how the value of K can be calculated and how this value can be employed to determine concentrations at equilibrium.

Although the general features of such calculations are probably familiar, it is worthwhile to review them here, emphasizing a number of significant points of particular relevance to later discussions.

The expression for the equilibrium constant can be written most conveniently for a general, chemically reversible, and homogeneous (occurring in a single phase) reaction,

$$bB + dD \rightleftharpoons mM + nN$$

as either

$$K_c = \frac{[M]^m[N]^n}{[B]^b\,[D]^d} \tag{3-1}$$

or

$$K_p = \frac{(P_M)^m(P_N)^n}{(P_B)^b(P_D)^d} \tag{3-2}$$

for use with concentrations in moles per liter or partial pressures in atmospheres for a gas phase reaction, respectively.

For a heterogeneous equilibrium—one which occurs in two phases, for example, at a gas–solid interface—the concentration of any pure species in the condensed phase is constant* and by convention is included in the K value. Thus for a gas–solid reaction such as $bB(s) + dD(g) \rightleftharpoons mM(s) + nN(g)$, the equilibrium constant would be simply $K = (P_N)^n/(P_D)^d$.

---

* Strictly speaking, the fundamental equations are set up in terms of activities, in which case the activities of pure solids or liquids are unity. The resulting equation is the same, however, as when these concentrations are simply assumed to be constant, and we need not make use of the formal activity concept until later.

At a given temperature the system will have a unique value of K_c or K_p, whose units depend on the stoichiometry. If the four coefficients in the general reaction are all unity, then K_c and K_p both would be dimensionless. In addition, for this special case they would have the same numerical value since the factor for conversion from atmospheres to moles per liter to be applied to each term would cancel. In any other case where the coefficients are not unity, or where there is a change in the number of moles in going from reactants to products, K_c and K_p will have different numerical values, even though they describe the same equilibrium state. For a process such as $B + D \rightleftharpoons M$, substitution of concentrations in moles per liter in the equilibrium expression shows that K_c has units of liters per mole, while K_p, with gas pressures in atmospheres, has units of reciprocal atmospheres.

It is frequently the practice to quote equilibrium constant values without units. This usually does not create any difficulty since in most problems the units of K do not enter into the calculation directly. It must be specified, however, whether the equilibrium constant is K_c or K_p.

Two types of calculation involving equilibrium constants are encountered: the calculation of K from experimentally determined concentrations at equilibrium and the somewhat more complex calculation of equilibrium composition from a known K value. For the latter determination we also must have some information about the concentration of at least some of the components either initially or at equilibrium. With this information and the stoichiometry of the process, an equation can be set up and solved for the unknown equilibrium concentration. This type of evaluation will be of primary importance when we see later how the equilibrium constant can be obtained from thermodynamic data. We will then be in a position to proceed via calculation alone from tabulated functions to composition at equilibrium.

The techniques involved in these two general problems are illustrated in the following examples. There are many variations, but all problems have the same essential features that require application of the same general methods of attack.

EXAMPLE 3-1. At a certain temperature the partial pressure of gases at equilibrium via $N_2(g) + 3 H_2(g) \rightleftharpoons 2 NH_3(g)$ are $P_{N_2} = 6.84$ atm, $P_{H_2} = 4.12$ atm, and $P_{NH_3} = 7.25$ atm. Calculate the equilibrium constant K_p.

$$K_p = \frac{(P_{NH_3})^2}{(P_{H_2})^3(P_{N_2})} = \frac{(7.25)^2}{(4.12)^3(6.84)}$$

$$K_p = \underline{0.110}$$

EXAMPLE 3-2. The equilibrium constant K_c for the mutarotation of glucose in aqueous solution, α-d-glucose $\rightleftharpoons \beta$-d-glucose, a transformation in structure giving rise to changes in optical properties, is 1.75. If the initial concentration of the α-form is 0.01 moles/liter, calculate the concentration of

β-d-glucose at equilibrium at the temperature at which K_c was determined. Let X equal moles per liter of β-d-glucose at equilibrium:

$$\alpha\text{-}d\text{-glucose} \quad \rightleftharpoons \quad \beta\text{-}d\text{-glucose}$$

	α-d-glucose	β-d-glucose
Initial concentration	0.01	0.00
Equilibrium concentration	0.01 $-$ X	X

Since the stoichiometry indicates a 1:1 mole ratio between product and reactant formed and consumed, if X moles/liter of the β-form are produced to attain equilibrium, the same concentration of reactant therefore must be consumed, and (0.01 $-$ X) moles/liter remain.

$$K_c = \frac{X}{0.01 - X} = 1.75$$
$$X = (0.01 - X)1.75 = \underline{0.00658 \text{ moles/liter}}$$

3-3 EQUILIBRIUM AND FREE ENERGY

In all processes considered in the discussion of thermodynamics, we invariably considered state-to-state transitions only in one direction. The change in the various state functions was calculated as the system was altered from an initial to a final state—from reactants to products.

To apply thermodynamics to chemical reactions that can occur to appreciable extents in both forward and reverse directions, the treatment is carried one step further. We examine how the thermodynamic functions change as the system goes from the "final" state *to* the "initial" state. If a value of $\Delta H = +10,000$ cal has been obtained for a system undergoing the process $state_1 \rightarrow state_2$, it follows directly that for the transition $state_2 \rightarrow state_1$, $\Delta H = -10,000$ cal. The enthalpy change just indicates the separation of the two states on the absolute enthalpy scale.

We can utilize an approach of this kind to the relationship between equilibrium and free energy. A reaction R $\rightarrow$ P with a positive ΔG under some specified conditions is thermodynamically nonspontaneous. If the system exhibits this property it must mean that the reverse process P $\rightarrow$ R has a negative free energy change and is therefore spontaneous. Thus if the reaction is chemically reversible to an observable extent, a positive ΔG indicates that the process will occur in reverse, *not* that "nothing will happen." This is illustrated in Fig. 3-1.

A nonzero value for ΔG thus indicates the possibility that the system can attain a state of lower free energy through the formation of more reactants or more products, as the case may be. Considering the free energy decrease as the driving force for chemical reactions, the system will move in a direction consistent with this tendency. There will be the "normal" reactant

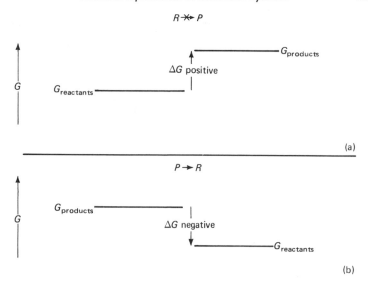

FIG. 3-1 *Schematic representation of free energy changes in the process* $R \to P$. *If* ΔG *for* $R \to P$ *is positive* (a), *the process is thermodynamically impossible, but the reverse process,* $P \to R$ *as shown in* (b), *has a negative* ΔG *and is therefore spontaneous.*

to product conversion or the opposite reaction. However, if the free energy difference is or becomes zero, there is no thermodynamic incentive for further change—no driving force for either the forward or reverse reactions—and consequently the system has reached equilibrium. Thus we arrive at the same conclusion reached in Sec. 2-4-2 on the basis of heat effects; A system that has $\Delta G = 0$ is in an equilibrium state.

Combination of this criterion with the relations developed earlier permits determination of the quantitative dependence of equilibrium conditions on free energy.

3-3-1   Relation Between $\Delta G°$ and K

The equation obtained previously,

$$\Delta G = \Delta G° + 2.303 \, RT \log Q \tag{2-37}$$

gives the free energy change in a system—the difference between $G_{products}$ and $G_{reactants}$—when the components of the system are present at activities whose ratio is given by Q. Under conditions of equilibrium, where $\Delta G = 0$, the equation becomes

$$0 = \Delta G° + 2.303 \, RT \log Q_{eq}$$
$$\Delta G° = -2.303 \, RT \log Q_{eq} \tag{3-3}$$

Since $\Delta G°$ has a fixed value at any one temperature, there can be only one value of Q that satisfies this condition; this has been specified in Eq. (3-3) as Q_{eq}. Recall that for a general process, which we now write as being chemically reversible, $bB + dD \rightleftharpoons mM + nN$, Q takes the form $(a_M)^m(a_N)^n/(a_B)^b(a_D)^d$. When substitution of appropriate concentration terms for activities is made, it is apparent that this is exactly the same expression as that of the equilibrium constant. We therefore can identify the value of Q at equilibrium with K:

$$K = Q_{eq} = \left[\frac{(a_M)^m(a_N)^n}{(a_B)^b(a_D)^d}\right]_{eq} \tag{3-4}$$

From the thermodynamic point of view the equilibrium constant is really the particular value of the activity quotient Q that results when $\Delta G = 0$. K therefore indicates the conditions under which the equilibrium state is reached at the temperature of interest.

Substitution of $K = Q$ in Eq. (3-3) yields one of the most important relations in thermodynamics;

$$\Delta G°_{T\,P} = -2.303\ RT \log K \tag{3-5}$$

The subscripts P and T have been put in as a reminder that the equations employed here are applicable to a reaction at constant temperature and pressure. The expression allows the calculation of the equilibrium constant from thermodynamic data—in particular, from the standard free energy change—or vice versa. It represents the fundamental connection between the experimentally measurable concentration of species in a real chemical process and the thermodynamic functions of the system.

Since both Q and K are represented by identical formal expressions, it is important to distinguish carefully between them:

Q = *RATIO OF ACTIVITIES OF PRODUCTS TO ACTIVITIES OF REACTANTS UNDER CONDITIONS OF INTEREST.*
K = *SPECIFIC VALUE OF Q WHEN THE SYSTEM IS IN EQUILIBRIUM AT A SPECIFIED TEMPERATURE.*

In molecular systems we invariably use nominal concentrations rather than activities. The expression for Q obtained by such substitution is usually designated, as for equilibrium constants as seen earlier, but with the subscripts p, x, and c for concentrations in pressure, mole fractions, and molarities. For the general purposes being discussed here, then, there are three alternative expressions:

$$Q_p = \frac{(P_M)^m(P_N)^n}{(P_B)^b(P_D)^d} \qquad Q_c = \frac{[M]^m[H]^n}{[B]^b[D]^d} \qquad Q_x = \frac{(X_M)^m(X_N)^n}{(X_B)^b(X_D)^d}$$

At a given temperature Q will have a specific numerical value for the system at equilibrium, K_p, K_c, or K_x.

We can set up Eqs. (2-37), (3-4), or (3-5) using whichever of these forms is convenient, but care must be taken to ensure that the standard states chosen to calculate $\Delta G°$ are consistent. For example, if we are using Q_p and therefore gas pressures in atmospheres, the standard free energies of formation employed to obtain $\Delta G°$ must be those in the gas phase at 1 atm. Particular attention to this requirement is necessary for reactions in solution where various standard state definitions may have been used for tabulated $\Delta G_f°$ values.

An instructive overall view of the changes in system composition as equilibrium is approached can be obtained in the following way. Substitution of the relation between $\Delta G°$ and K, Eq. (3-5), into the general expression Eq. (2-37) gives

$$\Delta G = -2.303 \, RT \log K + 2.303 \, RT \log Q \qquad (3-6)$$

The sign of ΔG at a particular temperature depends on the relative magnitude of Q and K. We can imagine two different nonequilibrium situations, that is, where K and Q are not equal.

If K is larger than Q, ΔG is negative and the process occurs spontaneously from left to right.* The physical significance of this situation is that the relative concentrations of product species present are smaller than at equilibrium. As the reaction proceeds, larger amounts of products are formed and the value of Q increases until $Q = K$ and equilibrium is established. In this sense Q is regarded as the instantaneous concentration ratio at any point during the approach to equilibrium.

A value of K less than that of Q is the second situation. Equation (3-6) indicates that here ΔG is positive, and hence the *reverse* reaction will be thermodynamically spontaneous. There consequently will be conversion of some products with a concomitant reduction in the value of Q. This continues until the equilibrium state is attained when $Q = K$.

To illustrate these points more clearly, consider a simple process $A \rightarrow B$ at 298 °K, and take $K_c = 25$. For this system Eq. (3-6) can be written

$$\Delta G = 2.303 \, RT \log \frac{Q}{K} = 1,364 \log \frac{Q}{25}$$

In the following tabulation free energy change values have been calculated using this expression for a number of representative Q values:

$Q =$	250	100	50	25	10	5	2.5
ΔG (cal) $=$	$+1364$	$+821$	$+411$	0	-543	-953	-1364

* The change in thermodynamic functions is always calculated for the process as written proceeding from left to right, reactants to products. The ΔG obtained from Eq. (3-6) will be the free energy difference $G_{products} - G_{reactants}$, regardless of whether the process occurs in the indicated direction.

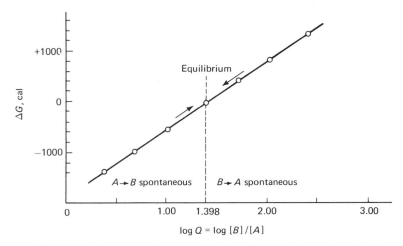

FIG. 3-2 The free energy change as a function of log Q for the reaction A → B with $K_c = 25$.

If ΔG is plotted as a function of log Q, we obtain the straight line graph given in Fig. 3-2. The point at log $Q = 1.398$ ($Q = 25$) represents the equilibrium state, where [B]/[A] = 25. For any case with a smaller relative concentration of B, which would be represented by points to the left of the equilibrium log Q value on the graph, the process A → B is spontaneous. Log Q values to the right of this point represent situations where the conversion B → A has a negative free energy change.

In each case the initial concentrations are such that Q and K are not identical, and hence either the forward or reverse reaction represents a path to lower free energy. The system accordingly moves in this direction with Q approaching its value at equilibrium and attains its lowest possible free energy at that temperature.

Chemical equilibrium is a dynamic situation. Reaction in both directions continues to occur in the equilibrium state. There is no inconsistency here, however, since the requirement of thermodynamics is solely that the concentrations remain constant once equilibrium is reached. It is irrelevant to the thermodynamic argument whether this is accomplished through equal rates of consumption and formation of the species present or simply because there is no further reaction. The former alternative, of course, is what actually happens.

One other important implication of Eq. (3-5) is that we can extract from it an indication of the physical significance of the *standard* free energy change.

It has been emphasized that the criterion for spontaneity is the value of ΔG, *not* $\Delta G°$. In addition, we also saw that the process to which $\Delta G°$ refers is often hypothetical. The fact that there is a quantitative relation between ΔG and K allows us not only to calculate the value of one from the other but also to establish a qualitative relation between the magnitude of $\Delta G°$ and the general characteristics of the system. From Eq. (3-5), if $\Delta G°$ is positive, log K must be negative and $K < 1$. Although quantitative interpretation of the magnitude of an equilibrium constant depends on the algebraic form of the activity ratio, a K value of less than 1 indicates that at equilibrium reactants predominate. The extent of the disproportion is determined by how small is the K value. Conversely, if $\Delta G°$ is negative, log K is positive and, therefore, K must be larger than unity. This indicates predominance of products at equilibrium. The case of $\Delta G° = 0$ has no particular significance other than indicating a K value of exactly 1.00, and in simple systems, for example, $A \rightarrow B$, identical concentrations of reactant product at equilibrium. Thus the significance of the $\Delta G°$ value is that at the temperature to which it applies it gives an indication as to whether products or reactants will be favored in the equilibrium state. It cannot be used directly to predict whether any process at nonstandard conditions will occur in one direction or the other. If the distinction between ΔG and $\Delta G°$ is not understood, quite erroneous conclusions about the thermodynamic spontaneity of reactions can be reached.

It is worthwhile to review here the different roles of these two important quantities as summarized in Table 3-2.

Although $\Delta G°$ is a fundamental characteristic of a process, we should take care not to conclude immediately that if $\Delta G°$ is fairly large and positive we would never obtain significant reactant to product conversion. From the tabulation above, it is apparent that in such circumstances at equilibrium the amount of product normally will be relatively small. As discussed in Sec. 2-5, however, the equation

$$\Delta G = \Delta G° + 2.303 \, RT \log Q \qquad (2\text{-}37)$$

TABLE 3-2. ΔG—THE FREE ENERGY CHANGE IN A PROCESS
UNDER CONDITIONS OF INTEREST

Magnitude of ΔG	*Significance*
$\Delta G < 0$	Process is thermodynamically spontaneous
$\Delta G > 0$	Process is thermodynamically nonspontaneous
$\Delta G = 0$	System is in equilibrium

$\Delta G°$—THE FREE ENERGY CHANGE IN A PROCESS WITH ALL SPECIES IN
STANDARD STATES

Magnitude of $\Delta G°$	*Significance*
$\Delta G° < 0$	$K > 1$—Products predominate at equilibrium
$\Delta G° > 0$	$K < 1$—Reactants predominate at equilibrium
$\Delta G° = 0$	$K = 1$—Equal distribution among products and reactants at equilibrium

clearly indicates that regardless of how large a positive $\Delta G°$ a system may exhibit, a negative value of ΔG always can be obtained. This can be accomplished by selection of a set of concentrations that yield a sufficiently large negative log Q term, although such a proposed process may sometimes be of theroretical interest only.

The central point is that the standard free energy changes are certainly indicative in a general way of whether a given reaction will occur to a significant extent as the system approaches equilibrium. Nevertheless, we must rely on a quantitative calculation of ΔG to make a thermodynamically definitive prediction in a particular set of circumstances.

3-3-2 Calculations Involving $\Delta G°$ and K

A number of useful types of calculation can be made by utilizing Eq. (3-5) relating $\Delta G°$ and the equilibrium constant. The following examples are representative of the general kind of information that can be obtained.

EXAMPLE 3-3.   In the enzyme-catalyzed biological conversion in aqueous solution D-mannose $\rightleftharpoons$ D-fructose, the equilibrium constant at 30 °C is 2.45. Calculate the standard free energy change in the process at this temperature.

$$\Delta G° = -2.303\ RT \log K = -2.303\ RT \log 2.45$$

$$\Delta G = -2.303 \times 1.987 \times 303 \times 0.3892$$

$$\Delta G° = \underline{-539\ \text{cal}}$$

EXAMPLE 3-4.   The free energy change at 298 °K in the process

$$2\ NO_2(g,\ 5\ \text{atm}) \rightleftharpoons N_2O_4(g,\ 5\ \text{atm})$$

is $-2{,}242$ cal. Calculate the equilibrium constant at this temperature.

$$\Delta G = \Delta G° + 2.303\ RT \log Q$$

$$\log Q = \log \frac{P_{N_2O_4}}{(P_{NO})^2} = \log \frac{5}{25} = -0.6990$$

$$-2{,}242 = \Delta G° + 2.303 \times 1.987 \times 298 \times (-0.6990)$$

$$\Delta G° = -2{,}242 + 953 = -1{,}289\ \text{cal}$$

$$\log K = \frac{-\Delta G°}{2.303\ RT} = \frac{+1{,}289}{2.303 \times 1.987 \times 298} = 0.9450$$

$$K = \underline{8.81}$$

EXAMPLE 3-5.   An organic compound exists in two isomeric forms, I and II. $\Delta G°$ at 298 for the isomerization I $\rightarrow$ II is $+500$ cal. Calculate the percent of form I in a mixture of the two isomers in equilibrium at 298 °K.

$$\log K = \frac{-500}{2.303 \times 1.987 \times 298} = -0.367$$

$$K = 0.429 = \frac{[\text{II}]}{[\text{I}]}$$

$$\text{Percent I} = \frac{[\text{I}]}{[\text{I}] + 0.429[\text{I}]} \times 100 = \frac{100}{1.429} = \underline{70 \text{ percent}}$$

EXAMPLE 3-6. In biological systems the equilibrium between thiols, RSH, and disulfides, RSSR, is of great importance since such compounds possess the ability to protect living systems against ionizing radiation. In an investigation* of the exchange equilibrium XSH + CSSC $\rightleftharpoons$ XSSC + SCH, under essentially physiological conditions in aqueous solution at 37 °C, when XSH = cysteine and CSSC = N,N'-diacetylcystamine, an equilibrium constant value of $K_c = 5.0$ was determined. Calculate the standard free energy change for the process. If an aqueous solution initially has 1.0×10^{-3} moles/liter each of XSH and CSSC and 2.0×10^{-3} moles/liter of products XSSC and CSH, what is the value of ΔG, in what direction will net reaction occur, and what will be the equilibrium concentrations?

$$\Delta G_{310} = -2.303 \, RT \log K \qquad \log K = \log 5.0 = 0.699$$

$$\Delta G_{310} = -2.303 \times 1.987 \times 3.10 \times 0.699 = \underline{-993 \text{ cal}}$$

$$Q = \frac{[\text{XSSC}][\text{CSH}]}{[\text{CSSC}][\text{XSH}]} = \frac{(2.0 \times 10^{-3})^2}{(1.0 \times 10^{-3})^2} = 4.0 \qquad \log Q = 0.602$$

$$\Delta G = \Delta G° + 2.303 \, RT \log Q$$

$$\Delta G = -993 + 2.303 \times 1.987 \times 310 \times 0.602$$

$$\Delta G = \underline{-138 \text{ cal}}$$

Therefore, net reaction proceeds from left to right.

Let x moles/liter be the additional amount of each product formed when the system comes to equilibrium. We thus have

	XSH	CSSC	XSSC	CSH
Initial concentrations	0.001	0.001	0.002	0.002
Equilibrium concentrations	$0.001 - x$	$0.001 - x$	$0.002 + x$	$0.002 + x$

$$K = \frac{[\text{XSSC}][\text{CSH}]}{[\text{XSH}][\text{CSSC}]} = 5.0 = \frac{(0.002 + x)^2}{(0.001 - x)^2}$$

In this particular case the necessity of solving a quadratic equation can be avoided by taking square roots of both sides of the equation to obtain

$$\frac{0.002 + x}{0.001 - x} = 2.23 \qquad x = 7 \times 10^{-5} \text{ mole/liter}$$

---

* L. Eldjarn and A. Phil, *J. Am. Chem. Soc.*, **79**, 4589 (1957).

Equilibrium concentrations:

$$[XSH] = [CSSC] = 0.001 - x = 9.3 \times 10^{-4} \text{ mole/liter}$$
$$[CSH] = [XSSC] = 0.002 + x = 2.07 \times 10^{-3} \text{ mole/liter}$$

3-3-3 Temperature Dependence of the Equilibrium Constant

In Sec. 2-4-6 the relation, Eq. (2-38), for the temperature dependence of the free energy change—standard or nonstandard—was developed. Since we have now seen how $\Delta G°$ and K are related, we can use virtually the same approach here to obtain an expression that allows calculation of K as a function of temperature.

Again the conventional derivation is based on differential calculus, but we can proceed satisfactorily as before, using algebra only.

To obtain the desired equation two fundamental equations for the standard free energy change can be utilized:

$$\Delta G° = -2.303 \, RT \log K \qquad (3\text{-}5)$$

$$\Delta G° = \Delta H° - T \, \Delta S° \qquad (2\text{-}23)$$

Combination of these two relations gives, after rearrangement,

$$2.303 \log K = -\frac{\Delta H°}{RT} + \frac{\Delta S°}{R} \qquad (3\text{-}7)$$

Writing this expression explicitly for two different temperatures T_1 and T_2 and subtracting the resulting equations, we get

$$2.303 \log K_1 = -\frac{\Delta H_1°}{RT_1} + \frac{\Delta S_1°}{R} \qquad 2.303 \log K_2 = -\frac{\Delta H_2°}{RT_2} + \frac{\Delta S_2°}{R}$$

$$2.303(\log K_2 - \log K_1) = +\frac{\Delta H_1°}{RT_1} - \frac{\Delta H_2°}{RT_2} - \frac{\Delta S_1°}{R} + \frac{\Delta S_2°}{R}$$

We again can simplify the relation by assuming that the variation of both $\Delta H°$ and $\Delta S°$ with temperature has a negligible effect on the K value. This assumption is usually valid for most systems, as we have seen, providing the temperature interval is not large. On this basis $\Delta H_1° = \Delta H_2° = \Delta H°$ and $\Delta S_1° = \Delta S_2° = \Delta S°$ and, therefore,

$$\log \frac{K_2}{K_1} = -\frac{\Delta H°}{2.303 \, R}\left[\frac{1}{T_2} - \frac{1}{T_1}\right] \qquad (3\text{-}8)$$

This important and versatile equation gives basically the temperature dependence of the equilibrium constant but has a number of other quite useful applications. It is known as the *van't Hoff equation*. If the value for the enthalpy change in the process is available, we can calculate K at a temperature of interest from a known value at one other temperature. Another application is to the case where the equilibrium constant is known at two or more temperatures. Here the data can be used in Eq. (3-8) to determine $\Delta H°$ for the process. The equation is equally well applicable to physical equilibria, and we will make use of it in this way later. Equations (3-5) and (3-8) together can be used to calculate the variation of the standard free energy change with temperature, thus providing an alternative to Eq. (2-38).

It will be noted that for the most frequently encountered kinds of thermodynamic calculations a number of somewhat different approaches can be made. For the fundamental properties of a system, it is very useful to have several relations available to be used under different circumstances where various combinations of data may be available.

EXAMPLE 3-7. (a) From the $\Delta G_f°$ and $\Delta H_f°$ values (Tables 1-2 and 2-2), for ammonia, calculate the following for the reaction $N_2(g) + 3 H_2(g) \rightleftharpoons 2 NH_3(g)$: the equilibrium constant K_p at 298 and 550 °K and $\Delta G_{550}°$, assuming $\Delta H°$ to be independent of temperature in this range. (b) In Example 3-1 a value of $K_p = 0.110$ was determined from observed pressures at equilibrium. To what temperature do these measurements apply?

(a) For the process as written,

$$\Delta G_{298}° = 2 \times \Delta G_{f,\mathrm{NH_3}(g)}° = -2 \times 3,976$$

$$= -7,952 \text{ cal} \quad \text{and} \quad \Delta H° = 2 \times \Delta H_{f,\mathrm{NH_3}(g)}°$$

$$\Delta H° = -2 \times 11,040 = -22,080$$

$$\log K_p = \frac{-\Delta G°}{2.303 \, RT} = \frac{+7,952}{2.303 \times 1.987 \times 298} = 5.831$$

$$K_{p,298} = 6.78 \times 10^5$$

$$\log \frac{K_2}{K_1} = -\frac{\Delta H°}{2.303 \, R} \left[\frac{1}{T_2} - \frac{1}{T_1} \right]$$

$$\log K_{550} - \log K_{298} = -\frac{-22,080}{2.303 \times 1.987} (1.82 - 3.36) \times 10^{-3}$$

$$\log K_{550} = +5.831 - 7.38 = -1.55$$

$$K_{550} = 2.8 \times 10^{-2}$$

$$\Delta G_{550}° = -2.303 \, RT \log K_{550}$$

$$= -2.303 \times 1.987 \times 550 \times (-1.55)$$

$$\Delta G_{550}° = +3,950 \text{ cal}$$

(b) $K = 0.110$ at temperature $= T$

$$\log K_T - \log K_{298} = -\frac{-22,080}{2.303 \, R} \left[\frac{1}{T} - \frac{1}{298} \right]$$

$$-0.959 - 5.831 = +4.817 \left(\frac{1}{T} - 3.36 \times 10^{-3} \right)$$

$$T = 513 \text{ °K}$$

EXAMPLE 3-8. Measurement of equilibrium gas pressures in the SO_2 + $\frac{1}{2}O_2 \rightleftharpoons SO_3$ system at temperatures of 834 and 921 °K yielded K values of 18.0 and 4.96, respectively, for the two temperatures. Calculate the average value of $\Delta H°$ over this temperature range.

$$\log \frac{4.96}{18.0} = -\frac{\Delta H°}{2.303 \times 1.987}\left[\frac{1}{921} - \frac{1}{834}\right]$$

$$-0.5602 = -\frac{\Delta H°}{4.576}(1.085 - 1.199) \times 10^{-3}$$

$$\Delta H° = -\frac{2.563}{0.114 \times 10^{-3}} = \underline{-22,500 \text{ cal}}$$

The last example illustrates a type of problem often encountered in evaluating enthalpy changes in both chemical and physical processes from equilibrium data. If experimental determinations have been made at a number of different temperatures, it is usual to obtain the ΔH value by graphical analysis rather than by solution of the van't Hoff equation for various pairs of T and K values. For purposes of graphical analysis, Eq. (3-7) is a more convenient form, and for this application it can be written simply as

$$\log K = \frac{-\Delta H°}{2.303\,R}\frac{1}{T} + \text{constant} \qquad (3\text{-}9)$$

This equation indicates that a plot of $\log K$ vs. $1/T$ will yield a straight line with the slope $-\Delta H°/2.303\,R$. The constant is, of course, $\Delta S°/R$, but it is usually not convenient to obtain this from the intercept of the graph. It can be calculated directly, if desired, once $\Delta H°$ is known. An advantage of the graphical approach is that any significant temperature dependence of the enthalpy change will be immediately apparent in a curvature of the plot. A true straight line will be obtained only if $\Delta H°$ is rigorously independent of temperature. If curvature is present we can obtain an average value and at the same time, the amount of variation in ΔH can be estimated.

EXAMPLE 3-9. For the equilibrium of the type $A + B \rightleftharpoons AB$ between pyridine, phenol, and their hydrogen-bonded complex, the following set of equilibrium constants have been determined:[*]

K_c	88	66	49
$T(°C)$	22	30	40

Obtain a value for $\Delta H°$ by graphical analysis.
The data calculated for the graph is as follows:

K	$\log K$	$T, °C$	$T, °K$	$1/T, °K \times 10^3$
88	1.9445	22	295	3.390
66	1.8195	30	303	3.300
49	1.6902	40	313	3.194

[*] A. K. Chandra and S. Banerjee, *J. Phys. Chem.*, **66**, 952 (1962).

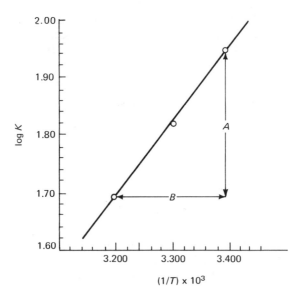

FIG. 3-3 Log K vs. 1/T plot for Example 3-9.

Data are plotted in Fig. 3-3, from which we obtain

$$\text{Slope} = \frac{A}{B} = \frac{1.955 - 1.695}{(3.400 - 3.200) \times 10^{-3}} = 1.30 \times 10^{3}$$

$$-\frac{\Delta H^{\circ}}{2.303\, R} = 1{,}300 \qquad \Delta H^{\circ} = -1{,}300 \times 4.576$$

$$\Delta H^{\circ} = -6{,}000 \text{ cal}$$

3-4 EQUILIBRIUM IN NONIDEAL SYSTEMS: THE ACTIVITY

In molecular systems deviations from ideal behavior leading to significant errors do not occur frequently, particularly for gases under normally encountered conditions. It is useful, however, to examine the relations developed to determine their limits of applicability. Treatment of the effects of nonideality is based on the free energy of a substance as a gas or vapor in equilibrium with the condensed phase. These effects become important when we move from the gas phase into solution, particularly when the solution contains ionic species. Thus, although refinements will not be required for gaseous systems as such, they are a necessary prerequisite for the later discussion on real solutions.

It was indicated briefly in Chapter 2 that substances exhibiting nonideal behavior can be dealt with using a corrected concentration. Up to this point the activity in such equations as

$$G = G° + 2.303 \, RT \log a \tag{2-34}$$

has been employed solely as a convenient symbol to represent the concentration of substances that behave ideally. To utilize this concept in handling nonideal systems we have to investigate the formal basis of the activity and its significance.

3-4-1   Real Gases

For an ideal gas we have seen that a change in pressure results in an alteration in the free energy per mole given by

$$\Delta G = 2.303 \, RT \log(P_2/P_1) \tag{2-26}$$

For a real gas the most convenient approach is to modify this equation. We replace P_2, which must be the pressure of an *ideal* gas if the relation is to be valid, by the product of the real gas pressure and a term correcting for nonideal behavior. We retain, however, the perfect gas at 1 atm as the standard state. Thus for a real gas the standard state is a hypothetical one, since ideal behavior will be followed quantitatively only in the limit of low pressures. To calculate the ΔG for a transition from the standard state to nonstandard state, P_1 in Eq. (2-26) becomes $P°$, the pressure in the standard state, and is set equal to 1 atm. It refers to the real gas, with free energy $G°$ per mole, behaving ideally at this pressure. The quantity substituted for P_2 is the thermodynamically effective pressure in the nonstandard state. It represents the pressure the gas would exhibit were it to behave ideally. The ratio of the corrected nonstandard state pressure to the pressure in the ideal gas state at 1 atm is termed the *activity a*,* a dimensionless quantity since the pressure units cancel. If we wished to use pressure units other than atm, $G°$ values must be recalculated to correspond to a different standard state where $P°$ is unity in the pressure units employed.

The correction factor that must be applied to the pressure is called the *activity coefficient* and is represented by γ. Thus

$$a = \frac{\gamma P}{P°} \equiv \gamma P \tag{3-10}$$

Let us examine this often-confusing concept in another way. We can determine experimentally from PVT measurements the difference in free energy between a real gas in two states characterized by different pressures.

---

* The corrected pressure is known as the *fugacity f*, and the activity is defined formally as $a = f/f°$, where f is the fugacity in the nonstandard state and $f°$ is that in the standard state chosen as 1 atm. Since $f° = 1$, the activity and fugacity are numerically equal; the treatment here, however, does not require explicit use of fugacities.

For the most straightforward case where the gas behaves ideally within acceptable experimental error at 1 atm but deviates significantly at higher pressures, by working in reverse with Eq. (2-26) and setting $P_1 = 1$ atm, we can calculate a numerical value for P_2. This quantity is the activity—the pressure that is consistent with the actual free energy difference between the two states at different pressures. The activity coefficient then can be calculated from Eq. (3-10). For a gas not behaving ideally at 1 atm pressure, the same kind of calculation can be performed utilizing additional measurements at low pressures where the activity and pressure are numerically equal.

On the basis of these arguments we can write for the free energy per mole of a real gas in some nonstandard state,

$$G = G^\circ + 2.303 \, RT \log a \tag{2-34}$$

This equation was encountered previously, but we now interpret a as the activity defined by Eq. (3-10). Substitution of this definition for a gives

$$G = G^\circ + 2.303 \, RT \log \gamma P \tag{3-11}$$

The activity coefficient γ will be unity only for a gas that behaves ideally.

The practical consideration of how this affects equilibrium calculations evidently turns on whether γ can be assumed to be unity. When this assumption is valid, a and p are numerically the same. The activity quotient Q, $(a_M)^m (a_N)^n / (a_B)^b (a_D)^d$, in

$$\Delta G = \Delta G^\circ + 2.303 \, RT \log Q \tag{2-36}$$

and equations derived from this relation, can be replaced properly by $Q_p = (P_M)^m (P_N)^n / (P_B)^b (P_D)^d$ (where we are referring throughout to the general reaction $bB + dD \rightleftharpoons nM + nN$ used earlier). In other words, the previous treatment is valid so long as the activity coefficients are unity. When $\gamma \neq 1$, the derivation of Eq. (2-36) and subsequent equations must be based on Eq. (3-11). The net effect is that each activity in Q must be replaced by the appropriate γP term.

In a gaseous reaction such as $2 \, SO_2 + O_2 \rightleftharpoons 2 \, SO_3$, for example, the activity quotient at equilibrium K would be written as

$$K = \frac{(P_{SO_3})^2}{(P_{SO_2})^2 (P_{O_2})} \frac{(\gamma_{SO_3})^2}{(\gamma_{SO_2})^2 (\gamma_{O_2})} = K_p K_\gamma$$

Note that K_γ, the conventional representation of the activity *coefficient* ratio, is *not* an equilibrium constant.

The circumstances under which such modifications must be employed depend on the accuracy required and the extent of nonideality. In general, deviations are greatest at high pressures and low temperatures (see Chapter 8). In any such cases, for example, in making accurate calculations from bomb calorimetry where high gas pressures are involved, it would be necessary to employ experimentally determined or calculated* activity coefficients. In normally encountered reactions in the laboratory or in biological systems, however, the behavior of gases can be considered to be ideal to a very good approximation. For example, at 0 °C and 1 atm, $\gamma_{N_2} = 0.9996$, and even at 150 atm the value has dropped by only 3 percent to 0.97. The purpose in examining gases has been primarily to introduce the concept of activity that we can make profitable use of in the condensed phase.

3-4-2 Nonideal Solutions

The techniques were outlined in Sec. 2-4-4 whereby the free energy of solution components can be evaluated indirectly by working with the substance in the vapor phase in equilibrium with the solution.

Looking first at the use of mole fractions, we obtained the equation

$$\Delta G = 2.303 \, RT \log \frac{X_2}{X_1} \qquad (2\text{-}29)$$

applicable to a solution component whose mole fraction changes from X_1 to X_2. Inherent in the derivation are the two assumptions that the vapor behaves as an ideal gas and that the solution is ideal. For real solutions the error in assuming that the vapor phase behaves ideally is almost always negligible, and attention therefore must focus on the condensed phase where nonideality will manifest itself as a departure for Raoult's law.† The approach is again one of making appropriate corrections to the nominal mole fraction of a solution component.

We again can define the activity of a solution component considered as the solvent as the ratio of its mole fraction, corrected for nonideality and in the nonstandard state, to that in the standard state, the latter being taken as the pure substance,‡ that is, at $X_{\text{solvent}} = 1.0$. Unlike the situation with real gases, we are dealing here with a physically realizable standard state. The correction term is again the activity coefficient γ, and

$$a = \frac{\gamma X}{X^\circ} \equiv \gamma X \quad (solvent) \qquad (3\text{-}12)$$

---

* Several methods are available for the estimation of activity coefficients of real gases from the critical constants and associated data. See, for example, K. S. Pitzer and L. Brewer, *Thermodynamics* (New York: McGraw-Hill, 1961, pp. 605–629).

† The nature of these deviations essentially involving a lack of simple and direct proportionality between vapor pressure and mole fraction in solution, is treated in detail in Chapter 9 in the general discussion of solution properties.

‡ Strictly speaking, the standard state is the pure substance under 1 atm pressure, but any pressure effects on the condensed phase can be neglected in all but very exacting work.

An expression for ΔG between the nonstandard and standard state can be obtained by using Eq. (2-29) as follows: The mole fractions comprising the X_2/X_1 ratio must be those of an ideal solution component. The term can be replaced by the activity since X_2/X_1 represents the ratio of mole fraction, corrected to be that of the ideally behaving component, to the mole fraction in the standard state where the "solution" is always ideal because it is just the pure solvent. Thus Eq. (2-29) becomes identical with the general expression $\Delta G = 2.303\,RT \log a$, wherein we can substitute the definition of activity and obtain

$$G = G° + 2.303\,RT \log \gamma X \qquad (3\text{-}13)$$

In any situation where we are concerned directly with the free energy of a solute, we can base the treatment on concentrations expressed as molarities. For an ideal solution the appropriate equation is

$$\Delta G = 2.303\,RT \log \frac{C_2}{C_1} \qquad (3\text{-}14)$$

This is an alternative to the previously discussed relation $G = G° + 2,303\,RT \log C$, Eq. (2-32). To obtain the analogous expression for a nonideal solute we proceed as before, replace C_2 by the activity coefficient-corrected molarity, use C_1 to represent the standard state (taken as $C° = 1.0$ mole/liter), and define the activity as the ratio of these concentrations:

$$a = \frac{\gamma C}{C°} \equiv \gamma C \quad (solute) \qquad (3\text{-}15)$$

The standard state again is now a hypothetical one since it is usually only at concentrations considerably less than 1 molar that a real solution will behave in an ideal manner. With these alterations Eq. (3-14) can be written in the form

$$G = G° + 2.303\,RT \log \gamma C \qquad (3\text{-}16)$$

As will be seen in subsequent discussions of solution properties, deviations from ideal behavior increase as the solution becomes more concentrated. For solutions of electrolytes, the effect of interionic forces gives rise to serious deviations and activity coefficients are quite significantly different from unity. Even in molecular systems, however, cases requiring the use of activity coefficients are encountered, although activity coefficients are sometimes unjustifiably ignored solely on the grounds that the solute is molecular.

For equilibrium calculations in solution, relations incorporating the activity quotient Q must be altered when necessary so that each activity is replaced by the product of the activity coefficient and the nominal concentration. For the process D-mannose $\rightleftharpoons$ D-fructose (see Example 3-3), which here is abbreviated as M $\rightleftharpoons$ F, the activity quotient would be written

$$Q = (a_F)/(a_M) = \frac{[F]\gamma_F}{[M]\gamma_M}$$

where $[i]$ and γ_i represent the concentration in moles per liter and the activity coefficient of the ith component.

To conclude the discussion of the techniques involved in the treatment of real systems, it is useful to review the basic idea that has been applied to various systems. The fundamental equations giving the dependence of thermodynamic functions on concentrations are derived on the assumption that the substance whose concentration terms are employed behaves ideally. The approach to real systems has been to use exactly the same equations but to use an "ideal" or thermodynamically effective concentration, the activity— which we obtain by correcting the nominal concentration. This is done in preference to developing an entirely new set of relations for real systems. The activity satisfies the requirement that the values substituted for concentrations—whether pressures, mole fractions, or molarities—must be those of a substance behaving ideally.

The rather detailed examination of standard state definitions may appear superficially to be more exhaustive than is consistent with the level of the general approach to physical chemistry in this book. Although an attempt continues to be made here to simplify many aspects of the subject, the very real difficulty of selection of standard states cannot be glossed over. If there is any uncertainty about the standard state on which data are based—or worse, if the whole idea of carefully defined standard states is ignored—then the answers obtained may very well be meaningless. The question is not whether the calculations are approximate but whether they are correct or incorrect. Although in some systems the problem can be very difficult and force some rather arbitrary choices of standard states, most cases require nothing more than an appreciation of the necessity of self-consistent data to avoid difficulty.

PROBLEMS

3-1. Assuming ideal gas behavior, derive an expression in terms of n, R, and T between K_c and K_p for the process

$$4\,HCl(g) + O_2(g) \rightleftharpoons 2\,H_2O(g) + 2\,Cl_2(g)$$

3-2. Using data from Table 2-2, calculate the standard free energy change at 298 °K for the reaction in Problem 3-1. Calculate a value for the equilibrium constant at this temperature. Does the K value correspond to a K_c or K_p? Explain.

3-3. Making reference to Table 3-1, calculate the concentration of $CH_3COOC_2H_5$ at equilibrium at 25 °C when the initial concentrations of ethanol and acetic acid are both 1 mole/liter.

3-4. Calculate the equilibrium constant at 298 °K for the following reactions:
(a) $CO(g) + \frac{1}{2} O_2(g) \rightleftharpoons CO_2(g)$
(b) $2 NO_2(g) \rightleftharpoons N_2O_4(g)$
(c) $N_2(g) + O_2(g) = 2 NO(g)$

3-5. In the gaseous system $2 HI \rightleftharpoons I_2 + H_2$, an equilibrium mixture was found to contain 0.5 atm each of H_2 and I_2 and 3.7 atm of HI. Find K_p and hence determine the partial pressure of HI after equilibrium has been attained in a reactor into which 1 atm each of H_2 and I_2 are placed at this temperature.

3-6. Determine the standard entropy change in the system discussed in Example 3-9.

3-7. Find the equilibrium constant for the following reaction at 50 °C where the standard free energy change is $+910$ cal:

$$CH_3ONO(g) + HCl(g) \rightleftharpoons CH_3OH(g) + NOCl(g)$$

3-8. Using the value of $\Delta H^\circ = +17.06$ kcal (assumed to be temperature independent) and $\Delta G^\circ_{298} = +9.72$ kcal, determine the temperature at which $K_p = 1.00$ for the system $2 NOCl(g) \rightleftharpoons 2 NO(g) + Cl_2(g)$.

3-9. Find the equilibrium pressure of $NH_3(g)$ and $HCl(g)$ at 250 °C (note!) that results from the decomposition (which is chemically reversible to a detectable extent) $NH_4Cl(s) \rightleftharpoons NH_3(g) + HCl(g)$, taking $\Delta H^\circ = +42.3$ kcal (assume this value to be temperature independent but see, nevertheless, Example 1-4) and $\Delta G^\circ_{298} = +21.98$ kcal.

3-10. Using standard enthalpies and free energies of formation for the compounds involved in the following process, evaluate ΔG°_{298}, K_{298}, K_{373}, ΔG°_{373} and ΔG_{373}:

$$CCl_4(g, 2 \text{ atm}) + 2H_2(g, 2 \text{ atm}) \rightleftharpoons 2 Cl_2 (g, 5 \text{ atm}) + CH_4(g, 5 \text{ atm})$$

Ionic Equilibrium I:
Acid-Base Systems

4

The approach taken here is similar to that employed in the discussion of molecular equilibrium: initially treating systems where, to a first approximation, nonideality can be ignored. Although it was indicated that in ionic solutions activity coefficients are an important consideration, there are many instances where quite satisfactory results can be obtained by using the uncorrected ion concentration.

In the general area of acid-base reactions there is a wide range of systems that can be dealt with adequately on the less rigorous basis. These systems include many processes of biological interest and importance.

Those aspects of acid-base equilibria where the Law of Mass Action approach to an equilibrium constant such as K_c can be utilized are examined in this chapter. It is assumed throughout that concentration and activity are identical and, therefore, that K_c satisfies previously derived relations for the thermodynamic equilibrium constant.

4-1 FUNDAMENTAL CONCEPTS

4-1-1 Acids and Bases

One of the most useful descriptions of acid-base systems is that based on the Bronsted–Lowry definitions:

*ACID: A SUBSTANCE THAT DONATES A PROTON IN A CHEMICAL
REACTION.
BASE: A SUBSTANCE THAT ACCEPTS A PROTON IN A CHEMICAL
REACTION.*

The interaction between an acid and a base is visualized as a *proton transfer
reaction* or *protolysis*: The acid donates, and the base accepts the proton
transferred from one species to another.

A thorough appreciation of this interpretation of acids and bases, as
opposed to a mechanism involving simple fragmentation of the molecule, is
important. Conventionally, the term "dissociation" constant of an acid
is used to refer to the reaction wherein the acid transfers a proton to another
molecule.

The idea of proton transfer also resolves the apparent conflict between
the fact that acids exist in ionic form in solution but are covalent compounds
in their pure state. They are bonded through sharing of electrons rather than
by electrostatic attraction between ions. Thus we cannot explain ionic
products in terms of the kind of dissociation reaction that characterizes an
ionic compound such as NaCl. Sodium chloride exists as positive and
negative ions in the crystal, and these pass into solution when interionic
attractive forces are counterbalanced by polar solvent molecules. For
covalent compounds, on the other hand, ions are produced as a result of
proton transfer.

In general it is convenient to make an overall classification into *strong*
and *weak* acids on the basis of the extent to which proton transfer occurs.
Hydrochloric acid is classified as a *strong acid* since, in aqueous solution,
HCl undergoes virtually complete proton transfer to water molecules:

$$HCl + H_2O \rightarrow H_3O^+ + Cl^-$$

This reaction proceeds so far to the right under most concentration conditions
that normally it is not treated as an equilibrium system. However, the process
may be influenced by the equilibrium that exists between water molecules
themselves, which will be encountered again in discussion of ionic equilibrium.
The representation of this proton transfer reaction between HCl and H_2O
is often abbreviated to $HCl \rightarrow H^+ + Cl^-$. We must not lose sight of the fact
that we are dealing with a reaction, not a dissociation and, equally important,
that the species giving rise to the acid characteristics of the solution is the
hydronium ion, H_3O^+. There is considerable experimental and theoretical
evidence that an isolated proton cannot exist in aqueous solution.

Similarly, a strong base such as the amide ion NH_2^- is so classified
because its proton acceptance reactions go practically to completion:

$$NH_2^- + H_2O \rightarrow NH_3 + OH^-$$

Weak acids and bases are characterized by proton transfer reactions that do not go to completion. They therefore constitute a system that can be treated in terms of equilibrium conditions. Classically, weak acids are represented by acetic acid whose protolysis reaction with water is

$$CH_3COOH + H_2O \rightleftharpoons H_3O^+ + CH_3COO^-$$

Here equilibrium is attained when only a small fraction of the acid has formed acetate and hydronium ions. Again the process can be written as $CH_3COOH \rightarrow H^+ + CH_3COO^-$, but the caution against misinterpretation mentioned for HCl applies equally well here.

The classic weak base is ammonia,

$$NH_3 + H_2O \rightleftharpoons NH_4^+ + OH^-$$

To represent the process as $NH_4OH \rightleftharpoons NH_4^+ + OH^-$ or to refer to an aqueous solution of ammonia and ammonium hydroxide is misleading since the "undissociated base" is NH_3 in water, not ionic NH_4OH that separates into a positive and negative ion.

Another result of the Bronsted–Lowry approach is the idea of a *conjugate acid-base pair*. In the CH_3COOH equilibrium it is apparent that the CH_3COO^- ion acts as a base by accepting a proton from H_3O^+ when the reaction proceeds from right to left. CH_3COOH and CH_3COO^- are referred to as conjugate acid-base pair. The general case can be represented as

$$\underset{\text{Acid}}{HA} + X \rightleftharpoons HX^+ + \underset{\text{Conjugate base}}{A^-}$$

Similarly, in the ammonia system the NH_4^+ ion acts as an acid, donating a proton to water, and the NH_3 and NH_4^+ constitute an acid-base pair. For bases the general scheme is

$$\underset{\text{Base}}{B} + HX \rightleftharpoons X^- + \underset{\text{Conjugate acid}}{BH^+}$$

A number of additional significant points emerge from a consideration of these example systems. First, a Bronsted–Lowry acid or base need not be a molecular species; ions are equally capable of participating in proton transfer reactions. Second, although we deal exclusively with aqueous solutions here, there is no such restriction in the definitions. In the equations illustrating the conjugate acid-base nomenclature, the second species in each case has been designated simply as X and XH. These examples also indicate that water has the capability of acting either as an acid or as a base. In the protolysis of acetic acid, water accepts a proton, while in the ammonia system, H_2O is a proton donor. Substances that can act in this way—can accept or donate a proton—are known as *amphoteric* or *amphiprotic* substances. The implications of amphoteric behavior of water are of widespread importance.

4-1-2 Proton Transfer in Pure Water

The most important result of the ability of water to act as an acid or base is the mutual proton transfer reaction that can occur in pure water—the so-called "self-ionization of water." The reaction is

$$H_2O + H_2O \rightleftharpoons H_3O^+ + OH^-$$

One water molecule is considered to be the acid. It donates a proton to the second molecule, the base. The process is chemically reversible, and the equilibrium position lies very much in favor of the "reactants." K_c for the process can be represented by the conventional expression

$$K_c = \frac{[H_3O^+][OH^-]}{[H_2O]^2}$$

The concentration of water is considered to remain essentially constant. It therefore is included in a modified equilibrium constant K_w, the *ionization constant for water*, defined as

$$K_w = [H_3O^+][OH^-] = 1 \times 10^{-14} \tag{4-1}$$

where the brackets indicate molar concentrations. The value quoted is approximately correct at temperatures close to 25 °C. K_w, like other equilibrium constants, is temperature dependent, but this is of minor importance here since we are concerned exclusively with solutions at room temperature. The value at body temperature is 3.13×10^{-14}. The equation also is often written as $[H^+][OH^-] = K_w$.

Equation (4-1) has a variety of important implications. In common with other equilibrium expressions, it does not restrict the *absolute values* of the hydronium and hydroxyl ion concentrations. It imposes the restriction that the *product* of $[H_3O^+]$ and $[OH^-]$ shall be 1×10^{-14} at room temperature. For pure water the concentrations of these two species will be equal. In every acid–base reaction in aqueous solution there is available at least this one relationship—the K_w expression—that holds regardless of the proportion of hydronium and hydroxyl ions and independent of the concentration of other species.*

Once the value of either $[H_3O^+]$ or $[OH^-]$ is known, the other can be calculated quite simply from Eq. (4-1). For example, if $[H_3O^+]$ is known to be 1.0×10^{-3}, we calculate that $[OH^-] = K_w/[H_3O^+] = 1 \times 10^{-14}/1 \times 10^{-3} = 1 \times 10^{-11}$ mole/liter. In complex equilibrium systems where several quantities may be unknown, the availability of Eq. (4-1) reduces by one the number of equations we must obtain to solve for unknown quantities.

---

* This holds true strictly only when *activities* of hydronium and hydroxyl ions are used. This point is discussed in Chapter 5.

4-1-3 The pH Scale

The H_3O^+ concentration varies over a considerable range. In the titration of a weak base with a strong acid, for example, a typical variation might be from 10^{-2} to 10^{-10} mole/liter. Any attempt to present a change of this magnitude graphically would not give a particularly useful graph. At best an accurate plot could be obtained for only two or three orders of magnitude. For this particular example it might be possible to have a reasonable presentation of values from 10^{-2} to 10^{-4}, but the variation between 10^{-4} and 10^{-10} likely would occupy the bottom 1 mm of the scale on the graph. However, if a logarithmic scale is employed, each unit becomes a tenfold variation and the range of values can be handled easily. It is on this basis and also for some electrochemical systems that it is convenient* to use a logarithmic expression for the hydronium ion concentration. The pH is defined by the relation

$$pH = -\log a_{H_3O^+} \approx -\log[H_3O^+] \qquad (4\text{-}2)$$

In dilute solutions the concentration in moles per liter is adequate, while activities (the nominal concentration corrected by the activity coefficient) must be employed for many intermediate concentrations. For very concentrated solutions the whole concept of pH becomes rather nebulous since it becomes impossible on a practical basis to determine the activity of hydronium ions accurately.

The pX type notation is used for several other related quantities such as

$$pOH = -\log[OH^-] \qquad (4\text{-}3)$$

$$pK = -\log K \qquad (4\text{-}4)$$

where in Eq. (4-4) K can be K_w or the equilibrium constant to be defined presently for an acid or base. This system of nomenclature is applied to the water equilibrium as follows:

$$[H_3O^+][OH^-] = K_w = 1.0 \times 10^{-14}$$
$$pH + pOH = pK_w = 14 \qquad (4\text{-}5)$$

By analogy with the interpretation of Eq. (4-1) we can state that pH and pOH can have any value, but their *sum* must be 14. One other point implied by these comments, but worthy of explicit statement, is that the pH

---

* Arguments put forth to justify the use of pH as a "convenience" are rather tenuous. Many calculations involving $[H_3O^+]$ expressed as an exponential can be carried out at least approximately by mental arithmetic, whereas it is certainly the exceptional student who can recite a set of logarithm tables from memory. However, the logarithm-based expression is generally accepted and will likely continue, and familiarity with its use therefore is essential.

TABLE 4-1.   RELATION OF pH TO ION CONCENTRATIONS IN
AQUEOUS SOLUTIONS

Solution classification	$[H_3O^+]^a$	$[OH^-]^a$	*pH*	*pOH*
Acid	1	10^{-14}	0	14
Acid	10^{-3}	10^{-11}	3	11
Neutral	10^{-7}	10^{-7}	7	7
Basic	10^{-10}	10^{-3}	10	3
Basic	10^{-14}	1	14	0

^{a} Moles per liter.

is *not* limited to the range 1 to 14, although this impression is often given. Such situations are not met very frequently, but it is quite possible to have $[H_3O^+]$ greater than 1 mole/liter and hence a negative value of pH and pOH greater than $+14$.

EXAMPLE 4-1.   Calculate the pH of solutions that are (1) 0.01 and (2) 4.36×10^{-3} molar* in hydronium ion.

(1) $pH = -\log[H_3O^+] = -\log(10^{-2}) = -(-2) = \underline{2}$
(2) $pH = -\log(4.36 \times 10^{-3}) = -(+0.64 - 3.00) = \underline{2.36}$

EXAMPLE 4-2.   Calculate the hydronium ion concentration in solutions with pH = (1) 9.00 and (2) 4.92.
(1) $\log[H_3O^+] = -9.00$ $[H_3O^+] = 1.0 \times 10^{-9}$
(2) $\log[H_3O^+] = -4.92 = +0.08 - 5.00$
    $[H_3O^+] = 1.20 \times 10^{-5}$

The classification of a given solution as acidic or basic depends on the relative concentrations of hydronium and hydroxyl ions. For pure water

$$[H_3O^+] = [OH^-] = \sqrt{K_w} = (1.0 \times 10^{-14})^{\frac{1}{2}}$$

$$[H_3O^+] = 1.0 \times 10^{-7} \text{ mole/liter}$$

In acid solution hydronium ions predominate. They can be quantitatively defined as solutions in which $[H_3O^+] > 1.0 \times 10^{-7}$ or pH < 7. Conversely, basic solutions have $[OH^-] > 1.0 \times 10^{-7}$ and pH > 7. The relations between pH and ion concentrations are summarized in Table 4-1.

4-1-4 Principle of Electroneutrality

In the equilibrium characterizing pure water, the stoichiometry requires that $[H_3O^+]$ and $[OH^-]$ have the same value. Hence it is possible to calculate this quantity from the K_w expression. More complex systems require on

---

* Except for calculations specifically concerned with acid–base titrations where another measure of concentration will be used, acid and base concentrations will be expressed in terms of moles per liter.

occasion, in addition to the various equilibrium expressions that may be applicable, a similar stoichiometric relationship that will permit evaluation of additional unknown concentrations. Although in some situations we can write such a relation by inspection of the stoichiometry, it is useful to have available a principle that is generally applicable.

We are dealing here with ionic solutes. The solution must always remain electroneutral regardless of the source or mechanism of production of its solute ions. This has the following stoichiometric implication:

> **PRINCIPLE OF ELECTRONEUTRALITY: IN ANY SOLUTION THE TOTAL CONCENTRATION OF POSITIVE AND NEGATIVE CHARGES MUST BE THE SAME.**

In pure water the condition is simply $[H_3O^+] = [OH^-]$. For an aqueous acetic acid solution, the equality is $[H_3O^+] = [CH_3COO^-] + [OH^-]$, while for $Mg(OH)_2$ as solute we would have $2[Mg^{2+}] + [H_3O^+] = [OH^-]$. In applying such relationships to ionic equilibrium problems, the procedure least prone to error consists in writing the equality initially for *all* ions present. Subsequently, from a knowledge of the characteristics of the system it may be possible to neglect certain terms in the summation. Any such modification, of course, introduces an assumption that should be checked when unknown concentrations have been computed. In general, simple cases can be handled without invoking this principle, but it is valuable in systems involving simultaneous equilibria.

4-2 AQUEOUS SOLUTIONS OF STRONG ACIDS AND BASES

As we have seen, the proton transfer reaction involving a strong acid or base proceeds to completion for practical purposes and thus is not susceptible to treatment as an equilibrium system. The main tasks for such systems therefore will be the calculation of pH or pOH on a simple stoichiometric basis and the application of the K_w condition to find $[OH^-]$ if $[H_3O^+]$ is known. We also will need to examine the conditions under which the pH of such solutions is directly influenced by the water equilibrium.

In all but the most dilute solutions, the contribution from the self-ionization of water, to $[H_3O^+]$ in a strong acid solution or to $[OH^-]$ in basic solutions, is quite negligible. For a general strong acid, then, the stoichiometry is adequately represented by

$$HX + H_2O \rightarrow H_3O^+ + X^-$$

Therefore, if the nominal concentration of HX is C_a moles/liter, $[H_3O^+]$ has the same value. The term *nominal concentration* will be used in the discussion of ionic equilibrium to indicate the moles of acid or base present per liter of solution without regard to the fraction that may undergo proton transfer.

For a strong acid the working relations are Eqs. (4-6) and (4-7)

$$[H_3O^+] = C_a \tag{4-6}$$

$$pH = -\log C_a \tag{4-7}$$

and for a strong base they are Eqs. (4-8) and (4-9).

$$[OH^-] = C_b \tag{4-8}$$

$$pOH = -\log C_b \tag{4-9}$$

In general the equations will be written only for the acid case. The analogous expression for a base can be written down directly by substituting pOH for pH, $[OH^-]$ for $[H_3O^+]$, etc.

Having obtained either the hydronium or hydroxyl ion concentration from the relevant equations above, the concentration of the other can be calculated from Eq. (4-1). In an acid solution, for example, we have $[OH^-] = K_w/[H_3O^+]$. In concentrated solutions the concentration is no longer an adequate representation of the activity, and activity coefficients must be incorporated into the equations. The techniques are discussed in Chapter 5.

We now have to consider the circumstances in which the amounts of H_3O^+ or OH^- ion produced from water itself will influence significantly the total concentration of that species. Dealing specifically with a strong acid, there always will be two sources of hydronium ion—the proton transfer from the acid and the self-ionization of water. The situation now will be characterized by both $[H_3O^+]$ and $[OH^-]$ being unknown quantities. Two equations therefore are required, the water equilibrium expression and the principle of electroneutrality, written for the strong acid HX as

$$K_w = [H_3O^+][OH^-] = 1 \times 10^{-14} \tag{4-1}$$
$$[H_3O^+] = [OH^-] + [X^-]$$

For a strong acid the concentration of the X^- ion is equal to the nominal concentration C_a, and

$$[H_3O^+] = [OH^-] + C_a$$

or

$$[OH^-] = [H_3O^+] - C_a$$

Substitution of this expression for $[OH^-]$ in Eq. (4-1) gives

$$[H_3O^+]([H_3O^+] - C_a) = K_w$$

$$[H_3O^+]^2 - C_a[H_3O^+] - K_w = 0$$

Applying the rule for the solution of a quadratic equation gives the working equation

$$[H_3O^+] = \frac{C_a \pm (C_a^2 + 4K_w)^{\frac{1}{2}}}{2} \tag{4-10}$$

A series of calculations carried out using Eq. (4-10) for various values of C_a indicates that this equation, rather than Eq. (4-6), is required for accurate results for $C_a < 1 \times 10^{-6}$ mole/liter. An analogous expression can be written for $[OH^-]$ in dilute base solutions.

EXAMPLE 4-3.   Calculate pH, $[OH^-]$, and pOH for a solution containing 2.75×10^{-2} mole/liter of HNO_3.

$$pH = -\log C_a = -\log(2.75 \times 10^{-2}) = -(+0.44 - 2.00) = \underline{1.56}$$

$$[OH^-] = \frac{K_w}{[H_3O^+]} = \frac{1 \times 10^{-14}}{2.75 \times 10^{-2}} = 3.64 \times 10^{-13} \text{ mole/liter}$$

$$pOH = -\log[OH^-] = -\log(3.64 \times 10^{-13}) \text{ mole/liter} = \underline{12.44}$$

(Check: pH + pOH = 1.56 + 12.44 = 14.)

EXAMPLE 4-4.   Calculate $[OH^-]$ in a 5×10^{-7} molar NaOH solution.

$$[OH^-] = \tfrac{1}{2}[C_b \pm (C_b + 4\,K_w)^{\frac{1}{2}}]$$

$$[OH^-] = \tfrac{1}{2}[5 \times 10^{-7} \pm (25 \times 10^{-14} + 4 \times 10^{-14})^{\frac{1}{2}}]$$

$$[OH^-] = \tfrac{1}{2}[5 \times 10^{-7} \pm (29 \times 10^{-14})^{\frac{1}{2}}]$$

$$[OH^-] = 5.19 \times 10^{-7} \text{ mole/liter}$$

(The negative square root gives a meaningless negative $[OH^-]$ value.)

4-3   SOLUTIONS OF WEAK ACIDS AND BASES

4-3-1   *Equilibrium and pH Calculations*

The equilibrium between a weak acid in water and its conjugate base and hydronium ions can be handled formally in the same way as the molecular equilibria discussed in Chapter 3. The relevant equations for the general weak acid are

$$HA + H_2O \rightleftharpoons H_3O^+ + A^-$$

$$K_c = \frac{[H_3O^+][A^-]}{[HA][H_2O]}$$

The amount of proton transfer that occurs is small, and the concentration of undissociated acid added initially will normally be less than 1 mole/liter (the concentration of water is 55.5 moles/liter). $[H_2O]$ will undergo a quite negligible change in the process. Hence it can be included in a modified equilibrium constant K_a, the *acid dissociation constant* (or *acid ionization constant*) defined by the equation

$$K_a = \frac{[H_3O^+][A^-]}{[HA]} \tag{4-11}$$

Table 4-2 lists values of K_a for some frequently encountered acids, and K_b,

TABLE 4-2.   ACID-BASE DISSOCIATION CONSTANTS[a]

Acid	Formula	T, $°C$	K_a	pK_a
Acetic	CH_3COOH	25	1.75×10^{-5}	4.76
Benzoic	C_6H_5COOH	25	6.46×10^{-5}	4.19
n-Butyric	*n*-C_3H_7COOH	20	1.54×10^{-5}	4.81
Carbonic	H_2CO_3	25 K_{1a}	4.30×10^{-7}	6.37
	HCO_3^-	25 K_{2a}	5.61×10^{-11}	10.25
Citric	$HOC(CH_2COOH)_2COOH$	18 K_{1a}	8.4×10^{-4}	3.08
	$HOC(CH_2COOH)_2COO^-$	18 K_{2a}	1.8×10^{-5}	4.74
	$HOC(CH_2COOH)(COO^-)_2$	18 K_{3a}	4.0×10^{-6}	5.40
Formic	$HCOOH$	20	1.77×10^{-4}	3.75
Hydrocyanic	HCN	25	4.93×10^{-10}	9.31
Hydrofluoric	HF	25	3.53×10^{-4}	3.45
Lactic	$CH_3CHOHCOOH$	100	8.4×10^{-4}	3.08
Nitrous	HNO_2	12.5	4.6×10^{-4}	3.37
Oxalic	$(COOH)_2$	25 K_{1a}	5.90×10^{-2}	1.23
	$HOOCCOO^-$	25 K_{2a}	6.40×10^{-5}	4.19
Phenol	C_6H_5OH	20	1.25×10^{-10}	9.89
Phosphoric	H_3PO_4	25 K_{1a}	7.5×10^{-3}	2.12
	$H_2PO_4^-$	25 K_{2a}	6.23×10^{-8}	7.19
	HPO_4^-	25 K_{3a}	1.7×10^{-12}	11.77
Sulfurous	H_2SO_3	18 K_{1a}	1.54×10^{-2}	1.81
	HSO_3^-	18 K_{2a}	1.02×10^{-7}	6.91
Hydrosulfuric	H_2S	18 K_{1a}	9.1×10^{-8}	7.07
	HS^-	18 K_{2a}	1.2×10^{-15}	14.92

Base	Formula	T, $°C$	K_b	pK_b
Acetamide	CH_3CONH_2	25	2.5×10^{-13}	12.60
Ammonia	NH_3	25	1.75×10^{-5}	4.76
Aniline	$C_6H_5NH_2$	25	3.82×10^{-10}	9.42
Caffeine	$C_8H_{10}O_2N_4$	40	4.1×10^{-14}	13.39
Calcium hydroxide	$Ca(OH)_2$	25 K_{1b}	3.74×10^{-3}	2.43
Diethyl amine	$(C_2H_5)_2NH$	25	9.6×10^{-4}	3.02
Dimethyl amine	$(CH_3)_2NH$	25	5.20×10^{-4}	3.28
Nicotine	$C_{10}H_{14}N_4$	25	7×10^{-7}	6.15
Thiourea	$CS(NH_2)_2$	25	1.1×10^{-15}	14.96
Silver hydroxide	$AgOH$	25	1.1×10^{-4}	3.96
Urea	$CO(NH_2)_2$	25	1.5×10^{-14}	13.82

[a] Data mostly from *Handbook of Chemistry and Physics*, 45th ed. (Cleveland: Chemical Rubber Co., 1964). Reprinted with permission.

defined as $[OH^-][BH^+]/[B]$, for some bases. It is important to appreciate that the concentrations in Eq. (4-11) are those at equilibrium. Thus, as it stands, this expression contains three unknown quantities. The necessary modifications to allow $[H_3O^+]$ to be calculated from K_a and the nominal concentration can be carried out as follows.

From the stoichiometry of the proton transfer reaction, $[H_3O^+] = [A^-]$ if we neglect contributions to the hydronium ion concentration from water itself. The concentration of undissociated acid at equilibrium will be the initial concentration less an amount, equal to $[H_3O^+]$, that has transferred a proton to water, and thus $[HA] = C_a - [H_3O^+]$. Substituted into Eq. (4-11), these equalities alter the relation to the form

$$K_a = \frac{[H_3O^+]^2}{C_a - [H_3O^+]} \tag{4-12}$$

This is a quadratic equation in hydronium ion concentration, but in the majority of cases the further simplification can be made that $C_a - [H_3O^+] \approx C_a$. Since we are dealing with weak electrolytes, the amount of acid that dissociates is generally quite small, and to a very good approximation the initial and equilibrium concentrations of the acid are the same. In solving numerical problems the most efficient procedure is to assume initially that $[HA]$ can be replaced by C_a. It often will be immediately apparent that this assumption would not be valid. In general if $C_a \leq 2,000\, K_a$ the assumption cannot be made without introducing unacceptable error. Normally, however, the simplifying assumption would be made initially. The value of the hydronium ion concentration obtained subsequently can be used to verify the assumption. In this form the K_a expression is $K_a = [H_3O^+]^2/C_a$, whence

$$[H_3O^+] = \sqrt{K_a C_a} \tag{4-13}$$

If $[H_3O^+]$ calculated on this basis is greater than about 5 percent of C_a, the error in the absolute value of $[H_3O^+]$ will be 2 percent or greater, and hence the value should be recalculated using Eq. (4-12). If that expression is recast in conventional quadratic form, and the solution formula is applied, the result is

$$[H_3O^+] = \frac{-K_a \pm (K_a^2 + 4\,C_a K_a)^{\frac{1}{2}}}{2} \tag{4-14}$$

For a basic solution $[OH^-]$ is obtained from the analogous expression with C_b and K_b. The ultimate refinement of including the contribution to $[H_3O^+]$ from water itself leads to a cubic equation, but the situation is not encountered frequently enough to justify inclusion here of the detailed algebraic discussion required.

Equation (4-13) is applicable to such a wide range of weak acid situation (or for weak bases, $[OH^-] = \sqrt{K_b C_b}$) that it is often written in a form that gives the pH directly:

$$pH = -\log(K_a C_a)^{\frac{1}{2}}$$
$$pH = \tfrac{1}{2}pK_a - \tfrac{1}{2}\log C_a \qquad (4\text{-}15)$$

Another conventional representation of ionic equilibrium data is through the value of the *degree of dissociation* α, which is defined for an acid as

$$\alpha = \frac{[H_3O^+]}{C_a} \qquad (4\text{-}16)$$

and is given as either a decimal fraction or a percentage. Of course, it is possible to write a series of equations, such as those presented in this section, in terms of α. This, however, is not necessary. Although some relations set up in this way are quite useful for routine calculations, they can be derived, when required, from the basic equations. The approach to be taken in problems involving the degree of ionization is to calculate $[H_3O^+]$ from C and α before or after application of the general equations.

The application of Eqs. (4-10) to (4-16) is illustrated in the following examples.

EXAMPLE 4-5.   Calculate the pH in a solution 1.0×10^{-3} molar in formic acid, HCOOH.

$$[H_3O^+] = \sqrt{K_a C_a} \quad \text{(assuming } C_a - [H_3O^+] \approx C_a)$$
$$[H_3O^+] = (1.77 \times 10^{-4} \times 10^{-3})^{\frac{1}{2}} = (17.7 \times 10^{-8})^{\frac{1}{2}} = 4.20 \times 10^{-4}$$

$[H_3O^+] > 0.05 C_a$ and, therefore, the use of the exact Eq. (4-14) is required.

$$[H_3O^+] = \tfrac{1}{2}[-K_a \pm (K_a^2 + 4 C_a K_a)^{\frac{1}{2}}]$$
$$[H_3O^+] = \tfrac{1}{2}[-1.77 \times 10^{-4} \pm (3.13 \times 10^{-8} + 4 \times 10^{-3} \times 1.77 \times 10^{-4})^{\frac{1}{2}}]$$
$$[H_3O^+] = \tfrac{1}{2}(-1.77 \times 10^{-4} \pm 8.60 \times 10^{-4}) = +3.42 \times 10^{-4}$$
$$pH = -\log(3.42 \times 10^{-4}) = -(-4.00 + 0.53) = \underline{3.47}$$

EXAMPLE 4-6.   Calculate pOH in a 0.01 molar solution of NH_3 in water.

$$[OH^-] = \sqrt{K_b C_b} \quad \text{(assuming negligible dissociation)}$$
$$[OH^-] = (1.75 \times 10^{-5} \times 1 \times 10^{-2})^{\frac{1}{2}} = 4.18 \times 10^{-4}$$
$$\frac{[OH^-]}{C_b} = 0.04 \quad \text{(therefore assumption is acceptable)}$$

EXAMPLE 4-7.   The degree of dissociation for benzoic acid, C_6H_5COOH, in 0.02 molar aqueous solution is 1.81 percent. Calculate the K_a value for this acid.

$$\alpha = [H_3O^+]/C \qquad [H_3O^+] = \alpha C = 0.2 \times 1.81 \times 10^{-2} = 3.62 \times 10^{-3}$$
$$K_a = \frac{[H_3O^+]^2}{C_a - [H_3O^+]} = \frac{(3.62 \times 10^{-3})^2}{200 \times 10^{-3} - 4 \times 10^{-3}} = \underline{6.5 \times 10^{-5}}$$

The standard free energy change in any proton transfer process is easily obtained from the general equation derived in Chapter 3;

$$\Delta G° = -2.303 \, RT \log K \tag{3-5}$$

For a weak acid, substitution of pK_a for the $-\log K$ term gives the more convenient expression

$$\Delta G° = 2.303 \, RT \, pK_a \tag{4-17}$$

Thus for acetic acid at 25 °C, the standard free energy change is

$$\Delta G° = 2.303 \times 1.987 \times 298 \times 4.76$$
$$\Delta G° = 6,487 \text{ cal}$$

This relatively large and positive value indicates that the occurrence of the proton transfer process

$$CH_3COOH + H_2O \rightleftharpoons CH_3COO^- + H_3O^+$$

to the extent that $a_{CH_3COO^-}$ and $a_{H_3O^+}$ are unity is thermodynamically nonspontaneous when a_{CH_3COOH} is unity. Using the interpretation of $\Delta G°$ developed in Chapter 3, at equilibrium the reactants in the equation as written are favored.

This approach represents an alternative rationalization of the experimental fact, manifested by an equilibrium constant considerably less than unity, that the proton transfer from acetic acid in all but extremely dilute solutions occurs to a rather limited extent.

4-3-2 Conjugate Acid and Base Pairs

Utilizing the equations derived, a very useful relationship between the dissociation constants of a conjugate pair of Bronsted–Lowry acids and bases can be obtained.

For a general acid the proton transfer reaction with water is

$$HA + H_2O \rightleftharpoons H_3O^+ + A^-$$

where in the reverse process A^- acts as a base. Thus an aqueous solution of the A^- ion, introduced, say, as the sodium salt, will be basic because of the proton transfer equilibrium

$$A^- + H_2O \rightleftharpoons HA + OH^-$$

We can write the equilibrium expressions for these two processes as

$$K_a = \frac{[H_3O^+][A^-]}{[HA]} \quad \text{and} \quad K_b = \frac{[HA][A^-]}{[A^-]}$$

Multiplying these equations, we get

$$K_a K_b = \frac{[H_3O^+][A^-][HA][OH^-]}{[HA][A^-]}$$

where all terms on the right-hand side cancel, except for $[H_3O^+]$ and $[OH^-]$. The product of these two concentrations is K_w, and therefore the final equation is

$$K_a K_b = K_w \tag{4-18}$$

Taking logarithms of both sides and converting to the pK notation yields an alternative form,

$$pK_a + pK_b = pK_w = 14 \tag{4-18'}$$

Either of these two equations provides a convenient method of calculating the dissociation constant for one member of a conjugate acid-base pair when that of the other is known. Thus for any acid or base listed in Table 4-2 we can obtain the pK_b or pK_a of the conjugate base or acid simply by subtracting the indicated pK value from 14. Thus pK_a for $CH_3COOH = 4.76$, and therefore $pK_{b,CH_3COO^-} = 14 - 4.76 = 9.24$. Alternatively, the numerical K value can be obtained from Eq. (4-18) as $K_b = 1 \times 10^{-14}/1.75 \times 10^{-5} = 5.7 \times 10^{-10}$.

4-3-3 Hydrolysis

In the context of acid-base reaction this term refers to a proton transfer reaction between an ion and water. It is important to appreciate the fact that although hydrolysis reactions are very often treated as a separate type of acid-base process, in terms of the Bronsted–Lowry concept they are simply protolysis reactions in which the acid or base happens to be an ion. The purpose of this discussion is to outline the manner in which the nomenclature and equations conventionally used when hydrolysis processes are treated independently are related to the general equations here.

Consider the weak base system

$$(I) \quad NH_3 + H_2O \rightleftharpoons NH_4^+ + OH^-$$

for which we can write the usual K_b expression. In a manner analogous to the weak acid case just discussed, in a solution of NH_4Cl (a strong electrolyte dissociating completely into NH_4^+ and Cl^-) we will observe the process

$$(II) \quad NH_4^+ + H_2O \rightleftharpoons NH_3 + H_3O^+$$

The equilibrium constant is given by

$$K_a = \frac{[NH_3][H_3O^+]}{[NH_4^+]}$$

Process (II) also qualifies as a hydrolysis reaction according to the definition given above. Considered from this viewpoint, the equilibrium constant would be designated K_{hyd} for the NH_4^+ ion but is identical to the normal K_a written as for any Bronsted–Lowry acid.

It should be recognized, then, that systems involving a process of this type can be treated in the same manner as other acid-base equilibria.

4-4 MULTIPLE ACID-BASE EQUILIBRIA

In this section two important applications of the general principles that have been developed for the treatment of acid-base equilibria are examined. Although these applications are to systems somewhat more complex than the cases examined thus far, the approach is the same. The general equilibrium expressions that apply are written considering the several equilibria individually. Subsequently, these equations are combined and the resulting general relation is modified on the basis of what is known or can be reasonably predicted about the case of interest.

4-4-1 Amphoteric Substances

In many acid-base situations where several proton transfer equilibria may occur simultaneously, one or more of the species present can act either as an acid or a base. We have already seen the important implications of the ability of water to behave in this way. Here we examine equations that are applicable to amphoteric species in general.

If the amphoteric substance is represented as XH, then by definition this species is capable of donating a proton and forming its conjugate base X^-. The negative ion is produced from the parent molecule on removal of the proton. XH can accept a proton, forming in the process the conjugate acid XH_2^+. In aqueous solution the two processes are

$$XH + H_2O \rightleftharpoons H_3O^+ + X^-$$
$$XH + H_2O \rightleftharpoons OH^- + XH_2^+$$

The associated equilibrium expressions are

$$K_a' = \frac{[H_3O^+][X^-]}{[XH]} \tag{4-11'}$$

$$K_b' = \frac{[OH^-][XH_2^+]}{[XH]} \tag{4-11''}$$

The K values are written as K_a' and K_b' to emphasize that they apply to the *same species* and to distinguish the relations derived here from those involving the K_a and K_b of a *conjugate pair*.

An expression for the hydronium ion concentration in terms of the dissociation constants and [XH] is desired and, therefore, the algebraic manipulations must permit elimination of the terms [OH⁻], [X⁻], and [XH₂⁺].

Since the solution is aqueous, the K_w condition applies and the concentration of hydroxyl ion can be replaced wherever it occurs by $K_w/[H_3O^+]$. The first use of this relation can be made by rewriting the K'_b expression, Eq. (4-11″), as

$$K'_b = \frac{K_w[XH_2^+]}{[H_3O^+][XH]} \tag{4-19}$$

The other relationship applicable to the solution is the principle of electroneutrality, which here takes the form

$$[H_3O^+] + [XH_2^+] = [X^-] + \frac{K_w}{[H_3O^+]} \tag{4-20}$$

in which [OH⁻] has been replaced by $K_w/[H_3O^+]$. The final step in the derivation consists in rearranging the two expressions for the dissociation constant, Eqs. (4-11′) and (4-11″), to give a relation for [X⁻] and [XH₂⁺], respectively, which can be substituted into Eq. (4-20) to yield

$$[H_3O^+] + \frac{K'_b[H_3O^+][XH]}{K_w} = \frac{K'_a[XH]}{[H_3O^+]} + \frac{K_w}{[H_3O^+]}$$

This can be rearranged to the form

$$[H_3O^+]^2\left(1 + \frac{K'_b[XH]}{K_w}\right) = K_w + K'_a[XH] \tag{4-21}$$

Although this equation can be solved for the hydronium ion concentration as it stands, in the majority of cases considerable simplification occurs since both $K'_a[XH]$ and $K'_b[XH]$ will be considerably larger than K_w. If this is the case then the right-hand side of Eq. (4-21) reduces to $K_a[XH]$. In addition, it follows that $K_b[XH]/K_w \gg 1$ and, therefore, the term in brackets becomes $K'_b[XH]/K_w$. With these simplifications Eq. (4-21) takes the form

$$[H_3O^+]^2 \frac{K'_b[XH]}{K_w} = K'_a[XH]$$

or

$$[H_3O^+] = \sqrt{\frac{K'_a K_w}{K'_b}} \tag{4-22}$$

which indicates that the acidity of the solution of an amphoteric substance is independent of its concentration as long as the value of [XH] is consistent with the assumptions made above. This behavior is verified by experimental pH determinations. The general nature of the expression is consistent with the fact that the species present in solution can produce *both* hydronium and hydroxyl ions. Since K_a' and K_b' indicate its tendency to undergo these processes, the resulting hydronium ion concentration is not unexpectedly a function of their ratio.

Equation (4-22) applies only to a solution where the amphoteric species is the only solute present. If there is an additional source of a conjugate acid or base, or both, the equation is no longer valid. The relation can also be written in the pX-type notation as

$$pH = 0.5\, pK_a' + 0.5\, pK_w - 0.5\, pK_b'$$

or

$$pH = 0.5\,(pK_a' - pK_b') + 7 \qquad (4\text{-}23)$$

This type of system is the first encounter with an important concept in acid-base equilibria. It demonstrates an essential pH-stability of solutions containing species (in this case the same substance) that can act as an acid or base. This behavior is also exhibited in a somewhat different manner by buffer solutions.

4-4-2 Polyprotic Acids and Bases

Many acids and bases are characterized by the ability to transfer more than one proton. The protons involved are invariably transferred singly in a series of protolysis reactions that are usually appreciably incomplete, that is, they can be treated by equilibrium techniques.

For the case of a polyprotic acid, the protons are donated successively from the parent acid molecule. The conjugate base formed in the first step acts as the acid in the next stage of the dissociation. It is through these species, as well as through the hydronium ion formed in each stage, that the several equilibria are interrelated.

To illustrate the treatment of polyprotic systems we examine phosphoric acid. H_3PO_4 is capable of donating all three protons and, hence, the acid is involved in three successive proton transfer equilibria. These can be written, along with the expression for the K_a of each step, as follows:

I. $H_3PO_4 + H_2O \rightleftharpoons H_3O^+ + H_2PO_4^-$ $K_{1a} = \dfrac{[H_3O^+][H_2PO_4^-]}{[H_3PO_4]}$

II. $H_2PO_4^- + H_2O \rightleftharpoons H_3O^+ + HPO_4^{2-}$ $K_{2a} = \dfrac{[H_3O^+][HPO_4^{2-}]}{[H_2PO_4^-]}$

III. $HPO_4^{2-} + H_2O \rightleftharpoons H_3O^+ + PO_4^{3-}$ $K_{3a} = \dfrac{[H_3O^+][PO_4^{3-}]}{[HPO_4^{2-}]}$

For a complete, generally valid equation yielding the hydronium ion concentration it is apparent that it would be necessary to solve the three equilibrium expressions, along with the electroneutrality expression. The tedious algebra involved can be avoided, however, by taking advantage of several properties of the system result from the large differences in the successive K values. For phosphoric acid, the dissociation constant values (from Table 4-2) are

$$K_{1a} = 7.5 \times 10^{-3}$$
$$K_{2a} = 6.2 \times 10^{-8}$$
$$K_{3a} = 1.7 \times 10^{-12}$$

This variation between successive K values is typical of polyprotic inorganic systems. It is readily explained on the basis of the increasing difficulty of successive proton removal because of the increasingly negative charge on the acid species.

The net result of the behavior is that each equilibrium may be considered independently. When the actual numerical values for a given system have been calculated, the validity of this approach can be demonstrated.

Consider the first proton transfer process. Since K is considerably larger than in processes II and III, the concentration of hydronium ions formed by these two latter species donating their proton to water is negligible. Thus we can write

$$[H_3O^+]_{tot} = [H_3O^+]_I + [H_3O^+]_{II} + [H_3O^+]_{III} \approx [H_3O^+]_I$$

where the subscripts indicate which process is responsible for the production of that particular concentration. In any equilibrium expression written for the system, the term $[H_3O^+]$ always will refer to the *total* concentration. The approximation being made here is that this can adequately be represented by the concentration arising from process I.

If $[H_3O^+]_{II}$ is considered to be negligible, then it is consistent to assume that the decrease in the concentration of the $H_2PO_4^-$ ion owing to proton transfer in process II also can be neglected, that is,

$$[H_2PO_4^-] = [H_2PO_4^-]_I - [H_3O^+]_{II} \approx [H_2PO_4^-]_I = [H_3O^+]_I$$

The last equality in this expression is based simply on the stoichiometry of process I. The equilibrium expression for this step, therefore, is

$$K_{1a} = \frac{[H_3O^+]^2}{C - [H_3O^+]} \qquad (4\text{-}12')$$

In actual calculations the same techniques are used as for the case of monoprotic acids, that is, it is assumed initially that the denominator in Eq. (4-12) can be represented adequately by C, and if recalculation is necessary the full equation is used.

In the second proton-donating stage the situation is simplified considerably by the assumptions already made. In the equilibrium equation

$$K_{2a} = \frac{[H_3O^+][HPO_4^{2-}]}{[H_2PO_4^-]}$$

$[H_3O^+]$ is again the *total* hydronium ion concentration, assumed to be equal to $[H_3O^+]_I$. As indicated, the consistent assumptions follow that $[H_3O^+]$ is the same as the total concentration of the $H_2PO_4^-$ ions and that loss of HPO_4^{2-} in process III produces a negligible decrease in its concentration. Consequently, the terms $[H_3O^+]$ and $[H_2PO_4^-]$ cancel and

$$K_{2a} = [HPO_4^{2-}] \qquad (4\text{-}24)$$

indicating that the concentration of the HPO_4^{2-} ion is independent of initial phosphoric acid concentration.

For the third proton transfer process the relevant relation is

$$K_{3a} = \frac{[H_3O^+][PO_4^{3-}]}{[HPO_4^{2-}]}$$

Here there is only one substitution to be made, that of K_{2a} for $[HPO_4^{2-}]$, giving

$$K_{3a}K_{2a} = [H_3O^+][PO_4^{3-}] \qquad (4\text{-}25)$$

From this the phosphate ion concentration can be determined from the K values and $[H_3O^+]$.

In Eqs. (4-12′), (4-24), and (4-25) we have three simple expressions that adequately describe what is superficially a rather complex system.

The various assumptions made in arriving at these equations can be verified by calculating the concentration of species present in a 1 molar solution of phosphoric acid.

For $[H_3O^+]$ we have from Eq. (4-12) that $[H_3O^+] = (7.5 \times 10^{-3})^{\frac{1}{2}} = 8.7 \times 10^{-2}$. This is greater than $0.05\,C$ and, therefore, the full quadratic equation is required:

$$[H_3O^+] = \tfrac{1}{2}[-K_{1a} \pm (K_{1a}^2 + 4C_aK_{1a})^{\frac{1}{2}}] = 8.25 \times 10^{-2} \text{ mole/liter}$$

From Eq. (4-24) we have directly that

$$[HPO_4^{2-}] = 6.2 \times 10^{-8} \text{ mole/liter}$$

and from Eq. (4-25) we have

$$[PO_4^{3-}] = \frac{K_{2a}K_{3a}}{[H_3O^+]} = 1.3 \times 10^{-18} \text{ mole/liter}$$

Since $[HPO_4^{2-}] = [H_3O^+]_{II}$ and $[PO_4^{3-}] = [H_3O^+]_{III}$, the first assumption can be verified by computing the total hydronium ion concentration:

$$[H_3O^+]_{tot} = (8.3 \times 10^{-2}) + (6.2 \times 10^{-8}) + (1.3 \times 10^{-18})$$
$$\approx 8.3 \times 10^{-2} \text{ mole/liter}$$

The consumption of $H_2PO_4^-$ in process II is negligible compared to the amount formed in the first step: 6.2×10^{-8} mole/liter vs. 8.3×10^{-2} mole/liter. Similarly, while the concentration of HPO_4^{2-} is decreased in process III by 1.3×10^{-18} mole/liter, this is a negligible fraction of the total concentration of that species, 6.2×10^{-8}.

The situation is analogous for polyprotic bases. For a solution containing PO_4^{3-} ions, introduced, say, as Na_3PO_4, we consider three successive proton-accepting reactions, each giving rise to OH^- ions in water, and assume that the total $[OH^-]$ is adequately given by their concentration produced in the first step.

Another polyprotic system, important in biological contexts, is H_2CO_3. In the two-step sequence characterizing this system, the hydronium ion concentration is obtained from Eq. (4-12') while that of CO_3^{2-} is obtained by an equation analogous to Eq. (4-24), indicating that $[CO_3^{2-}]$ is independent of the total acid concentration.

The equality in Eq. (4-24), of course, is applicable only to a solution containing initially the pure acid itself—all other sources of intermediate ions must be absent.

4-5 BUFFER SOLUTIONS

Buffer solutions are of great importance in biological systems and in other areas of chemistry, in addition to their role in the physical chemistry of ionic equilibria. Their distinctive features arise from what is generally referred to as the *common ion effect.*

As the name indicates, what is involved here is a solution wherein one ion is common to two processes—an ionic equilibrium and an ionic dissociation process.

The fundamental ideas can be illustrated by considering the classic acid-base common ion case, a mixture of sodium acetate and acetic acid. The two processes involved are

$$\text{(I)} \quad CH_3COOH + H_2O \underset{}{\overset{K_a}{\rightleftharpoons}} H_3O^+ + CH_3COO^-$$

$$\text{(II)} \quad CH_3COONa \xrightarrow[\text{``100 percent''}]{} Na^+ + CH_3COO^-$$

Process II, for practical purposes, is a complete dissociation, so that the introduction of, for example, 0.1 mole/liter of sodium acetate gives rise to

[CH_3COO^-] = 0.1. The acetate ion is *common* to the dissociation of the salt and to the acid's proton transfer equilibrium. The sodium ion does not participate in the reaction.

Consider the solution as containing initially only acetic acid. On a qualitative basis in terms of Le Châtelier's principle, the effect of the addition of sodium acetate (effectively acetate ion) will be to suppress the dissociation of the acid with a resulting decrease in [H_3O^+].

4-5-1   The Henderson–Hasselbalch Equation

To develop a quantitative expression for the hydronium ion concentration we first note that the addition of "extra" acetate ion will not affect the fundamental requirement imposed by the equilibrium expression

$$K_a = \frac{[H_3O^+][CH_3COO^-]}{[CH_3COOH]}$$

However, in contrast to a solution of pure acetic acid, the hydronium and acetate ion concentrations here will not be the same. Neglecting the self-ionization of water as a significant source of H_3O^+ (a valid assumption in almost all common ion situations), this species is produced only by the acid. Acetate ion comes from both the acid and the salt.

The solution contains both the acid CH_3COOH, at nominal concentration C_a, and its conjugate base CH_3COO^-, at concentration C_b. The terms in the equilibrium expression will be given by

$$[CH_3COO^-] = C_b + [H_3O^+] \approx C_b$$

$$[CH_3COOH] = C_a - [H_3O^+] \approx C_a$$

The indicated assumptions will almost invariably be valid for common ion solutions where the acid and base are present in concentrations of the same order of magnitude. The acid proton transfer proceeds only to a limited extent even in pure acid solution and will be appreciably suppressed in the presence of the common ion. The amount of acid dissociating, therefore, will be a negligible fraction of its total concentration. On the basis of similar arguments, the very small contribution to [CH_3COO^-] via proton transfer also can be neglected.

Making these substitutions in the K_a equation results in

$$K_a = [H_3O^+]\frac{C_b}{C_a}$$

or

$$[H_3O^+] = K_a\frac{C_a}{C_b} \tag{4-26}$$

Converting to pX notation, the equation becomes

$$-\log[H_3O^+] = -\log K_a - \log \frac{C_a}{C_b}$$

$$pH = pK_a + \log \frac{C_b}{C_a} \tag{4-27}$$

This very useful relation is known as the *Henderson–Hasselbalch equation*. The C_a/C_b ratio is conventionally inverted to give a $+\log C_b/C_a$ term. The analogous expression for a base and its conjugate acid is

$$pOH = pK_b + \log \frac{C_a}{C_b} \tag{4-28}$$

Before discussing the important properties of solutions of this type it is useful to examine in the following examples the mechanics of calculations that use the Henderson–Hasselbalch equation.

EXAMPLE 4-8. Calculate (1) the pH of a 0.1 molar aqueous solution of CH_3COOH and (2) the pH after addition of 0.1 mole/liter of CH_3COONa.
(1) $pH = 0.5 \, pK_a - 0.5 \log C_a$
     $pH = 0.5(4.76 + 1.00) = \underline{2.88}$

(2) $pH = pK_a + \log \dfrac{C_b}{C_a}$

$$pH = 4.76 + \log \frac{0.1}{0.1} = \underline{4.76}$$

EXAMPLE 4-9. Calculate the pOH of a solution 0.01 molar in NH_3 and 0.05 molar in NH_4Cl.

$$pOH = pK_b + \log \frac{C_a}{C_b} = 4.76 + \log \frac{0.05}{0.01} = 4.76 + 0.70 = \underline{5.46}$$

4-5-2 Properties of Buffer Solutions

The discussion of solutions containing a common ion thus far has dealt only with the mechanical aspects, illustrating the application of the basic principles of ionic equilibrium. Such solutions, however, have a property of considerable importance because they contain *both* an acid and a base. If a small amount of strong acid or base (effectively H_3O^+ or OH^-) is added to the solution, the addend can react with the base or acid already present. The change in solution pH is quite small since the product of the reaction accepts or donates protons to a rather limited extent. For this reason such solutions are known as *buffers*, defined formally as follows:

BUFFER: A SOLUTION THAT RESISTS A CHANGE IN pH ON THE ADDITION OF SMALL AMOUNTS OF STRONG ACID OR BASE.

The solution will only *resist* a change in its pH. There will always be some alteration in the value, even if this is quite small. Furthermore, no buffer can compensate for the addition of large amounts of acid or base.

To illustrate the changes that occur in actual solutions, examine again the situation dealt with in Example 4-8, a solution 0.1 molar in both acetic acid and acetate ions. At equilibrium the concentrations are

$$H_2O + CH_3COOH \rightleftharpoons \qquad H_3O^+ \qquad + \quad CH_3COO^-$$

0.1 mole/liter 1.75×10^{-5} mole/liter   0.1 mole/liter

If strong acid is added, the hydronium ions from this source react with acetate ions to produce more acetic acid. The hydroxyl ions from an added strong base would give rise to more acetate ion by reacting with acetic acid. Since the relative amounts of materials present per unit volume are controlled by the K_a value, there will be a consequent alteration in the hydronium ion concentration in either case.

Suppose 1×10^{-3} mole/liter of H_3O^+ is added via a strong acid such as HCl. The result would be an instantaneous concentration of H_3O^+ very much greater than the equilibrium value of about 2×10^{-5} mole/liter, and thus hydronium ions will react with CH_3COO^-. Assuming for the moment that the added acid is consumed completely in this way, the stoichiometry is

$$H_3O^+ \qquad + \quad CH_3COO^- \quad \rightarrow \quad CH_3COOH \quad + H_2O$$

0.001 mole/liter   0.001 mole/liter   0.001 mole/liter

The new concentrations will be

$$[CH_3COOH] = 0.100 + 0.001 = 0.101$$
$$[CH_3COO^-] = 0.100 - 0.001 = 0.099$$

The hydronium ion concentration then will be

$$pH = pK_a + \log \frac{C_b}{C_a} = 4.76 + \log \frac{0.099}{0.101}$$

$$pH = 4.76 + \log(0.980) = 4.76 - 0.0088$$

$$pH = \underline{4.75}$$

This result is consistent with the assumption that virtually all of the added hydronium ion reacts with acetate. The solution pH has changed by only 0.01 units despite the fact that approximately 50 times as much H_3O^+ as present initially was added. This behavior may be contrasted with that of pure water in the same circumstances. In pure water the pH is 7.0. On addition of 1×10^{-3} molar hydronium ions, we have simply $[H_3O^+] = 1 \times 10^{-3}$, and hence pH = 3. Thus the pH changes by four units as compared to the practically negligible alteration of 0.01 units for the buffer.

It was pointed out previously that a buffer is capable of accommodating only small amounts of acid or base added. It therefore is desirable to ascertain the quantitative limits of buffer capacity in general—the amount of acid or base that can be added to a specific system without a drastic change in pH.

In an extreme case, 0.1 molar acetic acid has 1.3×10^{-3} mole/liter of H_3O^+ and CH_3COO^- (both an acid and its conjugate base are present). It is obvious that relatively small amounts of added strong acid would overwhelm the minute concentration of available acetate ions, and the change in pH would be quite drastic. A fundamental consideration thus is the range of the C_b/C_a ratio in the buffer solution consistent with reasonable buffer capacity. For any particular system it is always possible to carry out a calculation such as that used in the illustration above, but it is more convenient to extract a number of principles applicable to buffers in general.

To accomplish this several variables must be considered—the pH of the buffer as determined by the K_a and C_b/C_a, the *change* in C_b/C_a per unit amount of H_3O^+ or OH^- added, and the resultant ΔpH. The most useful approach is to use the Henderson–Hasselbalch equation and plot a graph of pH values for various C_b/C_a ratios. For convenience, the *fraction in the base form*, $C_b/(C_b + C_a)$ is used rather than the ratio, since the latter varies from zero to infinity, while the fraction goes only from 0 to 1 and is therefore more readily presented in a graph.

Such a plot, applicable to any *total* buffer concentration, but calculated specifically for the acetic acid-sodium acetate system using Eq. (4-27), is shown in Fig. 4-1. A graph of identical form can be obtained for any buffer system by simply displacing the vertical axis so that the pH has the value of pK_a at the midpoint of the curve. At this point the fraction in the base form is 0.5; hence log C_b/C_a is zero and, from Eq. (4-27), pH = pK_a.

The most important property of buffers is their resistance to pH variation when strong acid or base is added. One approach to the specification of the usefulness of a particular buffer is through the ratio

$$\text{Buffer efficiency} = \frac{\text{change in fraction in base form}}{\text{change in pH}}$$

For a given addition of strong acid or base, resulting in a change in the C_b/C_a ratio, the smaller the pH change the greater the efficiency. For any buffer the efficiency can be viewed as the reciprocal of the rate of change of pH with the fraction in the base form. A hypothetical buffer of perfect efficiency would have a horizontal pH curve. The plot in Fig. 4-1 indicates that the efficiency is practically constant throughout the central portion of the plot, while at the ends the efficiency ratio decreases seriously. In general the useful buffer range is considered to be approximately in the range shown, where the fraction in the base form is between 0.1 and 0.9. This corresponds to pH = $pK_a \pm 1$. As is apparent from the graph, however, this range extends

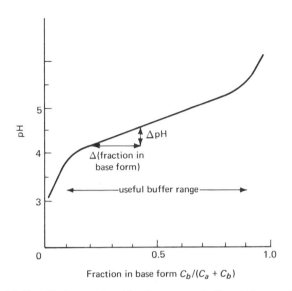

FIG. 4-1 The pH of an acetic acid-sodium acetate buffer solution as a function of the fraction of the total concentration in the base CH_3COO^- form.

beyond the portion of the plot that is essentially linear. In the region where the efficiency is more or less constant, pH = $pK_a \pm 0.5$. Thus a more appropriate delineation of the useful buffer range is perhaps a pK_a within 0.5 units of the desired pH.

Another measure of the usefulness of a buffer solution, closely related to the efficiency, is the buffer capacity. At a particular value of the "fraction in the base form" the buffer will have some value of the efficiency ratio. The larger the capacity, the larger the change in C_b/C_a that can be accommodated without moving into a region where the efficiency is low. It is apparent from the figure that for a given change in the fraction as base, the pH will change by the greatest amount if the buffer pH is initially near the end of the useful buffer range. Therefore, the maximum capacity with respect to addition of *either* strong acid *or* strong base will be at the midpoint where pH = pK_a.

One additional factor to be considered is the *total* capacity of the buffer. Although the efficiency and the capacity, in the sense just discussed, are both independent of the total acid plus base concentration, it is obvious that a solution that is 10^{-5} molar in acid and base would be a rather ineffective buffer. Only a very small amount of strong acid or base added would consume completely the appropriate member of the buffer pair and cause a large change in pH. It is usual, therefore, to make up buffer solutions to a total concentration in the range of 0.1 to 1 molar.

Many buffers, particularly those of biological interest, contain acid-base pairs that are part of a polyprotic system. For example, a mixture of

$NaHCO_3$ and Na_2CO_3 would form a buffer solution in which the principal equilibrium would be

$$\underset{\text{(acid, } C_a)}{HCO_3^-} + H_2O \rightleftharpoons H_3O^+ + \underset{\text{(base, } C_b)}{CO_3^{2-}}$$

where $pH = pK_a = 10.25$ when $C_a = C_b$.

In this and similar situations the effects of *other* equilibria involving the acid and conjugate base generally can be neglected. Here the bicarbonate ion could accept a proton from water to form H_2CO_3 and OH^-. The solution as constituted, however, effectively buffers not only against *added* OH^- but also against any hydroxyl ion that otherwise might be produced by the solution components themselves.

In general buffers maintaining the pH at a specific value within the range $pH = 3$–10 are most often encountered. Solutions of strong acids or bases also can act as buffers, according to the definition given previously, in solutions where $pH > 10$ or $pH < 3$. For example, a 0.10 molar solution of HCl has $pH = 1.00$. On addition of 10^{-2} mole/liter of additional hydronium ion, the pH is altered only to 0.96—a ΔpH of 0.04 units. Such buffers are infrequently encountered and are not important in biological systems.

4-5-3 Preparation of Buffer Solutions

For buffers with pH in the range 3 to 10, all methods of preparation have in common that either directly or indirectly they result in a solution containing an acid and its conjugate base in concentrations of the same order of magnitude. It is informative to examine the several methods available, not only from the practical point of view but also because these methods illustrate some further applications of acid-base equilibria.

Weak acid and its conjugate base. This is the most obvious approach where the two species are added directly to the solution in the correct proportions. The calculation technique is illustrated in the following example:

EXAMPLE 4-10. How many grams of sodium acetate should be added to 1 liter of 0.1 molar acetic acid to give a buffer solution with $pH = 4.90$?

$$pH = pK_a + \log \frac{C_b}{C_a}$$

$$4.90 = 4.76 + \log x$$

$$\log x = 4.90 - 4.76 = 0.14$$

$$\frac{C_b}{C_a} = 1.38$$

$$C_b = 1.38 \times C_a = 1.38 \times 0.1 = 0.138 \text{ mole/liter}$$

mol. wt. $CH_3COONa = 82.03$ gm

wt. of acetate required $= 82.03$ gm/mole $\times 0.138$ mole/liter $= \underline{11.32 \text{ gm/liter}}$

Partial neutralization of a weak acid or base. When such a species undergoes neutralization, its conjugate base is formed. If the conversion is carried out only partially, the solution will contain appreciable concentrations of both the acid and base. For example, if an acetic acid-sodium acetate buffer is required at pH $=$ pK_a, the solution can be obtained by neutralizing exactly one-half of the acid in a solution containing initially only acetic acid. This would result in an equal concentration of acid and CH_3COO^- ions, and hence the desired pH.

Partial neutralization of a polyprotic acid or base. The same technique is applicable here as outlined for monoprotic species. If the buffer is to be composed of an acid and base pair from the second or third stage of the proton transfer sequence, the neutralization must be carried far enough to produce these species. A buffer that is to contain, say, 0.05 mole/liter each of $H_2PO_4^-$ and HPO_4^{2-} could be prepared by adding 0.15 mole/liter of a strong base to a solution of 0.1 molar H_3PO_4. The first 0.1 mole/liter of OH^- would convert the system essentially to 0.1 molar $H_2PO_4^-$, and the remaining 0.05 mole/liter would reduce [$H_2PO_4^-$] to 0.05 mole/liter (one-half neutralized) and at the same time produce 0.05 molar HPO_4^{2-}.

EXAMPLE 4-11.   Calculate the number of moles of sodium hydroxide that must be added to 1 liter of 0.01 molar aqueous H_2CO_3 to produce a bicarbonate–carbonate buffer with pH $=$ 10.50. The equilibrium involved here is $HCO_3^- + H_2O \rightleftharpoons H_3O^+ + CO_3^{2-}$, and p$K_a$ for HCO_3^- is 10.25.

$$pH = pK_a + \log \frac{C_b}{C_a}$$

$$\log \frac{C_b}{C_a} = 10.50 - 10.25 = 0.25 \qquad \frac{C_b}{C_a} = 1.78$$

$$C_b + C_a = 0.01 \qquad C_b = 6.3 \times 10^{-3} \qquad C_a = 3.6 \times 10^{-3} \, \text{molar}$$

Base required to convert 0.01 molar H_2CO_3 to 0.01 molar $HCO_3^- = 10 \times 10^{-3}$ mole/liter. Base required to give 6.3×10^{-3} molar CO_3^{2-} and leave 3.6×10^{-3} molar $HCO_3^- = 6.3 \times 10^{-3}$ mole/liter. Total base required $= 16.3 \times 10^{-3}$ mole/liter.

4-6   ACID-BASE TITRATIONS

A useful consolidation of the principles developed here results from a consideration of pH changes during *titration*—the successive addition of accurately measured amounts of a solution of known concentration containing the neutralizing acid or base. Furthermore, there is the practical aspect of understanding a very useful and widely applicable analytical technique.

The primary requirement is the determination of solution pH during the course of a titration and particularly the observation of the point at which neutralization is complete by observing the more or less rapid pH alteration.

There are two fundamentally different methods of experimental measurement of pH: The electrochemical method, utilizing pH meters, involves determination of the electrical potential developed by hydronium ions with respect to certain reference electrodes. This method is discussed in detail in Chapter 7. The second method involves the use of acid-base indicators that are complex organic dyes whose color depends on the concentration of H_3O^+ in the solution. This technique is known as *colorimetric* determination. Although pH meters give accurate and reliable values, indicators nonetheless are quite useful and often more convenient for some applications.

4-6-1   Colorimetric Determination of pH

The high molecular organic molecules employed as indicators behave as weak acids or bases. Their electronic absorption spectrum in the visible region is altered significantly by the loss or gain of a proton. This results in the undissociated acid and its conjugate base having different colors or, in one form, being colorless.

Typical of the indicators of this type is phenolphthalein that in acid solution has the structure

In writing equilibrium equations for indicators, the usual representation of the acid and base form are HIn and In⁻ respectively. The equilibrium is

$$HIn + H_2O \rightleftharpoons H_3O^+ + In^-$$

Perhaps the most important aspect of this equilibrium is that since the indicator is present in very small total concentrations, the proton transfer by the indicator itself will have a negligible effect on $[H_3O^+]_{tot}$. It is the pH of the solution that will determine the relative concentrations of the two forms HIn and In⁻. At high H_3O^+ concentrations, the indicator equilibrium will be shifted in favor of the HIn form and its color therefore will be characteristic of acidic solutions. Conversely, in basic solutions $[H_3O^+]$ will be small and In⁻ will predominate and impart its color to the solution.

On a quantitative basis, equilibrium expression is

$$K_{In} = \frac{[H_3O^+][In^-]}{[HIn]}$$

where

$$[H_3O^+] = K_{In} \frac{[HIn]}{[In^-]} \tag{4-29}$$

It is also convenient to have Eq. (4-29) in a form giving pH directly. Converting to the pX type notation by taking logarithms, we have

$$pH = pK_{In} + \log \frac{[In^-]}{[HIn]} \tag{4-30}$$

The formal similarity between this and the Henderson–Hasselbalch equation is obvious, and from the algebraic and graphic analysis point of view the two are identical. It must be emphasized, however, that there is the fundamental difference between the two situations. In a buffer solution the ratio of base to acid determines the pH, while in the indicator equilibrium the pH of the solution determines the value of the $[In^-]/[HIn]$ ratio. From the practical point of view it also should be apparent that the amount of indicator added to the solution must be quite small. The essential difference between an indicator and buffer pair is the large difference in their total concentration.

Equation (4-30) indicates that for a particular indicator with its characteristic pK_{In} value, the relative amount of HIn and In$^-$ depends on the difference between the value of pK_{In} and pH. Since the two forms are colored differently, the ratio therefore determines the color observed visually.

When a solution containing two colored species is observed by the human eye, a given species usually must be present to the extent of at least about 10 percent of the total concentration before its effect on the color of the solution can be perceived. In a solution whose pH is varied over a wide range, the distribution of the total indicator concentration might vary from essentially 0 to 100 percent as HIn. As the fraction HIn varies from 0 to 10 percent, no color change is visible—the solution continually shows the characteristic color of the In$^-$ form. As the fraction then changes from 10 to 90 percent, the solution undergoes a gradual, observable color change. Finally, in the pH range where HIn is present in amounts greater than 90 percent of the total, the observable color again becomes independent of pH. The 10–90 percent range is approximate so that for convenience the pH range over which observable color change occurs can be represented by values of the $[In^-]/[HIn]$ ratio between 0.1 and 10 or $\log[In^-]/[HIn] = -1$ to $+1$. From these limits and Eq. (4-30) we have that

$$\Delta pH_{\text{color change}} = pK_{In} \pm 1$$

A clearer picture of the indicator equilibrium is obtained by examining a graph of the variation of pH as the equilibrium distribution shifts from the In$^-$ to HIn form. As in graphic presentations of the Henderson–Hasselbalch equation, it is also convenient to plot the fraction of the total indicator concentration in the base form rather than $[In^-]/[HIn]$.

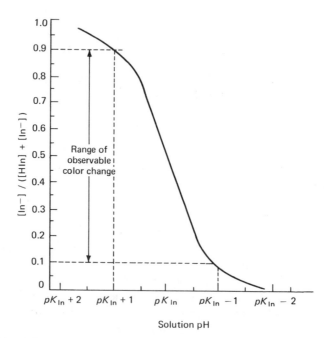

FIG. 4-2 *Fraction of an indicator in the base form as a function of solution* pH *expressed as values of* pK_{In}.

In Fig. 4-2 the fraction in the base form for an indicator is plotted as the *dependent variable* (since it depends on $[H_3O^+]$), vs. solution pH, given relative to the pK_{In} value rather than in absolute values. The graph therefore is applicable to any indicator by substitution for pK_{In}, $(pK_{In} - 1)$, etc. The numerical values are then absolute pH values for the system in question.

It is apparent from the discussion and the graph in Fig. 4-2, that the color of a single indicator in solution normally cannot be used to determine the exact value of the pH.* Visual observation can indicate that the pH is in one of three ranges:

pH > $(pK_{In} + 1)$ — solution exhibits basic color
pH = $(pK_{In} \pm 1)$ — solution exhibits a color intermediate between that associated with the acid and base forms
pH < $(pK_{In} - 1)$ — solution exhibits acidic color

---

* It is possible, however, to determine the pH more or less exactly by measuring the actual absorption spectrum of the solution in the visible region with a spectrophotometer. These measurements would allow determination of the absolute concentration of the indicator in both forms and, hence, pH could be calculated directly from Eq. (4-30). If instrumental methods for exact pH measurements are required, however, it is generally more convenient to determine pH electrochemically with a pH meter.

By testing the solution separately with a number of different indicators, each with a different pK_{In} and hence a different color change pH range, the pH can be determined to within at least 1 unit.

The pH ranges for a number of acid-base indicators are given in Table 4-3.

The most important application of indicator pH determinations is the detection of the point in titrations where the neutralization is complete. In many systems, some of which are examined in detail in the next section, there is a very rapid change in hydronium ion concentration just at this point and, hence, although the indicator changes color over a range of pH values, a sharp color alteration is observed.

4-6-2   Change in pH in Acid-Base Titrations

In practical calculations for acid-base titrations the concentration is conveniently expressed in terms of the number of *equivalents* of each species involved. The *gram equivalent weight* of an acid or base (often referred to simply as the *equivalent weight*, eq. wt.) is defined as

$$\text{Equivalent weight} = \frac{\text{gram molecular weight}}{\text{number of protons accepted or donated per molecule or ion}}$$

TABLE 4-3.   pH RANGE OF SOME ACID-BASE INDICATORS[a]

Indicator	pH range	Acid color	Base color
Thymol blue	1.2–2.8	red	yellow
2,4-Dinitrophenol	2.4–4.0	colorless	yellow
Methyl orange	3.1–4.4	red	orange
Bromophenol blue	3.0–4.6	yellow	blue–violet
Bromocresol green	4.0–5.6	yellow	blue
Methyl red	4.4–6.2	red	yellow
Chlorophenol red	5.4–6.8	yellow	red
Bromothymol blue	6.2–7.6	yellow	blue
Phenol red	6.4–8.0	yellow	red
Cresol red	7.2–8.8	yellow	red
Thymol blue	8.0–9.6	yellow	blue
Phenolphthalein	8.0–10.0	colorless	red
Thymolphthalein	9.4–10.6	colorless	blue
Alizarin yellow	10.0–12.0	yellow	lilac
Nitramine	11.0–13.0	colorless	orange brown
Trinitrobenzoic acid	12.0–13.4	colorless	orange red

[a] Data from I. M. Kolthoff and V. A. Stenger, *Volumetric Analysis*, Vol. 1 (New York: Interscience Publishers, 1942, pp. 92–93). Reprinted with permission.

Thus the equivalent weight of H_2CO_3 is one-half the molecular weight since two protons can be donated per acid molecule, while for the PO_4^{3-} ion, a base, the equivalent weight is one-third the molecular weight. For acids or bases, one equivalent* is therefore a *weight in grams of the species, equal to its equivalent weight,* and thus by definition represents a mole (Avogadro's number) of protons that can be accepted or donated. The concentration of the solution can be expressed as

NORMALITY IS THE NUMBER OF EQUIVALENTS PER LITER OF SOLUTION.

The advantage of the equivalent-based concentration is that it allows both mono- and polyprotic acids and bases to be treated using identical equations. Thus the product of normality and volume (in liters) will be the number of equivalents of the species present in the solution in question. Furthermore, for *all* systems, one equivalent of acid will require exactly one equivalent of base for neutralization. This is the origin of the very simple but useful relation

$$N_A V_A = N_B V_B \tag{4-31}$$

The normality of an "unknown" acid solution can be calculated if the volume of the acid required to neutralize a fixed volume of base of known normality is determined by titration. The terms $N_A V_A$ and $N_B V_B$ give, respectively, the numbers of equivalents of acid and base when the volume is given in liter units. In applying Eq. (4-31) to titration data the volume can be in any convenient units—usually milliliters—since the conversion factor will cancel on both sides of the equation.

To examine the manner in which the color change of an indicator can be used to detect the point at which Eq. (4-31) is applicable, it is necessary to determine how the pH will change during the process of titration.

This computation does not present any serious difficulty if it is appreciated that at each stage in the process the solution is one of the various types that have been dealt with previously. The best approach, then, is to consider at each point where a calculation is required what species are present in solution and what possible equilibria could be established among these species and therefore to determine the applicable equations.

Let us consider here two simple illustrative cases in some detail and then indicate how the same approach can be extended to more complex systems.

Strong acid—strong base. This is the simplest example of acid-base titrations. At all stages of the reaction the only equilibrium process that must

---

* Note that this definition applies *only* to acids and bases. The use of equivalents is essentially an experimental convenience, and the equivalent must be redefined for species involved in other than proton transfer reactions.

be considered is the self-ionization of water, and therefore the single expression applicable throughout is

$$[H_3O^+][OH^-] = K_w = 1.0 \times 10^{-14} \tag{4-1}$$

The stoichiometry of the neutralization process is equally simple:

$$H_3O^+ + OH^- \rightarrow 2\,H_2O$$

The other ions associated with the strong acid and base do not participate in the reaction. Thus, at the equivalence point* the solution will be neutral.

In a titration where the base is added to the acid, the initial hydronium ion concentration is progressively reduced until at the equivalence point $[H_3O^+] = [OH^-]$, following which the solution becomes more and more basic.

At any point during the titration before the equivalence point, $[H_3O^+]$ is determined simply by the amount of acid left unneutralized, whereas once this point has been passed OH^- will predominate and its concentration will be just that of the added "excess" base. In the region very near the equivalence point the hydronium ion concentration is reduced at some stage to less than 10^{-6} molar, for which situation Eq. (4-10) is required. However, the pH changes so rapidly with added base in this region that the calculation is not usually required.

To organize the calculations necessary to evaluate pH during the titration, it is convenient to use the number of equivalents of acid and base:

a = number of equivalents of acid initially present = $N_A V_A$

b = number of equivalents of base added = $N_B V_B$

where V_A is the initial volume of acid solution and V_B is the volume of base added at the point where the pH is to be determined. At any stage then, the total solution volume is $V_A + V_B$. The concentration of hydronium and hydroxyl ions before and after the equivalence point therefore can be determined from the equations:

$$\text{When } a > b \text{ (pre-end point)}, \quad [H_3O^+] = \frac{a-b}{V_{tot}} \tag{4-32}$$

$$\text{When } b > a \text{ (post-end point)}, \quad [OH^-] = \frac{b-a}{V_{tot}} \tag{4-33}$$

where the volumes must be expressed in liters.

* The equivalence point is where equivalent amounts of acid and base have been added to the solution. The "end point" represents the point in the titration where the indicator undergoes the observable sharp color change. The two are not necessarily identical, because of the error inherent in the pH range of color change of the indicator used. In most practical titrations, however, provided the indicator is properly chosen so that the pH change in the titration and for the indicator color change essentially coincide, the equivalence and end points are the same with a quite small experimental error. The terms will be used interchangeably here.

To illustrate the application of these working relations, consider the titration of 50 ml of 0.1 N HCl with 0.1 N NaOH:

1. *Initial pH* pH $= -\log C_a = -\log(0.1) = 1.0$.
2. *pH when 40 ml base have been added.*

$N_A V_A = a = 0.1$ eq./liter $\times$ (50/1,000) liter $= 0.005$ eq. of acid   ($V_A = 50$)

$N_B V_B = b = 0.1$ eq./liter $\times$ (40/1,000) liter $= 0.004$ eq. of base   ($V_B = 40$)

$$V_{tot} = \frac{90}{1,000} = 0.09 \text{ liter}$$

$$[H_3O^+] = \frac{0.005 - 0.004}{0.09} = 0.011 \text{ molar}$$

pH $= \underline{1.96}$

3. *pH when 50 ml of base have been added (equivalence point).*

$a = 0.005$ eq. of acid $b = 0.005$ eq. of base

Complete neutralization: $[H_3O^+] = [OH^-] = 1 \times 10^{-7}$ molar

$$pH = \underline{7.0}$$

4. *pH when 90 ml of base have been added.*

$a = 0.005$ eq. of acid

$b = 0.1$ eq./liter $\times$ (90/1,000) $= 0.009$ eq. of base

$$V_{tot} = \frac{50 + 90}{1,000} = 0.140 \text{ liter}$$

$$[OH^-] = \frac{0.009 - 0.005}{0.140} = 0.0286 \text{ molar}$$

pOH $= 1.54$ pH $= 14 - $ pOH $= \underline{12.5}$

When a large number of calculations of this type are carried out, the data required for a complete pH graph over the range from 0 to 100 ml of added base are obtained. The resulting curve is shown in Fig. 4-3, in which the points at which the pH was calculated above are indicated.

The most significant feature of this curve is the very sharp change in pH near the equivalence point. For this system any indicator whose $\Delta pH_{color\ change}$ lies completely within the pH $= 3$ to 10 range will exhibit an "instantaneous"

---

* By definition, C in this equation is the concentration of strong acid in moles per liter. For monoprotic acids and bases, normality and molarity are identical so that their concentration can be expressed correctly as the same numerical value in either moles per liter or equivalents per liter.

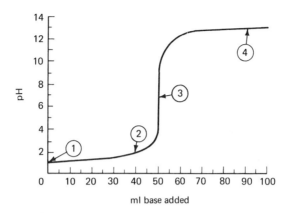

FIG. 4-3 *Variation in pH during the titration of 50 ml of 0.1 N HCl with 0.1 N NaOH. The calculation of the pH values at the points indicated is described in text.*

color change at the end point. In a practical titration in the laboratory, the addition of a very small amount of base solution produces the physical color change in the titration. The calculations such as outlined for this case indicate that the pH of the solution changes from 4.0 to 10.0 when the amount of base goes from 49.90 to 50.10 ml. The range of pH values over which a particular indicator changes color usually will be less than 6 pH units and, therefore, the volume of additional base required to pass through this point will, in practice, be considerably less than 0.2 ml.

Weak acid—strong base. The treatment to be developed here is identical to that applicable to the case of a weak base titrated with a strong acid. The working relations can be obtained by making the usual substitutions of OH^- for H_3O^+ etc. in the equations.

The essential difference between the weak acid–strong base situation and the case where both acid and base are completely dissociated is that the neutralization produces not only water but also the *conjugate base* of the acid being neutralized. The equilibrium between this weak base and H_2O and its role as a source of additional OH^- ions* therefore must be considered.

Since the weak acid is only slightly dissociated, the essential reaction will be between hydroxyl ions and the undissociated proton donor,

$$HA + OH^- \rightleftharpoons H_2O + A^-$$

* In terms of the Bronsted–Lowry definition, a conjugate base also is formed when a strong acid is neutralized, for example, Cl^- from HCl. Since the process goes virtually to completion, however, there is no detectable proton acceptance by the conjugate base. To state this in terms of Eq. (4-17), $K_aK_b = K_w$; since K_a for the strong acid is extremely large, K_b will be quite minute and the amount of proton transfer from water to the conjugate base is negligible.

Considering specifically the neutralization of CH_3COOH by NaOH, as the titration proceeds each molecule of acetic acid donating a proton to a hydroxyl ion will form an acetate ion and the concentration of this species increases. The solution becomes a buffer since both the acid and its conjugate base will be present in more or less comparable quantities. At the equivalence point the solution will be effectively the same as one obtained by dissolving the appropriate amount of CH_3COONa. Since CH_3COO^- is a weak base, the solution will not have pH $= 7$ but rather a higher, more basic pH, depending on acetate concentration. Further additions of strong base very rapidly give rise to hydroxyl ion concentrations significantly greater than those produced from the acetate ion hydrolysis. After a relatively small amount of excess base is added, the [OH^-] value is determined solely by that from the base. Thus the variation in pH is the same as that for the "excess base" region of the strong acid–strong base titration.

One further practical point is the importance of distinguishing between the *actual acidity* of the solution (the concentration of H_3O^+ ions under the given conditions) and the *titratable acidity* (the total concentration of protons available for transfer—existing either as free H_3O^+ or as undissociated acid). Thus in a 0.1 N acetic acid solution, the actual acidity is [H_3O^+] = 1.33 $\times$ 10^{-3} eq./liter, but the titratable acidity is 0.1 eq./liter—the total amount of acid per liter that can react with added OH^-.

The variation of pH as a function of volume of added base in the titration of 50 ml of 0.1 N CH_3COOH with 0.1 N NaOH is given in Fig. 4-4. At every point during the titration the system is one of the various types of solution that has been dealt with previously, and pH can be determined by use of the equation applicable to that particular case.

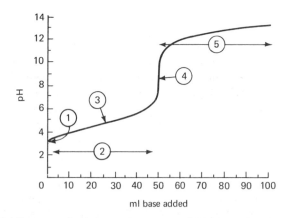

FIG. 4-4 *Variation of pH during the titration of 50 ml of 0.1 N CH₃COOH with 0.1 N NaOH. The calculation of the indicated pH values is described in text.*

The calculation of solution pH at the specific points indicated in the figure is carried out as follows:

(1) *Initial pH.* Before the addition of any base the solution is just that of a weak acid and, assuming $C - [H_3O^+] \approx C$, the pH is given by

$$pH = 0.5\ pK_a - 0.5\ \log C_a \qquad (4\text{-}15)$$

(2) *Buffer region.* The solution contains the acid and its conjugate base and the Henderson–Hasselbalch equation applies.

$$pH = pK_a + \log \frac{[CH_3COO^-]}{[CH_3COOH]} \qquad (4\text{-}27')$$

(3) *Midbuffer point.* Exactly one-half of the acid initially present has been neutralized, thus $[CH_3COOH] = [CH_3COO^-]$ and, from Eq. (4-27'), $pH = pK_a$.

(4) *Equivalence point.* All of the acid now has been converted to acetate ion, and the solution is essentially $0.05\ N\ CH_3COO^-$. The concentration is only one-half that of the initial C_a since the solution volume has been doubled. For this solution of weak base,

$$pOH = 0.5\ pK_b - 0.5\ \log C_b \qquad (4\text{-}15')$$

where K_b and C_b refer to the acetate ion. The pH is obtained from Eq. (4-5'): $pH = 14 - pOH$. The end point is at $pH = 8.72$.

(5) *Excess base region.* The hydroxyl ion concentration depends in this part of the titration only on the amount of strong base added. This will far outweigh the small contribution from proton acceptance by CH_3COO^-. $[OH^-]$ therefore can be calculated as in the excess base region for a strong acid–strong base situation, and $pH = 14 - pOH$.

There is one simplifying feature of the calculation involving Eq. (4-27'): It is necessary to calculate only the number of equivalents of acid and conjugate base present. The total volume of the solution will cancel in the C_b/C_a ratio. Also, the number of equivalents of acetate ion before the equivalence point will be equal to the number of equivalents of base added. Using a and b to represent, respectively, the equivalents of acetic acid present initially and the equivalents of strong base added, the ratio in Eq. (4-27) is given by

$$\frac{[CH_3COO^-]}{[CH_3COOH]} = \frac{b}{a - b} \qquad (4\text{-}34)$$

Examining the titration curve in Fig. 4-4, we find that the change in pH at the end point is not as large as in the HCl–NaOH system. The choice of indicator is more restricted, and the color change pH range must lie completely within the approximate range $pH = 7$ to 10. From Table 4-3 it is apparent that there are several indicators in this incomplete list that would be satisfactory.

Weak acid—weak base. This system is complicated by the simultaneous production of both a conjugate acid and a conjugate base. For example, in the CH_3COOH—NH_3 reaction,

$$CH_3COOH + NH_3 \rightleftharpoons NH_4^+ + CH_3COO^-$$

both products are capable of participating in proton transfer—the NH_4^+ ion as an acid and CH_3COO^- as a base. The system will exhibit buffer solution properties before and after the equivalence point.

The net result is that the pH change at the end point occurs over a rather small range and is not particularly sharp. The accurate detection of this point with indicators therefore is rather difficult. The solution pH when all acid has been neutralized depends on the relative magnitudes of the dissociation constant for the conjugate acid and base. In general such systems behave very much as if they were a solution containing a single amphoteric species, and thus by an approach identical to that employed to derive Eq. (4-22) the hydronium ion concentration is given by

$$[H_3O^+] = \sqrt{K_a K_w / K_b} \qquad (4\text{-}22')$$

In the acetic acid–ammonia system K_a (for NH_4^+, the conjugate acid to NH_3) and K_b (for CH_3COO^-) happen to be the same and, hence, $[H_3O^+] = (K_w)^{\frac{1}{2}} = 1 \times 10^{-7}$ mole/liter. Note that the formation of a neutral solution at the end point is not a general characteristic of such systems.

Polyprotic acid—strong base. In the earlier discussion of polyprotic systems, it was demonstrated that considerable simplification could be effected when each stage of the overall process was treated independently. The same approach is useful in considering the titration of such species. In the titration of H_3PO_4 with sodium hydroxide the system can be considered to comprise three sequential neutralization reactions:

$$H_3PO_4 + OH^- \rightleftharpoons H_2O + H_2PO_4^-$$

$$H_2PO_4^- + OH^- \rightleftharpoons H_2O + HPO_4^{2-}$$

$$HPO_4^{2-} + OH^- \rightleftharpoons H_2O + PO_4^{3-}$$

To construct the titration curve we can consider each of these processes as the neutralization of a weak acid by a strong base and proceed as before. The complete titration curve then is obtained by plotting the three component curves together. Such a plot for the phosphoric acid system is given in Fig. 4-5. The data for the various points indicated in the figure are summarized in Table 4-4.

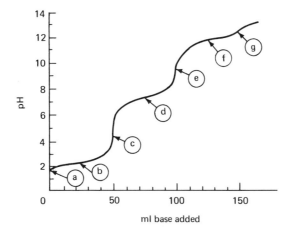

FIG. 4-5 *Variation of pH during the titration of 50 ml of 0.1 molar* H_3PO_4 *with 0.1 molar NaOH. The calculation of the indicated pH values is described in Table 4-4.*

It will be noted from the shape of the titration curve that it is progressively more difficult to detect each successive end point as the solution becomes increasingly basic and the pH change associated with completion of each neutralization becomes smaller. At the last equivalence point the change in pH is so small that it cannot normally be determined colorimetrically.

TABLE 4-4

Point	ml base added	pH	Principal species determining pH	Solution type
a	0	1.60	H_3PO_4	initially present weak acid solution
b	25	2.12	$[H_3PO_4] = [H_2PO_4^-]$	midbuffer point (1) $pH = pK_{1a}$
c	50	4.66	$H_2PO_4^-$	equivalence point (1) weak base solution
d	75	7.19	$[H_2PO_4^-] = [HPO_4^{2-}]$	midbuffer point (2) $pH = pK_{2a}$
e	100	9.48	HPO_4^{2-}	equivalence point (2) weak base solution
f	125	11.8	$[HPO_4^{2-}] = [PO_4^{3-}]$	midbuffer point (3) $pH = pK_{3a}$
g	150	12.3	PO_4^{3-}	equivalence point (3) "weak" base solution

4-7 ACID-BASE EQUILIBRIA IN BIOLOGICAL SYSTEMS

The pH of human blood plasma is maintained within the rather narrow limits of 7.4 ± 0.1. Variations in the hydronium ion concentration outside these limits—that is, outside the range $[H_3O^+] = 3.2 \times 10^{-8}$ to 5×10^{-8} mole/liter—are indicative of serious disruption of the normal physiological acid-base balance. These conditions are known as *acidosis*, pH < 7.4, and *alkalosis*, pH > 7.4.

The maintenance of a more or less constant pH is the result of the presence in the blood of a number of conjugate acid-base pairs that effectively compensate for production or consumption of excess acid. In effect, extra-cellular fluid constitutes a very efficient buffer system. One of the most important component buffers is the H_2CO_3 system. This acid is essentially dissolved carbon dioxide. It is influenced, therefore, not only by metabolic processes that form or remove H_3O^+ but also by CO_2 removal through respiration and the net excretion of hydronium ions by the kidney.

The relation of the carbonic acid system to the chemistry of respiration and the maintenance of acid-base balance in the body is an example of an elementary biological application of the principles developed in this chapter.

Amino acids and proteins also are intimately related to acid-base balance and metabolic functions. It is instructive to examine, therefore, the approach that can be taken for the quantitative treatment of the equilibria associated with these species.

The overall physiological acid-base system is extremely complex, but the discussion here is intended to indicate how this system can be treated by a logical extension of the principles, equations, and calculation techniques developed for simple cases. In this way the relevance of fundamental acid-base equilibrium theory, developed here primarily in the context of physical chemistry, may be apparent when the complex equilibria that obtain in living organisms are encountered in standard works in biochemistry and physiology.

4-7-1 The Bicarbonate System as a Physiological Buffer

In the discussion in Sec. 4-4-2, it was indicated that carbonic acid, H_2CO_3, undergoes two successive proton transfer reactions giving rise to bicarbonate, HCO_3^-, and carbonate, CO_3^{2-}, ions. Each process constitutes an equilibrium with a characteristic K_a value.

There is one additional process involving this system that has not yet been considered—that in which carbonic acid in aqueous solution is in equilibrium with gaseous carbon dioxide and liquid water:

$$CO_2(g) + H_2O(l) \rightleftharpoons H_2CO_3(aq)$$

The equation indicates that carbonic acid is essentially dissolved CO_2. Only a very small fraction of the dissolved gas, however, actually exists in solution in the H_2CO_3 form, but the total concentration of dissolved carbon dioxide is available for reaction, so that we can legitimately equate $[CO_2]_{dissolved}$ and $[H_2CO_3]$. For this equilibrium, if the concentration of water is considered to be constant, the following expression can be written:

$$K = \frac{[H_2CO_3(aq)]}{[CO_2(g)]}$$

$$K' = \frac{[H_2CO_3]}{P_{CO_2}} \tag{4-35}$$

If the solution involved is blood, the value of K' is 3×10^{-5} for CO_2 pressure in torr and acid concentration in moles per liter. Rearranging Eq. (4-35) and inserting the numerical value of K', we get

$$[H_2CO_3] = (3 \times 10^{-5})P_{CO_2} \tag{4-36}$$

The equation allows calculation of the concentration of carbonic acid in the blood in equilibrium with a given partial pressure of CO_2.

Conventionally, in biological systems the concentrations of acid-base species are expressed as millimoles (mM) or millequivalents (meq) per liter of solution. To facilitate references to previous discussions in this chapter, the units of mole or equivalents per liter will be retained. The conversion from mole to millimole can be made simply by multiplication by 10^3. For example, if in Eq. (4-36) the concentration of H_2CO_3 in millimoles per liter were required, then

$$[H_2CO_3]_{mM/liter} = (3 \times 10^{-2})P_{CO_2}$$

The several equilibrium processes involved in this system may be summarized as follows:

$$
\begin{array}{ccc}
H_2CO_3 + H_2O & \overset{K_{1a}}{\rightleftharpoons} & HCO_3^- + H_3O^+ \\
\updownarrow {\scriptstyle K'} & & + \\
CO_2(g) & & H_2O \\
+ & & \updownarrow {\scriptstyle K_{2a}} \\
H_2O(l) & & CO_3^{2-} \\
& & + \\
& & H_3O^+
\end{array}
$$

Since the value for each of the equilibrium constants is known, the appropriate equations can be set up to calculate the concentrations of the various species involved. As seen in the case of polyprotic acids, it is generally possible to treat each step in successive equilibria independently. For this particular sequence it is most helpful to base the calculations primarily on the equilibrium between carbonic acid and bicarbonate ion. Blood pH is 7.4 and is maintained at this level by both the carbonic acid itself and by a number of other physiological buffer pairs. The calculations are carried out conveniently using the Henderson–Hasselbalch equation, which for this case is

$$pH = pK_{1a} + \log \frac{[HCO_3^-]}{[H_2CO_3]} \qquad (4\text{-}27'')$$

Rearranging this equation and substituting for the acid concentration at equilibrium, we get the product of K' and P_{CO_2} from Eq. (4-36) as

$$pH = pK_{1a} + \log[HCO_3^-] - \log(3\,P_{CO_2} \times 10^{-5}) \qquad (4\text{-}37)$$

The CO_2 dissolved in the blood is in equilibrium with gaseous carbon dioxide in alveolar air in the lungs, in which the partial pressure of CO_2 remains constant under normal conditions at about 40 torr. Substitution of this value in Eq. (4-37) yields

$$7.4 = 6.1 + \log[HCO_3^-] - \log(1.2 \times 10^{-3})$$

$$\log[HCO_3^-] = -1.62$$

$$[HCO_3^-] = 24 \times 10^{-3} \text{ mole/liter}$$

The CO_3^{2-} ion concentration is now calculable directly from the equilibrium expression for

$$HCO_3^- + H_2O \rightleftharpoons H_3O^+ + CO_3^{2-}$$

$$K_{2a} = \frac{[H_3O^+][CO_3^{2-}]}{[HCO_3^-]} = 6 \times 10^{-11}$$

Substitution of the hydronium and bicarbonate ion concentrations in the system allows calculation of $[CO_3^{2-}]$. The value is 3.6×10^{-5} mole/liter. Thus the concentration of this species is very much smaller than that of either H_2CO_3 or HCO_3^-, that is, 0.04 mM/liter compared with 1.2 and 24 mM/liter. In biological systems, therefore, it is usual to neglect the participation of CO_3^{2-} ions. Their concentration will not influence significantly the fundamental property of the overall equilibrium—the relation between solution pH and the $[HCO_3^-]/[CO_3^{2-}]$ ratio. If for some reason a large concentration of CO_3^{2-} exists in solution, it may then become a significant factor and would have to be taken into account in calculating $[HCO_3^-]$.

The normal physiological situation as it relates to the bicarbonate system can be summarized in the following equilibrium sequence:

$$CO_2 + H_2O \underset{}{\overset{}{\rightleftharpoons}} \quad H_2CO_3 \quad \overset{H_2O}{\rightleftharpoons} \quad HCO_3^-$$

40 torr $\qquad$ 1.2×10^{-3} mole/liter $\qquad$ 24×10^{-3} mole/liter

$$+ \qquad H_3O^+$$

$$4 \times 10^{-8} \text{ mole/liter}$$

There are many processes occurring in the body that influence pH and hence the carbonic acid–bicarbonate ion concentration ratio. The role of this buffer system can be demonstrated by considering how its equilibrium position is influenced by the functions of the lungs and kidneys.

In respiration the rate of ventilation determines the partial pressure of CO_2 in equilibrium with the blood, which in turn regulates $[CO_2]_{dissolved}$ and therefore $[H_2CO_3]$. Increased ventilation will reduce P_{CO_2} and, therefore, tend to shift the equilibrium to lower $[H_3O^+]$ (higher pH) values. Conversely, a decreased ventilation rate tends to lower the pH through the reverse sequence. Since the respiratory apparatus is controlled in part by the pH, either of these deviations results in the appropriate response of a lower or higher ventilation rate.

In renal regulation through the kidney the balance is maintained by excretion of either HCO_3^- or hydronium ions. Thus the pH of the urine varies, depending on the requirements for removal of excess acid or base.

The bicarbonate–carbonic acid system is only one of the buffers in the blood. It is important, however, since regardless of their fundamental cause the changes that occur in the composite system are manifested in this component buffer by alterations that can be readily determined and used to monitor the overall process.

4-7-2 Equilibria Involving Amino Acids and Proteins

Proteins are organic macromolecules of great complexity. They play a vital role in the chemistry of living organisms. The fundamental units from which they are constructed are the amino acids, with the general structure

NH_2 is known as the *amino* group and COOH as the *carboxyl group*. For a given amino acid, R represents a specific group of atoms. A few examples of the nature of this substituent in amino acids occurring in proteins are given in the following tabulation:

R	Amino Acid
—H	Glycine
—CH(CH$_3$)$_2$	Valine
—CH$_2$—C$_6$H$_5$	Phenylalanine
—CH$_2$—COOH	Glutamic acid
—CH$_2$SH	Cysteine
—(CH$_2$)$_4$—NH$_2$	Lysine

Amino acid units are joined in the protein molecules by the *peptide linkage*. This chemical bond is formed by the elimination of H_2O between a hydrogen atom of $-NH_2$ of one amino acid and OH from the carboxyl group of a second molecule:

$$
\begin{array}{ccc}
\text{R} & \text{O} & \text{H} \quad \text{R} \\
| & \diagup\!\!\diagup & \diagdown \quad | \\
H_2N-CH-C & + & N-CH-COOH \rightarrow \\
& \diagdown & \diagup \\
& \underline{[OH\ \ H]} &
\end{array}
$$

$$
\begin{array}{c}
\text{O} \\
\diagup\!\!\diagup \\
H_2N-CH-C \qquad \text{R} \qquad\qquad + H_2O \\
\diagdown \qquad | \\
N-CH-COOH \\
\diagup \\
H
\end{array}
$$

The terminal $-NH_2$ and $-COOH$ groups of this molecule both can be used subsequently for peptide linkages, and through a series of such reactions the protein molecule is built up. At each stage amino acids with differing R groups can be incorporated. The result is a class of complex molecules that can perform a variety of biological functions and is involved in a wide range of decomposition and synthetic reactions in the body.

Only one aspect of these processes is examined here: the behavior of proteins as acids and bases and how the proton transfer equilibria involving these systems can be treated.

The simplest amino acid, glycine, is a convenient example. This molecule,

$$
\begin{array}{ccc}
\text{H} & \text{H} & \text{O} \\
\diagdown & | & \diagup\!\!\diagup \\
\text{N}-\text{C}-\text{C} \\
\diagup & | & \diagdown \\
\text{H} & \text{H} & \text{O}-\text{H}
\end{array}
$$

can be considered from two points of view: (1) as a substituted ammonia, $X-NH_2$, which by analogy with ammonia itself should behave as a weak base,

$$X-NH_2 + H_2O \rightleftharpoons X-NH_3^+ + OH^-$$

and (2) as a substituted acetic acid, $Y-CH_2COOH$, for which weak acid behavior,

$$Y-CH_2COOH + H_2O \rightleftharpoons Y-CH_2COO^- + H_3O^+$$

would be expected. Both these reactions are actually observed. Amino acids are therefore amphoteric, but unlike species of this kind discussed previously the acid and base functions are associated with separate groups on the molecule. It is possible, then, for the acid to exist in a form that would result from a proton transfer occurring within the molecule itself:

$$H_2\overset{\frown}{N}\text{—}CH_2\text{—}COOH \rightarrow H_3N^+\text{—}CH_2\text{—}COO^-$$

In the pure crystalline state amino acids exist in this doubly charged form known as the *zwitterion* or *dipolar ion.*

When the amino acid is placed in solution there are two possible proton transfer reactions:

$$\overset{+}{H_3N}\text{—}CH_2\text{—}COO^- + H_2O \overset{K'_a}{\rightleftharpoons} H_2N\text{—}CH_2\text{—}COO^- + H_3O^+$$

$$\overset{+}{H_3N}\text{—}CH_2\text{—}COO^- + H_2O \overset{K'_b}{\rightleftharpoons} \overset{+}{H_3N}\text{—}CH_2COOH + OH^-$$

These are identical to the two processes written in Sec. 4-4-1 for the general amphoteric species XH, and thus the hydronium ion concentration is given by

$$[H_3O^+] = \sqrt{\frac{K'_a K_w}{K'_b}} \tag{4-22}$$

Conventionally, the equilibrium constants for amino acids are given as K_1 and K_2, where $K_2 = K'_a$ and K_1 is the acid dissociation constant K_a for the positively charged species H_3N^+—CH_2—COOH. K_1 is equal to K_w/K'_b since the species to which it refers is the conjugate acid of the zwitterion. On this basis Eq. (4-22) becomes

$$[H_3O^+] = \sqrt{K_1 K_2} \tag{4-38}$$

It is evident that this system of nomenclature treats the cation (positively charged species) form of the amino as a diprotic acid. The K_1 and K_2 values are its successive acid dissociation constants:

$$\overset{+}{H_3N}\text{—}CH_2\text{—}COOH + H_2O \overset{K_1}{\rightleftharpoons} \overset{+}{H_3N}\text{—}CH_2COO^- + H_3O^+$$

$$\overset{+}{H_3N}\text{—}CH_2\text{—}COO^- + H_2O \overset{K_2}{\rightleftharpoons} H_2N\text{—}CH_2\text{—}COO^- + H_3O^+$$

This description of the interrelation between the various equilibrium constants that can be written for amino acids also points up the care that must be taken to ensure that one knows precisely to what processes K values listed in various tabulations apply.

In more complex amino acids, where the side chain R contains a functional group that also can be involved in proton transfer, there will be an additional equilibrium and, therefore, three K values for the molecule. Again the convention is to list these data for successive proton-donating steps, beginning with the most acidic form. For glutamic acid the three species and the K values for their proton-donating reactions are

$$
\overset{+}{H_3}N—CH—COOH \qquad \overset{+}{H_3}N—CH—COO^- \qquad \overset{+}{H_2}N—CH—COO^-
$$
$$
\quad\quad | \qquad\qquad\qquad\qquad | \qquad\qquad\qquad\quad\; | \quad |
$$
$$
CH_2CH_2COOH \qquad\quad CH_2CH_2COOH \qquad\quad H \;\; CH_2CH_2COO^-
$$
$$
\qquad K_1 \qquad\qquad\qquad\qquad K_2 \qquad\qquad\qquad\qquad K_3
$$

In each species the proton which is transferred is indicated as **H**.

In biological systems we often are concerned with the form in which a particular amino acid will exist at a given pH. Although the amino acids and proteins themselves exert an influence on the overall pH, the most illustrative approach is to consider the fraction of the amino acid that exists in the cation, zwitterion, and anion forms in solutions of various hydronium ion concentrations.

Again using glycine as an example, at high pH the amino acid will exist primarily in the anion form, $H_2N—CH_2—COO^-$. In acidic (low pH) solutions, the $H_3N^+—CH_2—COOH$ form is more important. Thus if we consider that in a given solution of amino acid the pH is altered by addition of a strong acid or a strong base, the anion and cation forms can be thought of as arising via neutralization of the base and acid functions of the zwitterion as

$$
H_2N—CH_2—COO^- \;\underset{\text{at high pH}}{\overset{OH^-}{\longleftarrow}}\; \overset{+}{H_3}N—CH_2—COO^- \;\overset{H_3O^+}{\longrightarrow}\; \overset{+}{H_3}N—CH_2—COOH
$$
$$
\qquad\qquad\qquad\qquad\qquad\qquad\qquad\qquad\qquad\qquad\qquad\qquad\qquad \text{at low pH}
$$

At the specific pH given by Eq. (4-38), the acid is present as the zwitterion itself along with small amounts of the cation and anion in equal concentrations. This pH, characteristic of each amino acid, is called the *isoelectric point* (IEP). The acid has no *net charge* at this pH since it exists as the dipolar species (mainly) and as *equal* concentrations of the positively and negatively charged forms. Therefore, the amino acid will *undergo no net migration in an electric field.**

This is a very important property of amino acids and proteins since it is a fundamental consideration in a method of separating proteins by means of differences in their rate of migration in an electric field—a technique known as *electrophoresis*. This phenomenon is discussed in Chapter 10.

To return to the problem of finding the concentration of the species present as a function of pH, two different situations can be considered.

---

* For exact considerations it is necessary to distinguish between the isoelectric point as defined by the absence of net migration and the *isoionic point*—the pH at which the substance in question has no net charge. The two may be slightly different if the cation and anion forms, present in small concentrations along with the zwitterion, migrate at different rates because of the presence of other ions in solution or because of ions adsorbed on the surface of the protein molecule. The present treatment neglects any such effects.

If the pH is greater than that at the IEP, the solution will contain primarily the zwitterion and anion—an acid and its conjugate base. It therefore is a buffer to which the Henderson–Hasselbalch equation can be applied in the form

$$pH = pK_2 + \log \frac{[H_2N\!-\!CH_2\!-\!COO^-]}{[H_3N^+\!-\!CH_2\!-\!COO^-]}$$

To see why K_2 is the appropriate equilibrium constant to employ in this case, consider the specific process to which it applies. The reaction is that in which the zwitterion acts as an acid producing $H_2N\!-\!CH_2\!-\!COO^-$ as its conjugate base.

If pH is less than the IEP, the two forms present are the zwitterion and the cation, and again the Henderson–Hasselbalch equation applies:

$$pH = pK_1 + \log \frac{[H_3N^+\!-\!CH_2\!-\!COO^-]}{[H_3N^+\!-\!CH_2\!-\!COOH]}$$

To be able to use the K_1 value in this equation, $\overset{+}{H_3N}\!-\!CH_2\!-\!COOH$ is considered as the acid member of the buffer pair and the zwitterion as its conjugate base. Thus, in the general form of the Henderson–Hasselbalch equation, $pK_a \equiv pK_1$ and C_b/C_a is the ratio of concentration of zwitterion to that of the cation.

These two equations are not the only way in which the pH relation can be set up for the two situations. They are convenient, however, since they utilize directly the ordinarily tabulated equilibrium constant values.

In practical calculations the techniques are identical to those discussed previously for simple systems. The only additional requirement for a case such as the glycine solution is to establish initially whether the pH of the solution is greater or less than the IEP. This determines which of the two forms of the Henderson–Hasselbalch equation given above is applicable to the system. For example, the idea of a midbuffer point, previously encountered in the titration of a weak acid with a strong base is equally applicable here. For a solution containing 2×10^{-3} mole/liter of glycine to which is added 1×10^{-3} mole/liter of strong base, exactly one-half of the amino acid would be converted to the anion form and, therefore, $[H_2N\!-\!CH_2\!-\!COO^-] = [H_3N^+\!-\!CH_2\!-\!COO^-]$. From the equation applicable at pH > IEP, we have pH = pK_2 = 9.7.

In more complex amino acids and in proteins constructed from these species, exactly the same kind of considerations can be applied by including additional equilibria, as required, for the various functional groups present. In the protein molecule the main amino and carboxyl groups of the constituent amino acids of course will be used in the formation of the peptide linkage, but the final protein will have large numbers of proton transfer groups incorporated as side chains, in addition to the terminal —NH$_2$ and —COOH groups.

If we again consider the change in form of the protein molecule with increasing pH, beginning in a very acidic solution wherein the protein exists in its most acidic form, it is apparent that the "titration curve" will be complex. The dissociation constant for successive proton transfers as the acidity of the solution is decreased will be very close together, and no sharp changes can be observed. A series of proton transfer reactions could be represented as

$$A \xrightarrow{-H_3O^+} B \xrightarrow{-H_3O^+} C \rightarrow etc.$$

When the dissociation constants are appreciably different, the conversion from A to B is essentially complete before a significant B → C reaction occurs. On the other hand, if the dissociation constants are close together, as is the case with proteins, then long before the A → B reaction is complete appreciable conversion of B to C will have occurred. Thus there will be no precise point at which the pH will be drastically altered. In these circumstances the simple approach represented by the two Henderson–Hasselbalch equations written above is inadequate.

In any protein solution, as the pH continues to increase we eventually reach a stage where the dissociation has proceeded to an extent where the total number of positive and negative charges on the molecule is the same— the isoelectric point. This will be a unique value for a given protein. In such cases, where a hundred or more acid-base groups may be present, the situation is not very susceptible to exact calculation and we rely more on experimental determinations.

PROBLEMS

4-1. Calculate the pH of a solution of HCl at concentrations in moles per liter of (a) 3.50, (b) 0.100, (c) 1.00 × 10^{-3}, and (d) 4.76 × 10^{-1}.

4-2. Calculate the hydronium and hydroxyl ion concentrations in a solution with pH of (a) 3.00, (b) 8.93, and (c) 11.72.

4-3. Calculate the concentration of hydronium ions, pH, and α in a solution of *n*-butyric acid of concentration (a) 0.5 and (b) 0.005.

4-4. A weak acid is 1.75 percent ionized in 0.1 molar solution. What is the value of [H$_3$O$^+$] and K_a. Calculate the pH of the solution and the pK_a for the acid.

4-5. Given that K_a of isovaleric acid, C$_4$H$_8$COOH, is 1.7 × 10^{-5}, calculate the pOH in a solution that is 0.05 molar in sodium isovalerate, C$_4$H$_8$COONa.

4-6. Calculate the pH and concentration of the S^{2-} ion in a 0.0200 normal solution of H$_2$S in water.

4-7. Determine the concentration of the NH_4^+ ion in a solution containing initially 0.1 M NH_3 and 0.05 M NaOH.

4-8. What weight of NH_4Cl in grams is required to be added to 500 ml of 0.1 M NH_3 to give a buffer solution with pH $= 8.75$?

4-9. Calculate the standard free energy change in the proton transfer reaction between (a) benzoic acid and (b) carbonate ion and water.

4-10. What nominal concentration of benzoic acid is required to give a solution in which the acid is 2 percent ionized?

4-11. Calculate the concentration of hydroxyl ions in an aqueous solution containing 15.00 gm/liter $NaNO_2$.

4-12. The amino acid valine has $pK_1 = 2.29$ and $pK_2 = 9.72$. Construct an appropriate graph to show how the composition of the system changes with solution pH. On the same graph, draw three curves to show the percent of the total acid concentration that exists in the cation, zwitterion, and anion form as a function of pH from 1 to 14. Calculate the isoelectric point for this acid. Determine if the curve representing the concentration of the zwitterion is consistent with this calculation.

4-13. Calculate the pH of a solution obtained by mixing 50 ml of 0.02 M NH_3, 25 ml of 0.03 M HCl, and 0.01 M NH_4Cl. (Caution: determine the extent of any proton transfer reactions before applying any equilibrium calculations.)

4-14. pK_{In} for bromcresol purple is 6.2. Spectrophotometric analysis of a solution containing this indicator indicated that 63.50 percent of the indicator was in the base form. What is the pH of this solution?

4-15. How effectively could the end points in the titration of citric acid with a strong base be determined by the use of indicators? Draw an approximate titration curve to illustrate the answer.

4-16. Calculate the weight in grams of solid NaH_2PO_4 and Na_2HPO_4 that must be added to 1 liter of water to have a total salt concentration of 0.75 mole/liter and a pH of 7.40.

4-17. For the following indicators calculate the percent in the base form at the pH indicated: (a) bromophenol blue ($pK_{In} = 4.0$), pH of 3.50, and (b) cresol red ($pK_{In} = 8.1$), pH of 9.50.

4-18. Calculate the approximate pH in an aqueous solution of $NaHSO_3$.

4-19. 100 ml of a solution 0.5 molar in $C_6H_5O^-$ is titrated with 0.5 molar HCl. Calculate the pH of the solution when 0, 20, 50, 70, 98, 100, 102, and 180 ml of acid have been added and present the data in a conventional pH vs. milliliters of acid added titration curve. Suggest two indicators that might be suitable for this titration.

Ionic Equilibrium II:
Ion Activities

5

It was indicated in Chapter 3 that the effects of deviations from ideal behavior were most pronounced in ionic solutions. Activities rather than concentrations therefore must be utilized frequently for equilibrium and related calculations in these systems.

This chapter deals initially with one example of the unique properties of ionic solutions and subsequently with some of the theories and equations proposed for the quantitative treatment of these effects in general. With these techniques available, some of the acid-base systems discussed in the previous chapter can be reconsidered to determine the conditions under which concentration-based equations must be modified and how these changes can be made.

Any discussion of ionic equilibrium would be incomplete without an examination of the equilibria characterizing solutions of slightly soluble salts. The principal features of this type of equilibrium will be examined and used to compare the application of the law of mass action and the thermodynamic approach utilizing activities.

5-1 CONCENTRATION EFFECTS IN IONIC SOLUTIONS

Substances that form ions when dissolved in water yield solutions that conduct electric current. Ion formation can result from complete or partial dissociation or from proton transfer.

There are a number of aspects of ionic solutions that could be used as a basis for the discussion of their fundamental properties.* Measurements of the efficiency with which ionic solutions conduct electricity, however, yield some of the most illustrative data. It will not be necessary to make extensive use of these data in later discussions, but they are of great value in elucidating the general problem of quantitative treatment of ions in solution and how these effects become more important as the concentration increases.

5-1-1 Determination of Equivalent Conductance

The passage of an electric current through a conductor is governed by Ohm's law:

$$I = E/R \qquad (5\text{-}1)$$

where I is the current in *amperes*, E is the electromotive force in *volts*, and R is the resistance in *ohms*. R depends on the physical dimensions of the conductor, the length l, cross-sectional area A, and the inherent resistance of the material to the flow of current. The latter property is expressed as the *resistivity* ρ, in ohm centimeters, which is equal to R for a conductor with $A = 1$ cm^2 and $l = 1$ cm. The resistance of a particular conductor is given by

$$R = \rho l/A \qquad (5\text{-}2)$$

For this discussion it is more convenient to think in terms of the ability of a solution to conduct electricity rather than of its resistance. Thus use is made of the *conductance*, which is defined as the reciprocal of R and which has units of mho (or ohm^{-1}). Like R, the conductance depends on the physical dimensions of the conductor so that in practice data are presented in terms of the *specific conductance*,

$$\kappa = 1/\rho \qquad (5\text{-}3)$$

in units of mho/cm. κ is independent of the size of the conductor for a given material.

In determining κ for an ionic solution there are problems not encountered with metallic conductors. When direct current is passed through the solution, *electrolysis*—a chemical reaction resulting from loss or gain of electrons by the ions—will occur. This process is discussed in Chapter 7. Even if the extent of the electrochemical reaction is reduced by utilizing small currents, there still occurs a net reduction in the concentration of ions of opposite sign near each electrode. This causes the solution to exhibit a resistance varying with the applied voltage. The effects can be overcome by measuring the resistance of the solution to alternating current in a *conductivity*

---

* The colligative properties of electrolyte solutions, for example, can be employed to obtain activity coefficient data. This technique is discussed in Chapter 9.

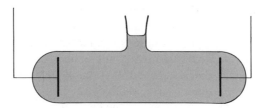

FIG. 5-1 Schematic diagram of a simple conductivity cell.

cell. This apparatus is a cylindrical cell with fixed electrodes, usually of disk shape, with known area. A simple conductivity cell is shown in Fig. 5-1. The resistance is measured by means of an alternating current Wheatstone bridge that operates essentially by balancing the resistance of the cell against that of a calibrated standard. It is imperative that the water used as solvent be virtually free from ionic impurities. Especially purified *conductivity water* is used for this purpose. Since the conductance is also temperature sensitive, accurate work also requires thermostating the apparatus.

For a simple cell, the resistivity, and hence the specific conductance, can be calculated by use of the measured values of l (the distance between the electrodes) and A (their area). Thus, from Eqs. (5-2) and (5-3) we have

$$\kappa = \left(\frac{l}{A}\right)\frac{1}{R} \tag{5-4}$$

It is more usual, however, to measure the resistance of the conductivity cell when it contains a solution of known κ (potassium chloride solutions are most often used) and from Eq. (5-4) to calculate the effective value of l/A, the *cell constant*, for that particular apparatus. Thus the working equation for conductivity measurements is

$$\kappa = \text{cell constant}/R \tag{5-5}$$

The specific conductance of any given solution depends on the concentration of ions and their charge. To facilitate comparisons between different solutions, an *equivalent conductance*, Λ is defined as the conductance of a solution containing one equivalent each of positive and negative ions measured by electrodes that are 1 cm apart. For an ion, an equivalent is that weight equal to the gram ionic weight divided by its charge. Thus one equivalent represents the number of ions required to carry Avogadro's number, a mole, of positive or negative charges. For the electrolyte as a whole the equivalent is that weight equal to the gram formula weight divided by the total number of positive *or* negative charges.

The equivalent and specific conductances are related by

$$\Lambda = \frac{\kappa}{C'} = \frac{1000\,\kappa}{C} \tag{5-6}$$

where C' is the concentration of electrolyte in equivalents per cubic centimeter and C is the number of equivalents per liter of solution. Λ has units of mho square centimeters per equivalent. The following example illustrates the use of these equations in obtaining the equivalent conductance from experimental data:

> EXAMPLE 5-1. A solution containing 1.115 gm/liter of Cu_2SO_4 was determined to have a resistance of 2,140 ohm in a conductivity cell with cell constant $= 1.78$ cm^{-1}. Calculate Λ of the solution.
>
> $$\text{Gram equivalent weight of } Cu_2SO_4 = \frac{\text{formula wt.}}{2} = \frac{223}{2} = 111.5 \text{ gm}$$
>
> $$C = \frac{1.115 \text{ gm/liter}}{111.5 \text{ gm/eq.}} = 0.01 \text{ eq./liter}$$
>
> $$\kappa = \text{cell constant}/R = 1.78/2{,}140 = 0.829 \times 10^{-3}$$
>
> $$\Lambda = \frac{10^3 \kappa}{C} = \frac{10^3 \times 0.829 \times 10^{-3}}{0.01} = 83 \text{ mho cm}^2/\text{eq.}$$

5-1-2 Concentration Dependence of Λ

If the specific conductance of a given solution were directly proportional to the concentration of the electrolytes, then the *equivalent* conductance should be the same at all concentrations. Thus once the value of Λ for a particular solution were determined at any one concentration convenient for experimental measurements, the specific conductance at any other concentration would be calculable from Eq. (5-6). For the solution in Example 5-1, if Λ were independent of concentration, the specific conductance, in mho per centimeter, would be 0.0083 when $C = 0.1$ and 0.083 when $C = 1$. The values actually observed are 0.0051 and 0.0293 at these concentrations. That is, the equivalent conductance decreases from 83 mho cm^2/eq. at $C = 0.01$ to 29.3 mho cm^2/eq. when $C = 1$ eq./liter.

A decrease in the equivalent conductance with increasing concentration is a characteristic of all ionic solutions. For *strong electrolytes*, graphical analysis shows that the equivalent conductance is approximately a linear function of $\sqrt{C}$ over a fairly wide range of C values. Thus the behavior can be represented by the empirical relation

$$\Lambda = \Lambda_0 + m\sqrt{C} \tag{5-7}$$

where Λ_0 is the equivalent conductance at infinite dilution—the intercept of the straight line plot with the Λ axis—and m is the slope of the line (an experimental constant characteristic of the solute in question).

The second class of electrolytes, *weak electrolytes*, exhibit a Λ vs. $\sqrt{C}$

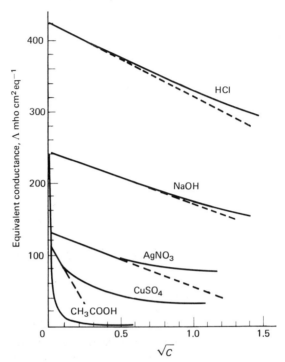

FIG. 5-2 *Equivalent conductance of electrolytes in aqueous solution as a function of $\sqrt{C}$. Adapted from W. J. Moore, Physical Chemistry, 3rd ed., Prentice-Hall, Inc., Englewood Cliffs, N.J., 1962, p. 329.*

plot that is not linear except in very dilute solutions where accurate determinations of the conductance are extremely difficult if not impossible to make.

Representative equivalent conductance vs. $\sqrt{C}$ curves for several electrolytes are shown in Fig. 5-2.

The theory proposed initially by Arrhenius suggested that solutions of both classes of electrolytes contained both dissociated and undissociated species in equilibrium. According to the law of mass action, as the concentration increases, the amount of dissociation is reduced. This decrease in the fraction of the solute existing in the ionic form then would cause a lowering of the equivalent conductance.

It is now known that this explanation applies only to weak electrolytes—those solutes exhibiting conductance curves that rise sharply at very low concentrations. The acetic acid curve in Fig. 5-2 is typical of their behavior. The equivalent conductance of weak electrolytes is directly proportional to the fraction that exists in the dissociated form. At infinite dilution this fraction must be one. Arrhenius related the degree of dissociation α at a particular concentration to the ratio of Λ in that solution to the equivalent conductance at infinite dilution as

$$\alpha = \Lambda/\Lambda_0 \tag{5-8}$$

If the weak electrolyte is an acid, this quantity is identical to that previously defined (Sec. 4-3-1) by

$$\alpha = \frac{[H_3O^+]}{C} \tag{4-16}$$

If a value of Λ_0 is known,* Eq. (5-8) represents an alternative method of calculating α from experimental data and hence of obtaining a value for the acid or base dissociation constant.

The primary interest in equivalent conductance data in the context of this discussion, however, is its significance for strong electrolytes in terms of "nonideal" behavior of ions in solution. These solutes are completely dissociated, and the Arrhenius theory is inapplicable. We now turn our attention to a somewhat different explanation of the behavior of the equivalent conductance of strong electrolytes and examine how these effects can be treated quantitatively by using techniques applicable to situations other than conductance determinations.

5-2 ACTIVITY COEFFICIENTS OF IONS

The most satisfactory approach to the problem of rationalizing strong electrolyte behavior in solution is the Debye–Huckel theory. Its essential features are that the strong electrolyte is 100 percent dissociated and that the decrease in equivalent conductance arises from electrostatic interaction between charged solute particles.

The effect of these interactions is to introduce an additional term that must be included when the free energy of an ion at concentration C is evaluated. The expression for the free energy per mole of a solute, in general, was seen previously to be given by

$$G = G^\circ + 2.303 \, RT \log C \tag{2-32}$$

This equation must now be modified by the addition of a G' term due to ionic interactions, such as

$$G = G^\circ + 2.303 \, RT \log C + G' \tag{5-9}$$

* As mentioned previously, it is not possible to extrapolate the Λ vs. $\sqrt{C}$ curve for weak electrolytes because of the large uncertainty in Λ values at small C. The equivalent conductance of a solution, however, is the sum of the equivalent conductances of the individual ions, λ_1, λ_2, etc. This rule applies strictly only at infinite dilution, where it takes the form $\Lambda_0 = \lambda_+^\circ + \lambda_-^\circ$ for a binary electrolyte. Thus the value of Λ_0 for a weak electrolyte can be found by determining λ° values for its component ions when these are present as part of a strong electrolyte whose equivalent conductance curve can easily be extrapolated.

Since the magnitude of this interaction is different at each concentration, it is appropriate to express the resulting free energy in the same formalism as the other concentration-dependent free energies, that is,

$$G' = 2.303 \ RT \log \gamma \qquad (5\text{-}10)$$

where γ is a parameter dependent on concentration. Substitution of this expression for G' into Eq. (5-9) and collection of the terms containing $2.303 \ RT$ gives

$$G = G° + 2.303 \ RT(\log C + \log \gamma)$$

or

$$G = G° + 2.303 \ RT \log \gamma C \qquad (3\text{-}16)$$

It is apparent, then, that we can identify the concentration-dependent parameter γ as the *activity coefficient*, since the equation obtained is identical to that derived previously for the free energy per mole of a solute at concentration C, corrected for nonideal behavior by the γ coefficient.

Although this approach is in no way a "derivation" of the defining equation for activity coefficients, it is useful in that it does indicate, at least qualitatively, the connection between the nonideal correction factor and the free energy that arises from electrostatic interaction between ions in solution.

The Debye–Huckel theory gives an expression for γ, derived on the basis of a rigorous consideration of the nature of the electrostatic forces in solution. The extent of these interactions depends not only on the concentration of solute ions but also on their charge. It therefore is useful to employ a measure of concentration that takes these charges into account and can be used in any expression for the concentration dependence of γ. The required quantity is an indication of the total charge concentration in the solution and is called the *ionic strength μ*. The defining equation is

$$\mu = \tfrac{1}{2} \sum_i C_i Z_i^2 \qquad (5\text{-}11)$$

where C_i is the concentration in moles per liter of the ith ion, Z_i is its charge, and $\sum_i$ indicates that the CZ^2 term for *all* ions in the solution are to be included. The ionic strength is one-half this total.

Considerable difficulty in subsequent calculations using μ values is avoided if it is appreciated that this quantity is a gross property of the solution—not of an individual ion. Thus when we wish to calculate the activity coefficient of a given ion, we first determine the ionic strength *of the solution*, via. Eq. (5-11), including the CZ^2 term not only for the ion of interest but also for all other ions in solution.

Before examining the method of calculating γ, it is helpful to illustrate how Eq. (5-11) is applied. For 0.01 molar KCl the ionic strength is

$$\mu = \tfrac{1}{2}(C_{K^+} Z_{K^+}^2 + C_{Cl^-} Z_{Cl^-}^2) = \tfrac{1}{2}[0.01(+1)^2 + 0.01(-1)^2] = 0.01$$

Thus for a simple electrolyte with singly charged ions in a 1:1 ratio, a *uniunivalent electrolyte*, μ and C are identical. On the other hand, these quantities will be different for more complex species. A solution 0.01 molar in K_2SO_4 would have

$$\mu = \tfrac{1}{2}(C_{K^+} Z_{K^+}^2 + C_{SO_4^{2-}} Z_{SO_4^{2-}}^2)$$

From the stoichiometry of the dissociation process, $C_{K^+} = 0.02$ and $C_{SO_4^{2-}} = 0.01$, and the ionic strength is therefore

$$\mu = \tfrac{1}{2}[0.02(+1)^2 + 0.01(-2)^2] = 0.03$$

The usefulness of the ionic strength is that at low concentrations the activity coefficient of a given ion depends only on its charge and the value of μ for the solution. γ is independent of the nature of the ions present.

It is not possible to isolate a single positive or negative ion and measure its activity coefficient. In practice we can determine only the *mean activity coefficient* $\gamma_\pm$, defined for a binary electrolyte as the geometric mean of the activity coefficients of the individual ions,

$$\gamma_\pm = \sqrt{\gamma_+\gamma_-} \tag{5-12}$$

The subscripts $+$ and $-$ indicate only whether the ion is positively or negatively charged and do not necessarily mean that it carries a single charge.

The essential result of the Debye–Huckel theory is the equation that gives the value of the mean activity coefficient in terms of the ionic strength. The general Debye–Huckel equation is

$$-\log \gamma_\pm = 1.83 \times 10^6 [Z_+ Z_-]\left(\frac{\mu \rho_0}{\epsilon^3 T^3}\right)^{\frac{1}{2}}$$

where Z_+ and Z_- are the magnitudes of the charges on the positive and negative ions respectively, ρ_0 is the solvent density, and ϵ is the dielectric constant—a dimensionless number that is a measure of the polarity of the solvent. For water at 298 °K, ρ_0 is 0.997 and $\epsilon = 78.54$. Substitution of these values gives the working equation for aqueous solutions at 25 °C;

$$-\log \gamma_\pm = 0.509[Z_+ Z_-]\sqrt{\mu} \tag{5-13}$$

Although it is not possible experimentally to measure the quantity, it nonetheless is quite useful to *calculate* the activity coefficient of an individual ion. This can be done by using a modified form of Eq. (5-13):

$$-\log \gamma = 0.509 \, Z^2 \sqrt{\mu} \tag{5-14}$$

As with the equation for $\gamma_\pm$, this relation applies only to an aqueous solution at 25 °C.

Equations (5-13) and (5-14) are valid for solutions where μ is less than about 0.01.

EXAMPLE 5-2. Calculate the activity coefficient and activity of the barium ion in a solution (1) 2×10^{-3} molar in $BaCl_2$ and (2) 2×10^{-3} molar in $BaCl_2$ and 0.01 molar in $NaNO_3$.

(1) $\quad \mu = \frac{1}{2}[C_{Ba^{2+}}Z_{Ba^{2+}}^2 + C_{Cl^-}Z_{Cl^-}^2] = \frac{1}{2}[0.002(+2)^2 + 0.004(-1)^2]$

$$\mu = 6 \times 10^{-3}$$

$-\log \gamma_+ = 0.509 \, Z^2 \sqrt{\mu} = 0.509(2)^2(6 \times 10^{-3})^{\frac{1}{2}} = 0.159$

$\gamma_+ = 0.69 \quad a_{Ba^{2+}} = [Ba^{2+}]\gamma_+ = 0.69 \times 2 \times 10^{-3} = \underline{1.4 \times 10^{-3}}$

(2) $\quad \mu = \frac{1}{2}[C_{Ba^{2+}}Z_{Ba^{2+}}^2 + C_{Cl^-}Z_{Cl^-}^2 + C_{Na^+}Z_{Na^+}^2 + C_{NO_3}Z_{NO_3}^2]$

$\mu = \frac{1}{2}[0.002(2)^2 + 0.004(-1)^2 + 0.01(1)^2 + 0.01(-1)^2]$

$$\mu = 1.6 \times 10^{-2}$$

$-\log \gamma_+ = 0.509 \, Z^2 \sqrt{\mu} = 0.509(2)^2(1.6 \times 10^{-2})^{\frac{1}{2}} = 0.258$

$\gamma_+ = 0.55 \quad a_{Ba^{2+}} = [Ba^{2+}]\gamma_+ = 0.55 \times 2 \times 10^{-3} = \underline{1.1 \times 10^{-3}}$

concentrated ~~dilute~~ soln

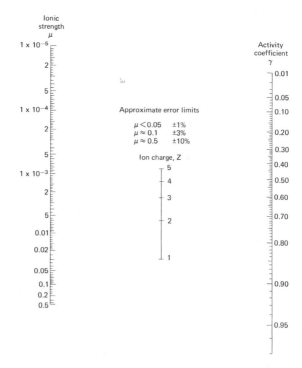

FIG. 5-3 Nomograph for the estimation of ion activity coefficients. To obtain γ, align the value of the solution ionic strength with the charge on the ion and read from the extrapolation of the line joining these two points, the value of the activity coefficient on the right-hand axis. Reprinted by permission from J. N. Butler, Ionic Equilibrium, Addison-Wesley Publishing Company, Inc., Reading, Mass., 1964, p. 473.

For more concentrated solutions two difficulties are encountered: The situation is no longer adequately described by the Debye–Huckel equation and, in addition, the activity coefficient varies from ion to ion of the same charge for a particular ionic strength. Several empirical equations extending the range of applicability of Eq. (5-13), by including additional terms with adjustable parameters have been suggested. For many calculations, however, it is sufficient to have a reasonably accurate estimate of a generally applicable value for ions in a solution of given ionic strength. One such general relation developed by Davies is given here in the form

$$-\log \gamma = 0.509 \, Z^2 \, \frac{\sqrt{\mu}}{1 + \sqrt{\mu}} - 0.2\sqrt{\mu} \qquad (5\text{-}15)$$

which is applicable to aqueous solutions at 25 °C. A nomograph for the determination of γ from μ and Z based on this equation has been devised by J. N. Butler and is shown in Fig. 5-3, along with the approximate error limits at various concentrations. Except for very dilute solutions where the Debye–Huckel equation, Eq. (5-13) or (5-14), gives more accurate values that can be obtained by use of the nomograph, the inherent errors in the modified equation are such that taking values directly from the nomograph does not introduce any significant new error. Thus this chart is very convenient for calculations requiring the use of ion activity coefficients.

5-3 pH CALCULATIONS USING THE ACTIVITY

The general principle to be followed here is that in any situation where the ionic strength of the solution is such that the assumption of $\gamma = 1$ leads to unacceptable errors, the concentration terms must be replaced by the activity—the product of the concentration in moles or equivalents per liter and the activity coefficient.

5-3-1 pH° of Strong Acids and Bases

The fundamental definition of pH has been given previously as

$$\text{pH} = -\log a_{\text{H}_3\text{O}^+} \approx -\log[\text{H}_3\text{O}^+] \qquad (4\text{-}2)$$

Incorporating the activity coefficient, we get

$$\text{pH}° = -\log \gamma_+[\text{H}_3\text{O}^+]$$

or

$$\text{pH}° = -\log C_a - \log \gamma_+ \qquad (5\text{-}16)$$

where the superscript ° indicates explicitly that the pH is based strictly on activities. Since the activity coefficient usually will be less than unity (except in very concentrated solutions that must be dealt with on a semiempirical basis), the effect of the $-\log \gamma_{\pm}$ will be to increase $pH°$, that is, the effective concentration of hydronium ions is lower than the nominal concentration.

The values of $pH°$ and pH are significantly different at concentration levels not normally thought of as being concentrated solutions. For example, in aqueous HCl solution, where $\mu = C$, the nomograph indicates that even in a solution that would not be considered as "concentrated" ($C = 0.01$ mole/liter) the activity coefficient of the hydronium ion is 0.90. The value of pH for this case is 2.0; the exact value is

$$pH° = -\log(10^{-2}) - \log 0.90 = \underline{2.05}$$

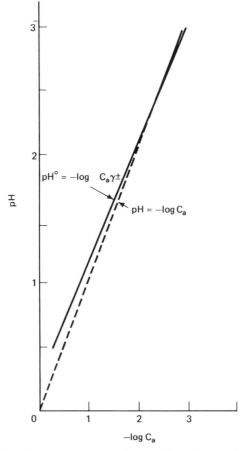

FIG. 5-4 *pH values, approximate and exact, for a solution of a uniunivalent strong acid in water.*

Thus pH is 2.5 percent in error compared with pH°. Since the definitions are in terms of logarithms, however, the assumption that concentration and activity are identical in this case is in error by 10 percent, that is, 0.90×10^{-2} compared with 1.00×10^{-2}.

A graph of pH and pH°, calculated as $-\log C_a$ and $-\log(C_a\gamma_\pm)$ respectively, as a function of $-\log C_a$ is shown in Fig. 5-4. This plot is valid for any uniunivalent strong acid present alone in aqueous solution. When other solutes are present, it is necessary to calculate the ionic strength and then obtain the activity coefficient from Fig. 5-3.

EXAMPLE 5-3. Calculate the pH° of a solution that contains HNO_3 at a concentration of 1.0×10^{-3} mole/liter and $Mg(NO_3)_2$ at 2.0×10^{-3} mole/liter

$$C_{NO_3^-} = 0.001 \text{ (from } HNO_3) + 0.004 \text{ [from } Mg(NO_3)_2] = 0.005$$
$$\mu = \tfrac{1}{2}[0.002(2)^2 + 0.005(-1)^2 + 0.001(+1)^2] = 0.007$$
$$\gamma_+ = 0.915 \qquad \log \gamma_+ = -0.00386 \quad \text{(for } H_3O^+ \text{ ion)}$$
$$pH° = -\log C_a - \log \gamma_+ = -(-3) + 0.0386 = \underline{3.04}$$

5-3-2 pH° in Weak Acid and Buffer Solutions

Beginning with the simplest system of a weak acid in aqueous solution, the equilibrium constant can be written as

$$K_a = \frac{a_{H_3O^+}a_{A^-}}{a_{HA}} = \frac{[H_3O^+][A^-]}{[HA]}\frac{\gamma_{H_3O^+}\gamma_{A^-}}{\gamma_{HA}} \qquad (5\text{-}17)$$

To facilitate the discussion of this equation there are a number of simplifications that can be made. The product of the activity coefficients of the hydronium and A^- ion is the square of the mean activity coefficient, $\gamma_\pm^2$, as defined by Eq. (5-12). From the stoichiometry, $[H_3O^+] = [A^-]$, except in very dilute solution, and thus $[H_3O^+][A^-]$ can be replaced by $[H_3O^+]^2$. The term [HA] refers to the concentration of undissociated acid at equilibrium—a molecular solute—and, therefore, γ_{HA} can be taken as unity with negligible error. Thus the product in the denominator of Eq. (5-17) becomes $C_a - [H_3O^+]$, and the whole equation can be rewritten as

$$K_a = \frac{[H_3O^+]^2\gamma_\pm^2}{C_a - [H_3O^+]} \qquad (5\text{-}18)$$

If it is assumed that the amount of dissociation is a negligible fraction of C_a, the denominator reduces to C_a and

$$K_a = ([H_3O^+]\gamma_\pm)^2/C_a$$
$$[H_3O^+]\gamma_\pm = a_{H_3O^+} = \sqrt{K_aC_a} \qquad (5\text{-}19)$$

It therefore is apparent that the value of $\sqrt{K_a C_a}$ is *not* identical with the hydronium ion concentration as assumed in Eq. (4-13) but rather is equal to the activity of this species. Furthermore, considering the exact definition of pH° as $-\log a_{H_3O^+}$, we have from Eq. (5-19) that

$$pH° = \tfrac{1}{2}pK_a - \tfrac{1}{2}\log C_a \qquad (4\text{-}15')$$

This is the same equation obtained previously by interpreting pH values in terms of $[H_3O^+]$ by assuming activity coefficients of unity. When activity coefficients are incorporated we see that, in fact, Eq. (4-15) gives the value of the "true" pH, which has been designated as pH°. This equality, of course, requires that the K_a value used be appropriate to a solution in which the ionic strength is zero—where the activity coefficients are exactly unity. The values for acid and base dissociations given in Table 4-2 are those obtained under these conditions, and therefore their use in Eq. (4-15′) is justified.

When the assumption that the equilibrium acid concentration can be represented by C_a alone is no longer justified, Eq. (5-18) must be solved in full. Again, however,

$$\frac{-K_a \pm \sqrt{K_a^2 + 4C_a K_a}}{2}$$

[which was equated with $[H_3O^+]$ in Eq. (4-13)] gives the value $a_{H_3O^+}$.

In practice, then, a solution of weak acid (or base) can be treated using the same formal equations, whether or not the activity coefficients are being considered. The results can be interpreted as pH or pH° as required for a given problem. It should be noted carefully, however, that these equations give *exactly* the value of the hydronium ion activity. The "inaccuracy" in the interpretation of these equations in Chapter 4 was that the relations were assumed to give values of the concentration when in fact they yield activities.

This convenient applicability of the same formalism to both types of situation does not extend to all cases. In a buffer solution containing an acid and its conjugate base, the concentration of hydronium ions and base ion will not be the same. In addition, the common ion will be introduced along with a positively charged ion, for example, Na^+; the concentration of this ion also must be considered in calculating the ionic strength. Equation (5-17) again applies, but not the equality $[H_3O^+] = [A^-]$. One simplifying feature noted in earlier discussions of buffers is that [HA] invariably can be replaced by C_a. Equation (5-17) is thus written for the buffer solution as

$$K_a = [H_3O^+]\gamma_\pm [A^-]\gamma_\pm / C_a$$

and thus

$$pH° = pK_a + \log \frac{[A^-]\gamma_\pm}{C_a} \qquad (5\text{-}20)$$

This is identical to the Henderson–Hasselbalch equation when $\gamma_\pm = 1$.

EXAMPLE 5-4.   Calculate the pH° of a solution 0.1 molar in formic acid, HCOOH, and sodium formate, HCOONa. If 0.1 mole/liter of $BaCl_2$ is added, what is the resulting change in pH°?

In the first solution, since $C_a = C_b$, $[H_3O^+] \approx K_a \approx 10^{-4}$ and, therefore, this species will make a negligible contribution to μ.

$$\mu = C_{HCOONa} = 0.1 \quad \text{(uniunivalent electrolyte)}$$

$$Z = +1 \qquad \gamma_\pm = 0.77 \quad \text{(Fig. 5-3)}$$

$$pH^\circ = pK_a + \log \frac{0.1 \times 0.77}{0.1} = 3.70 + \log 0.77 = \underline{3.59}$$

On addition of $BaCl_2$,

$$\mu = \tfrac{1}{2}[C_{Na^+}Z^2_{Na^+} + C_{HCOO^-}Z^2_{HCOO^-} + C_{Ba^{2+}}Z^2_{Ba^{2+}} + C_{Cl^-}Z^2_{Cl^-}]$$

$$\mu = \tfrac{1}{2}[0.1(1)^2 + 0.1(-1)^2 + 0.1(2)^2 + 0.2(-1)^2] = 0.4$$

$$Z = +1 \qquad \gamma_\pm = 0.69 \quad \text{(Fig. 5-3)}$$

$$pH^\circ = 3.70 + \log \frac{0.1 \times 0.69}{0.1} = \underline{3.54} \qquad \underline{\Delta pH = 0.05}$$

This example illustrates an important point. The pH° of a solution can be altered by the presence of ions that are not involved in proton transfer but nonetheless are influential because of their effect on the ionic strength.

Another case of interest, where the introduction of activity coefficients gives rise to a new equation, is where the weak acid is an ionic species in its undissociated form. For a general acid of this type XH^+—for example, NH_4^+—the equilibrium expression is

$$K_a = \frac{[X][H_3O^+]\gamma_\pm}{[XH^+]\gamma_\pm}$$

Although $[XH^+]$ and $[H_3O^+]$ will be quite different, the activity coefficients of these ions will be the same, since γ depends only on the value of μ for the solution and the charge on the ion. Since a value for pH° is required, however, the term $\gamma_\pm$—that is, $\gamma_{H_3O^+}$—must remain in the expression. If $\gamma_\pm$ values were canceled, the result would be an exact expression for the *concentration* of the hydronium ion—not the activity as required. From the stoichiometry, $[X] = [H_3O^+]$ and, consequently, the convenient substitution of $a_{H_3O^+}$ for $[X]$ can be made.

If it is assumed that $[XH^+] = C_a$, the equilibrium relation is altered, on making the substitutions listed, to the form

$$K_a = \frac{a_{H_3O^+}[H_3O^+]\gamma_\pm}{\gamma_\pm C_a \gamma_\pm} = \frac{(a_{H_3O^+})^2}{C_a(\gamma_\pm)^2}$$

$$a_{H_3O^+} = \gamma_\pm\sqrt{K_a C_a} \tag{5-21}$$

$$pH^\circ = \tfrac{1}{2}pK_a - \tfrac{1}{2}\log C_a - \log \gamma_\pm \tag{5-22}$$

An important application of equations such as (4-15') and (5-22) is in the experimental determination of acid dissociation constants. In order that the K_a values obtained be those defined by Eq. (5-17), that is, a true thermodynamic equilibrium constant defined in terms of activities, the method of determining pH must be such that, in fact, a value for pH° is obtained from the measurements. The most common method for precise determination of hydronium ion activities is that involving electromotive force measurements in electrochemical cells. This method is discussed in Chapter 7.

5-4 SLIGHTLY SOLUBLE SALTS

Among compounds that are completely dissociated in ionic form in solution there is a group of ionic salts that dissolve to only a very limited extent and thus give rise to solutions where the concentration of ions is quite minute.

In this type of situation the equilibrium between the ions in solution and the solid, undissolved salt can be treated adequately in terms of the law of mass action. On this basis a modified equilibrium constant conveniently related to the solubility of the salt can be obtained. When the solution in which the ions must dissolve has a significant ionic strength, however, the activity coefficients of the ions from the slightly soluble material will be less than unity. Their effective concentration will be altered, with a resulting change in the solubility. The general principles developed previously now can be applied to both these situations.

5-4-1 Solubility Products

The behavior of slightly soluble salt solutions can be represented, for the general salt A_mB_n, by the equilibrium

$$A_mB_n(s) \rightleftharpoons mA^{+n}(aq) + nB^{-m}(aq)$$

for which the equilibrium expression in terms of concentrations is

$$K = \frac{[A^{+n}]^m[B^{-m}]^n}{[A_mB_n]}$$

The term $[A_mB_n]$, the concentration of the solid salt, depends only on the density of the substance and, therefore, it can be included in the equilibrium constant,

$$K[A_mB_n] = K_{sp} = [A^{+n}]^m[B^{-m}]^n \tag{5-23}$$

where the modified equilibrium constant K_{sp} is the *solubility product*. If a logarithmic expression is more convenient,

$$pK_{sp} = -\log K_{sp} = -m\log[A^{+n}] - n\log[B^{-m}] \qquad (5\text{-}24)$$

Equation (5-23) or (5-24) can be employed so long as the concentration of ions in solution is relatively small and neither A^{+n} nor B^{-m} participate in any further reactions. In pure water or a solution containing no common ion the *solubility* of the salt $\mathscr{S}$ can be calculated from the K_{sp} value and the stoichiometry of the dissociation process. $\mathscr{S}$ may be expressed in moles per liter or grams per liter, but if the latter units are employed the values must be converted to molarity before substitution into the equations developed here. For the $A_m B_n$ case,

$$\mathscr{S} = \frac{[A^{+n}]}{m} = \frac{[B^{-m}]}{n} \qquad (5\text{-}25)$$

Substitution from Eq. (5-25) for the ion concentration in the K_{sp} expression gives

$$K_{sp} = (m\mathscr{S})^m (n\mathscr{S})^n \qquad (5\text{-}26)$$

For the specific case of MgF_2 (neglecting hydrolysis effects involving the F^- ion to be discussed below), Eqs. (5-25) and (5-26) take the form

$$\mathscr{S} = [Mg^{2+}] = \frac{[Cl^-]}{2} \qquad K_{sp} = (\mathscr{S})(2\mathscr{S})^2$$

where

$$\mathscr{S} = \left(\frac{K_{sp}}{4}\right)^{\frac{1}{3}}$$

In practice Eq. (5-26) is used to calculate K_{sp} values from experimental solubility measurements or, conversely, to determine solubilities from tabulated solubility product data.

EXAMPLE 5-5.   The solubility of $Mg(OH)_2$ in water is 9.1×10^{-3} gm/liter. Calculate the solubility product.

$$\mathscr{S} = [Mg^{2+}] = [OH^-]/2 \qquad K_{sp} = (\mathscr{S})(2\mathscr{S})^2 = 4\mathscr{S}^3$$

$$\mathscr{S} = \frac{9.1 \times 10^{-3} \text{ gm/liter}}{58.33 \text{ gm/mole}} = 1.56 \times 10^{-4} \text{ mole/liter}$$

$$K_{sp} = 4(1.56 \times 10^{-4})^3 = \underline{1.52 \times 10^{-11}}$$

There are two features of Eq. (5-23) that have important implications. The value of K_{sp} establishes an upper limit on the product of the concentrations of the ions of a slightly soluble salt that can exist in equilibrium in solution. Any solution containing these ions will be stable only if the concentration product does not exceed K_{sp}. If this does occur, precipitation of the solid salt results. Furthermore, the equation applies to *any* solution, regardless of the source of the ions. For example, the K_{sp} of silver chloride is 1.8×10^{-10}. This imposes the condition that the maximum concentration of Ag^+ and Cl^- ions that can coexist in aqueous solution must be those that have $[Ag^+][Cl^-] \leq 1.8 \times 10^{-10}$. This restriction applies to both a solution of the salt itself, where $[Ag^+]$ and $[Cl^-]$ are identical, and to a solution where the two ions are introduced from different sources, for example, NaCl and $AgNO_3$.

It therefore is possible to have a common ion effect quite analogous to that discussed in connection with acid-base systems. Using the silver chloride case to illustrate the general principles involved, in a solution obtained by addition of excess AgCl the following equilibrium occurs:

$$AgCl(s) \rightleftharpoons Ag^+(aq) + Cl^-(aq)$$

and $[Ag^+] = [Cl^-] = \sqrt{K_{sp}} = 1.33 \times 10^{-5}$ mole/liter. On the addition of more Cl^- ions to the system—for example, via HCl—the K_{sp} equality still applies. The result is that the concentration of silver ion and therefore the solubility of the salt are reduced. Conversely, if the concentration of Ag^+ is lowered by the addition of NH_3 which reacts to form the complex ion $Ag(NH_3)_2^+$, the Cl^- ion concentration and thus $\mathscr{S}$ increase to maintain $[Ag^+][Cl^-] = K_{sp}$. Such techniques are important in analytical chemistry. The introduction of other ions in solution, however, also effects the ionic strength, and frequently this has the result that Eq. (5-23) is no longer applicable since ion activities are required. The effect of the common ion, however, will be far more important, and thus to a first approximation the changes in solubility can be determined by considering only concentrations.

EXAMPLE 5-6. Calculate the solubility in moles per liter of $AgBrO_3$ ($K_{sp} = 5.0 \times 10^{-5}$) in a solution containing 0.020 moles of $AgNO_3$ per liter.

$$K_{sp} = [Ag^+][BrO_3^-] \qquad \mathscr{S} = [BrO_3^-]$$
$$[Ag^+] = \mathscr{S} + 0.020 \approx 0.020$$
$$\mathscr{S} = K_{sp}/0.020 = 5.0 \times 10^{-5}/0.02 = \underline{2.5 \times 10^{-3} \text{ moles/liter}}$$

The treatment of the common ion effect and the calculation of salt solubility becomes more difficult when the process responsible for the establishment of the concentration of one of the ions is an equilibrium rather than a reaction that for practical purposes is chemically irreversible. An important caution in dealing with K_{sp} problems in general is to recognize the possibility of additional equilibria.

The simple equations developed so far are inadequate for such cases, although often the additional effects are quite minor. The approach to be taken in these circumstances may be illustrated by considering the case of a slightly soluble salt of a weak acid. The equilibria involved can be represented as follows:

$$MA \xrightleftharpoons{K_{sp}} M^+ + A^-$$
$$+$$
$$H_2O \xrightleftharpoons{K_b} HA + OH^-$$

The concentration of the A^- ion is governed not only by the solubility of the salt MA but also by the extent to which A^- undergoes hydrolysis, as indicated by the magnitude of the K_b value. The result is that the value of $[A^-]$ must satisfy both the equilibrium expressions

$$K_{sp} = [M^+][A^-] \quad \text{and} \quad K_b = [HA][OH^-]/[A^-]$$

The solubility of the salt is equal to $[M^+]$ and, therefore, $[A^-] = K_{sp}/\mathscr{S}$. The general relation for the solubility is a cubic equation in $\mathscr{S}$, but to determine the solubility of the salt in a buffer solution in which the concentration of OH^- ions is fixed, the equation can be easily obtained. The concentration of HA produced by hydrolysis will be the difference between the total A^- concentration arising from the solution of the salt and the concentration at equilibrium, that is, $[HA] = \mathscr{S} - [A^-] = \mathscr{S} - K_{sp}/\mathscr{S}$. From the hydrolysis equilibrium expression we therefore have

$$[A^-] = \frac{(\mathscr{S} - K_{sp}/\mathscr{S})[OH^-]}{K_b}$$

Equating this with the expression $[A^-] = K_{sp}/\mathscr{S}$ obtained from the solubility relation, we get

$$K_b K_{sp}/\mathscr{S} = (\mathscr{S} - K_{sp}/\mathscr{S})[OH^-]$$

which rearranges to the more convenient form

$$\mathscr{S}^2 = K_{sp} + \frac{K_b K_{sp}}{[OH^-]} \tag{5-27}$$

The solubility therefore can be calculated when the value of $[OH^-]$ for the system (usually through pH) along with K_{sp} and K_b are known.

5-4-2 Thermodynamic Solubility Product

If the equilibrium between the dissolved ions and the solid salt is considered thermodynamically, the solubility product is defined in terms of ion activities and is designated K_{sp}°. To simplify the algebra, the general salt AB, which dissociates to A^+ and B^- ions will be used. The treatment can be easily extended to the more general $A_m B_n$ case.

The defining equation for K_{sp}° of the salt AB is

$$K_{sp}^\circ = (a_{A^+})(a_{B^-}) \tag{5-28}$$

The relation between K_{sp}° and K_{sp} as defined by Eq. (5-23) can be obtained by substitution of the product of ion concentration and activity coefficient for each activity:

$$K_{sp}^{\circ} = [A^+][B^-]\gamma_{\pm}^2 = K_{sp}\gamma_{\pm}^2 \qquad (5\text{-}29)$$

When the mean activity coefficient is unity, K_{sp} and K_{sp}° are identical. Thus K_{sp}° can be interpreted as the solubility product in a solution of zero ionic strength.

In terms of the solubility, Eq. (5-29) can be written as

$$K_{sp}^{\circ} = \mathscr{S}^2\gamma_{\pm}^2$$

or

$$\mathscr{S} = \frac{\sqrt{K_{sp}^{\circ}}}{\gamma_{\pm}} \qquad (5\text{-}30)$$

since for the salt AB, $\mathscr{S} = [A^+] = [B^-]$. The form of Eq. (5-30) indicates that as the mean activity coefficient decreases, as μ increases, there is an *increase* in solubility.

This behavior represents another manifestation of the ability of ions that do not participate chemically in a process to affect the equilibrium concentrations through their influence on the ionic strength. For example, the solubility of AgCl in aqueous solution is increased by the addition of $NaNO_3$, not through a direct chemical interaction but because of the increase in μ and the resulting lowering of the mean activity coefficient of the silver and chloride ions.

K_{sp}° can be determined experimentally by measuring the actual solubility of the salt in question as a function of the ionic strength. By extrapolation, the value of $\mathscr{S}^{\circ}$, the solubility at $\mu = 0$ and therefore at $\gamma_{\pm} = 1.00$, is obtained and the thermodynamic solubility product can be calculated from Eq. (5-30).

To facilitate the handling of the experimental data, Eq. (5-30) may be modified in the following manner. Taking logarithms of both sides gives

$$\log \mathscr{S} = \frac{\log K_{sp}^{\circ}}{2} - \log \gamma_{\pm} \qquad (5\text{-}31)$$

In dilute solution the Debye–Huckel law, Eq. (5-13), is applicable. The term $-\log \gamma_{\pm}$ can be replaced by $0.509Z_+Z_-\sqrt{\mu}$, which for the salt AB is just $0.509\sqrt{\mu}$. Making this substitution transforms Eq. (5-31) to

$$\log \mathscr{S} = \frac{\log K_{sp}^{\circ}}{2} + 0.509\sqrt{\mu} \qquad (5\text{-}32)$$

A plot of $\log \mathscr{S}$ vs. $\sqrt{\mu}$ therefore will give a straight line with a slope equal to 0.509 and an intercept equal to $\frac{1}{2} \log K_{sp}^{\circ}$. Figure 5-5 shows a plot of this equation for a typical slightly soluble salt.

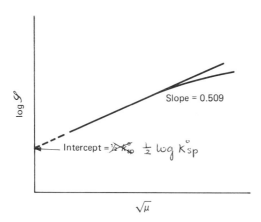

FIG. 5-5 Plot of Eq. (5-32) for a typical slightly soluble salt.

Table 5-1 lists the thermodynamic solubility product for a number of slightly soluble salts, including a number that are salts of weak acids. For solutions in which the only source of ions is the dissolution of the salt itself and K_{sp}° is not large, the activity coefficients can be taken as unity. Thus K_{sp}° would be equivalent to K_{sp}, and solubilities of acceptable accuracy can be determined from Eqs. (5-25) and (5-26). For salts whose solubilities are large and consequently of themselves give solutions of high ionic strength, or for any solution containing other ions such that μ is large, it is necessary to use Eq. (5-30) for accurate evaluation of salt solubility.

TABLE 5-1.   THERMODYNAMIC SOLUBILITY PRODUCTS
FOR SLIGHTLY SOLUBLE SALTS AT 25 °C[a]

Salt	K_{sp}°	pK_{sp}°	Salt	K_{sp}°	pK_{sp}°
CH_3COOAg[b]	4.0×10^{-3}	2.40	Ag_2SO_4[b]	1.6×10^{-5}	4.80
$AgBrO_3$	5.25×10^{-5}	4.28	BaF_2[b]	1.7×10^{-6}	5.76
$SrSO_4$[b]	6.8×10^{-7}	6.55	MgF_2[b]	5.25×10^{-9}	8.18
$AgIO_3$[b]	3.0×10^{-8}	7.52	SrF_2[b]	2.9×10^{-9}	8.54
$PbSO_4$[b]	1.6×10^{-8}	7.80	CaF_2[b]	4.0×10^{-11}	10.40
$AgCl$	1.8×10^{-10}	9.75	$Mg(OH)_2$	1.8×10^{-11}	10.74
$BaSO_4$[b]	1.1×10^{-10}	9.96	$Pb(IO_3)_2$[b]	2.6×10^{-13}	12.59
$AgBr$	5.25×10^{-13}	12.28	Hg_2Cl_2[b]	1.3×10^{-18}	17.88
AgI	8.3×10^{-17}	16.08			

[a] pK_{sp}° data from J. N. Butler, *Ionic Equilibrium* (Reading, Massachusetts: Addison-Wesley Publishing Company, Inc., 1964, p. 195). Reprinted with permission.
[b] Salts of weak acids.

EXAMPLE 5-7.   Calculate the solubility in moles per liter of AgBr in 0.1 molar $NaNO_3$.

$$[Na^+] = [NO_3^-] = 0.1$$

$$[Ag^+] = [Br^-] \approx \sqrt{K_{sp}^\circ} = (5.13 \times 10^{-13})^{\frac{1}{2}} \approx 10^{-6}$$

(These ions therefore make a negligible contribution to μ.)

$$\mu = c_{NaNO_3} = 0.1 \qquad Z = \pm 1 \qquad \gamma_\pm = 0.76 \quad \text{(Fig. 5-3)}$$

$$\mathscr{S} = \frac{\sqrt{K_{sp}^\circ}}{\gamma_\pm} = \frac{(51.3 \times 10^{-14})^{\frac{1}{2}}}{0.76} = \frac{7.16 \times 10^{-7}}{0.76}$$

$$\mathscr{S} = 9.4 \times 10^{-7} \text{ mole/liter}$$

Although the Debye–Huckel theory, whose applicability is limited to dilute solutions, was utilized in obtaining a function suitable for extrapolation to zero ionic strength, Eq. (5-30) is not subject to this limitation. Thus if a K_{sp}° value is obtained by solubility measurements at very small values of the ionic strength, or a tabulated value is available, solubility measurements in more concentrated solutions can be used to measure actual values of the activity coefficient. This procedure, along with electrochemical methods discussed in Chapter 7, represents the main source of available data on $\gamma_\pm$ values. As seen in this chapter, such data are a necessary prerequisite for accurate calculations in many systems in which ionic species are present.

PROBLEMS

5-1. The equivalent conductance of a 0.425 N aqueous solution of NaOH is found to be 200 mho cm²/eq. Calculate the cell constant of a conductivity cell in which the solution exhibits a resistance of 7.88 ohms.

5-2. The equivalent conductance at infinite dilution Λ_0 for acetic acid is 390.7 mho cm²/eq. Find the equivalent conductance in a 0.125 N solution of this acid, making use of the K_a value for CH_3COOH listed in Table 4-2.

5-3. Using the relation $[H_3O^+] = \sqrt{K_a C_a}/\gamma_\pm$, calculate the hydronium ion concentration in a 0.1 molar solution of acetic acid in (a) distilled water, (b) 0.05 molar NaCl, and (c) 0.02 molar $MgSO_4$.

5-4. Calculate pH° for 1.00×10^{-3} molar HCl in a solution containing 0.001, 0.005, 0.01, 0.05, 1, 0.2 and 0.5 mole/liter of NaCl. Present the data in tabular form and plot a graph of pH° vs. log concentration of NaCl.

5-5. Calculate the solubility of silver acetate in a buffer solution whose pH is 5.50.

5-6. Neglecting hydrolysis effects and nonunity activity coefficients, calculate the solubility of SrF_2 in water ($K_{sp} = 3 \times 10^{-9}$).

5-7. Using the data in Problem 5-6, calculate the solubility of SrF_2 in a solution containing 0.01 mole/liter of the F^- ion.

5-8. The solubility of $PbSO_4$ in a solution containing 1.65 gm of $Pb(NO_3)_2$ per 100 ml water is 2.12×10^{-7} mole/liter. Neglecting hydrolysis, calculate the K_{sp} of $PbSO_4$.

5-9. Using Eq. (5-30) and silver chloride as the slightly soluble salt, calculate several representative values of the solubility and plot the results in a graph such as that shown in Fig. 5-5. Calculate and plot on the same graph a curve of the solubility determined on the assumption that $\gamma_{\pm}$ remains at 1.00 throughout.

Chemical Kinetics

6

The study of chemical kinetics is involved with detailed examination of the rate at which chemical reactions take place and the mechanism of product formation. Concern with rate and mechanism in kinetics indicates the essential difference between the kinetic approach and the thermodynamic approach.

Previous chapters emphasized that thermodynamics considers a reaction as a transition from initial state to final state. Thermodynamic examination tells us whether or not a process is spontaneous but nothing about the time interval involved. To acquire the latter information we must look to a kinetic treatment.

The same considerations apply to the intimate details of the route by which reactants are transformed to products. For example, in the reaction

$$H_2 + O_2 \rightarrow H_2O$$

the changes in various thermodynamic functions can be evaluated as the hydrogen–oxygen system undergoes the indicated process. It is immaterial whether H_2 and O_2 molecules first dissociate into atoms before forming water or whether the product is formed directly in an encounter between reactant molecules. Insights of this nature are sought in a kinetic investigation of the reaction.

Suppose that thermodynamic calculations or exploratory experiments indicate that a particular reaction occurs. What is the utility of the kinetic approach? The techniques of chemical kinetics allow initially a determination of the overall rate of conversion of reactants to products. The dependence of this rate on such parameters as reactant concentration or temperature also can be expressed in convenient mathematical form.

The acquisition of this information, however, is just the first step toward a complete understanding of the reaction. This overall picture often conceals the real complexity of the process. Most chemical reactions occur through a sequence of steps, the *mechanism of the reaction*, rather than as a single chemical event. Kinetics also provides the techniques to determine the nature of the mechanism.

Detailed discussion of the methods by which the mechanism of a reaction may be elucidated is beyond the scope of this book. It is important, however, to realize that even the most complex mechanism is nothing more than an assembly of single step processes. The essential techniques of obtaining and using the *rate equations*, which relate rate to concentration, temperature, etc., can be applied directly to each individual step. By an extension of these techniques the entire mechanism can be handled and overall rate expressions can be derived.

Although for students in the life sciences primary interest is in the kinetics of biologically important reactions, these systems are not simple. It is necessary to examine and understand initially the basic ideas of kinetics as they apply to some of the more straightforward chemical systems. This information, along with an introductory indication of how the techniques can be applied to other systems, should provide the necessary background for a detailed study of biochemical kinetics at a more advanced level. Thus the discussion begins with an introduction of some fundamental concepts in chemical kinetics, along with the mathematical apparatus for the quantitative handling of kinetic data.

6-1 RATES OF CHEMICAL REACTIONS

6-1-1 Concept of Reaction Rate

The study of chemical kinetics centers around determination and mathematical expression of the *rate* of the reaction. This important term may be defined as follows:

> *REACTION RATE: THE INCREASE IN CONCENTRATION OF PRODUCTS OR DECREASE IN CONCENTRATION OF REACTANTS IN A CONSTANT VOLUME SYSTEM PER UNIT TIME.*

In a general way it can be said that the rate of a reaction is proportional to the concentration of reactants. That is, the larger the number of reacting species present per unit volume in the reactor, the more reaction will occur in a given time—the larger will be the rate.

This fact, combined with the definition of rate, indicates that we can satisfactorily view the rate as the amount of reaction in a given period. It is somewhat more difficult, however, to state precisely the numerical value of the rate since it evidently will *vary continuously with time*. As the reaction proceeds the concentration of reactants changes. Since the rate is dependent on this parameter, it too must undergo a concomitant alteration.

As an illustration, a very simple reaction can be used. A single reactant molecule A decomposes to give a single product species B whose concentration can be measured conveniently as the reaction proceeds. The same example process also will be employed in subsequent discussions. If a series of concentration measurements are made, beginning with [A] = 1.0 mole/liter, it might be found that the concentration of B (initially zero) increases with time, as shown in Fig. 6-1. The concentration is given in moles per liter. The reaction could be occurring in the gas phase or in solution, but for present purposes it is not necessary to make this specification.

Using these data, how can the rate of the reaction be calculated? Since it is the concentration of B that has been measured, the most straightforward manner of expressing the rate would be in terms of the amount of B formed in a given time. The curve in Fig. 6-1 shows that the rate of production of B is 0.6 mole/liter in 200 sec, or 0.85 mole/liter in 350 sec. More commonly, some fixed time interval is used and the data are expressed more formally in terms of the definition of rate given above. Thus, taking 100 sec as "unit

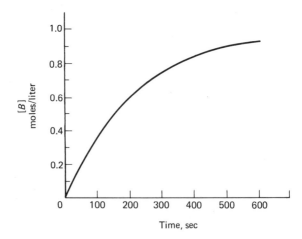

FIG. 6-1 Amount of product B formed in the reaction A → B as a function of reaction time in seconds.

time," the rate of this reaction could be represented as

Rate = moles per liter of B formed/100 sec.

When this calculation is carried out using the data in the figure, it is immediately apparent that the rate will have a different value during each 100-sec interval. Some of the values obtained in this way are listed in Table 6-1.

As reactant A is consumed, the rate at which it undergoes further decomposition decreases. This is reflected in the declining rate values in the table. For example, after 500 sec, since 0.9 mole/liter of B has formed, only 0.1 mole/liter of A remains unreacted. Hence the rate in the interval between 500 and 600 sec is one-tenth that in the first 100-sec period where the concentration of A was initially ten times as large.

This behavior of the reaction rate also illustrates the fundamental problem that the smaller the unit of time employed in defining the rate, the larger is the number of rate values necessary to describe the system completely. Some method of calculating the rate at a given point in time therefore must be used rather than attempting to give all rates or a gross rate over some large time interval.

The mechanics of the rate determinations resulting in the rates quoted in Table 6-1 are illustrated in Fig. 6-2. By drawing a line through two points on the curve a value of the ratio $\Delta[B]/\Delta t$ is obtained. As in the previous use of the symbol Δ, we are dealing here with a finite change in the variable, and the terms in brackets are to be interpreted as concentration in moles per liter unless otherwise indicated.

The rate is given by the slope of the line joining the two points on the curve. These points represent the concentration of B at two particular times of reaction. As the interval between the points is made smaller and smaller, this line whose slope gives the rate approximates more and more closely the actual curve drawn through the experimental points. When two points along this curve are just infinitesimally separated, the line becomes in effect a tangent to the curve. Although *mechanically* we would still measure

TABLE 6-1. RATES OF FORMATION OF
PRODUCT B IN THE REACTION A → B,
WITH INITIAL CONCENTRATION OF A = 1.0
MOLE/LITER

Time interval, sec	Rate of formation of B, mole/liter/100 sec
0–100	0.369
100–200	0.233
200–300	0.147
300–400	0.093
400–500	0.068
500–600	0.0369

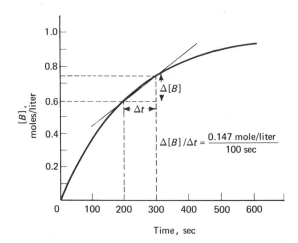

FIG. 6-2 *Amount of product B formed in the reaction* A → B *as a function of reaction time in seconds. Illustrated is the mechanical determination of the rate* Δ[B]/Δt *from the curve for the time interval* t = 200 *to* 300 *seconds.*

Δ[B]/Δt from the line, this would have the same value as $d[B]/dt$—the change in the concentration of B during the infinitesimal time interval dt. This procedure is shown in Fig. 6-3. The result of such a determination gives what can be viewed as the *instantaneous rate*, the rate of the reaction at just one point in time where we take the tangent to the curve. On this basis the formal definition of rate for this reaction can be written as

$$\text{Rate} = d[B]/dt \tag{6-1}$$

Since [B] increases with time, the numerical value of $d[B]/dt$ will be a positive number. Although the rate so delineated is for an infinitesimal time period, it is customary to express the rate in terms of a finite unit time, such as seconds or minutes. Thus the units of rate as defined by this equation are moles per liter second, or, using the unit designation system almost invariably employed in kinetic discussions, moles liter^{-1} sec^{-1}.

Frequently it is useful to express the rate in terms of reactant consumption. In the reaction A → B, 1 mole of A is consumed per mole of B formed. The rate of consumption of A will be numerically equal to $d[B]/dt$ but will have the opposite sign since [A] decreases with time. Consequently, the tangent to the curve of [A] vs. t will have a negative slope. This gives an alternative expression for the rate:

$$\text{Rate} = -d[A]/dt \tag{6-2}$$

In practice it is not possible to determine the amount of B formed at every instant of time. We are restricted to obtaining values of [B] at finite time

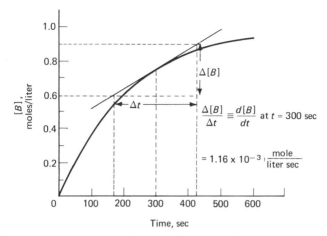

FIG. 6-3 Amount of product B formed in the reaction A → B as a function of reaction time in seconds. Illustrated is the technique of determining the "instantaneous rate" at t = 300 seconds by drawing a tangent to the curve and measuring its slope Δ[B]/Δt.

intervals, drawing a curve through the experimental points, and taking a tangent at some time of interest, as shown in Fig. 6.3. Taking a tangent to the curve is evidently a procedure that can lead to considerable error. Although it remains a valid method to which we will refer again shortly, it is necessary to develop other methods of data handling less subject to error. These methods, however, accomplish no more than is done in taking the tangent.

The other complication associated with a constantly changing rate is the difficulty in calculating *from* the rate the amount of product formed in a given finite time period. To calculate the amount of B formed in the first x sec of reaction using the expression

$$\text{Amount of reaction} = \text{rate} \times \text{time}$$

the problem that arises is analogous to that encountered in the calculation of PV work done or heat absorbed in a thermodynamic system. If the rate were constant with time the equation could be applied directly. Since this is not the case, the rate vs. time curve takes the form shown in Fig. 6-4 and the amount of reaction must be determined as the area under the curve between $t = 0$ and $t = x$ sec. This area represents the summation of products of time interval × applicable rate over the Δt of interest. Using the methematical formalism developed previously, the amount of reaction $\Delta[B]$ can be represented by the integration

$$\Delta \text{B} = \int_{t=0}^{t=x} (\text{instantaneous rate}) \, dt \tag{6-3}$$

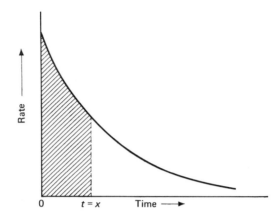

FIG. 6-4 *Variation of rate with time (arbitrary units) for a reaction such as* A → B. *The shaded area represents the amount of reaction from* t = 0 *to* t = x *seconds.*

The term in parentheses represents an expression (whose form will be obtained presently) for the rate as a function of time.

In summary, there are two basic problems in a quantitative treatment of reaction rate: How can the rate be evaluated from a measurement of the amount of product formed or reactant consumed or knowing the variation of rate with time and how can the amount of product formed after some time period be determined.

Here the simplest type of reaction has been utilized to outline the main features of the kinetic approach. For each general type of reaction it is possible to develop two forms of the basic rate law or expression to solve these two closely related problems.

Before examining the equations specific to general classifications of chemical reactions, some basic principles and terms used in kinetic descriptions of chemical reactions will be examined.

6-1-2 Rate Constant, Order, and Molecularity

To obtain the necessary relations to treat reaction rates quantitatively, the first requirement is a formal expression of the idea that the rate is proportional to the concentration of reactants.

Consider a general reaction in which a moles of reactant A combine with b moles of B to yield products that need not be specified for the moment. Written in the conventional manner the process is

$$aA + bB \rightarrow \text{products}$$

The rate is proportional to the concentration as

$$\text{Rate} \propto [A]^m[B]^n$$

$$\text{Rate} = k[A]^m[B]^n \qquad (6\text{-}4)$$

The proportionality factor k is the *rate constant*, which will have a unique value for any particular process at a given temperature. The powers to which the concentrations are raised, m and n, may or may not be the same as the coefficients a and b in the reaction equation giving the stoichiometry of the process. There is a very fundamental distinction to be made here. The reaction equation indicates the numbers of moles of A and B that will combine, while Eq. (6-4) represents the actual, experimentally determined dependence of rate on concentration. The reason for this distinction is the difference between the overall reaction and the individual steps in the mechanism. The stoichiometric reaction equation usually represents the net result of a number of intermediate processes. Since each of these will have a rate dependent on the concentrations of the species involved—including perhaps reactive intermediates that do not appear among the final products—the overall rate may have a quite different dependence on [A] and [B] than indicated by the stoichiometry.

Consistent with this argument is the fact that for a single step process $m = a$ and $n = b$. Thus, if the reaction actually proceeds by "a" molecules of A coming together with "b" molecules of B in one event to yield the observed product, the rate is correctly represented by

$$\text{Rate} = k[A]^a[B]^b \qquad (6\text{-}5)$$

One seldom encounters more than two species reacting in a true single-step process to form a final product, so Eq. (6-4) is much more frequently encountered.

The sum of the terms m and n in Eq. (6-4) represents an *experimentally observed* quantity—the actual dependence on concentration. It is a basic characteristic of the process and is known as the *reaction order*.

$$\text{Reaction order} = m + n \qquad (6\text{-}6)$$

It is important that the order of the reaction be correctly viewed as an experimental quantity.

Emphasizing this characteristic of reaction order is the fact that order can be determined directly from experimental measurements by one of the methods discussed below without knowledge of the mechanism of the reaction. Conversely, it *cannot* be evaluated by an examination of the stoichiometry of the process.

The sum of m and n can have values of 0, $\frac{1}{2}$, 1, $\frac{3}{2}$, 2, and 3 (and in a few instances other fractional values). The most commonly met situations are *first order* ($m + n = 1$, rate dependent on the first power of one reactant concentration) and *second order* ($m + n = 2$).

There also are many reactions whose stoichiometry is adequately represented by an equation such as $aA + bB \rightarrow$ products but whose rate is dependent on concentrations not in the simple way expressed by Eq. (6-4) but rather by a complex algebraic expression. In such circumstances the concept of overall order of the reaction becomes meaningless. It is often convenient, however, to invoke the concept of reaction order with respect to one particular reactant of interest—for example regarding a reaction as "second order in X" when the term $[X]^2$ appears in the rate equation.

Particularly since the point is not clearly made in older kinetic treatments, the order of a reaction must be clearly distinguished from another property, the *molecularity*, which is applicable only to a single-step process—as compared to the order of the overall reaction—and which is defined as follows.

> ***MOLECULARITY: THE NUMBER OF SPECIES (MOLECULES ATOMS, FREE RADICALS, OR IONS) TAKING PART IN A SINGLE-STEP PROCESS.***

Unlike reaction order, the molecularity only can be a small whole number, most often 1 (*unimolecular*) or 2 (*bimolecular*). To determine this quantity we require not just the dependence of rate on concentration but also information on the detailed nature of the process. Thus, if it is suggested that a reaction is bimolecular, there must be evidence to indicate that the products arise from the coming together of *two* reactant species and that this encounter leads *directly* to the formation of product.

Although molecularity is generally confined to a concerted process, it is occasionally valid (in a case where the rate of an overall reaction is determined by the rate of one particular step) to consider the molecularity of this process as that of the overall reaction.

The tendency to confuse the terms "order" and "molecularity" arises in part from the fact that they often are numerically the same. The reaction

$$2 \, HI \rightarrow H_2 + I_2$$

for example, is second order and bimolecular, but note that the implications of these two properties are not necessarily synonymous. Here the experimental rate data indicate a dependence of the rate of $[HI]^2$, and investigations of the mechanism of the reaction show that it is a single-step process in which H_2 and I_2 result directly from an encounter between 2 HI molecules.

On the other hand, the reaction whose stoichiometry is represented by

$$2 \, NO + O_2 \rightarrow 2 \, NO_2$$

is third order but bimolecular. There are numerous examples of reactions that have the same order and molecularity as well as those for which these parameters are not the same.

The most convenient general classification of chemical reactions is on the basis of order, which as noted earlier can vary from zero to three, including nonintegral values. Subsequent discussions deal primarily with the two types of reaction most often encountered: first and second order processes.

6-1-3 First Order Reactions

Any reaction qualifies as first order if it is determined experimentally that the rate depends on the first power of the concentration of one reactant. Methods available to establish order are discussed in a subsequent section.

To make use here of the example reaction $A \rightarrow B$ dealt with in the introductory discussion, it must be assumed that it has been established previously that the reaction is first order. In Sec. 6-2 the procedure for the example reaction is examined. The point is that before the process can legitimately be discussed as a first order reaction, this fact must be established experimentally. It does not follow automatically from the stoichiometric indication that "1 mole of A yields 1 mole of B."

The general rate equation can be written for this first order process on the basis of Eqs. (6-2) and (6-4) as

$$\text{Rate} = -d[A]/dt = d[B]/dt = k_1[A] \tag{6-7}$$

Since in practice the course of a reaction can be followed by measuring, as convenient, either [A] or [B] as a function of time, it is desirable to have available a form of the rate equation in which either type of data can be used directly. This type of approach is even more valuable in reactions of higher order.

The conventional method of accomplishing this modification to the rate expression is to employ the *transformation variable x*. For the reaction $A \rightarrow B$, x is defined as

$$x = [B] \text{ formed in time } t = [A] \text{ consumed in time } t$$

The initial concentration of A, $[A]_0$, is designated as a. Making these substitutions in Eq. (6-7), we get

$$\frac{dx}{dt} = k_1(a - x) \tag{6-8}$$

This relation is known as the *first order differential rate law*. It allows calculation of the rate of the reaction for a given concentration of reactant at any value of x if the rate constant is known. Conversely, if we know

dx/dt at some value of $(a - x)$, k_1 can be evaluated. The use of the subscript 1 specifies that the rate constant is for a first order reaction.

In the discussion of the example reaction, it was determined in Fig. 6-3 that the rate at $t = 300$ sec was 1.16×10^{-3} mole liter^{-1} sec^{-1}. The value of x at this point can be read from the y axis as $x = [B]_{t=300} = 0.749$ mole/liter. As indicated before, $a = 1.0$ mole/liter. From this information the value of k_1 can be determined as follows: For convenience Eq. (6-8) is rearranged to the form

$$k_1 = \frac{1}{(a - x)} \frac{dx}{dt}$$

(6-9)

and substitution of the appropriate values gives

$$k_1 = \frac{1}{(1.0 - 0.749)} 1.16 \times 10^{-3}$$

$$k_1 = \frac{1.16 \times 10^{-3}}{0.251} = 4.606 \times 10^{-3} \text{ sec}^{-1*}$$

The value of the rate constant is a fundamental property of the reaction at the temperature in question. While dx/dt varies with time, k_1, the proportionality factor, remains unchanged. Since this datum is now available it is possible to use the rate law to find dx/dt at any other time of interest.

Initial rate: $[A] = a = 1.0$ mole/liter $(a - x) = 1.0$ mole/liter

$$\frac{dx}{dt} = k_1(a - x) = 4.6^* \times 1.0 \times 10^{-3}$$

$$\frac{dx}{dt} = 4.6 \times 10^{-3} \text{ mole liter}^{-1} \text{ sec}^{-1} \qquad (t = 0)$$

Rate at $[A] = 0.05$: $(a - x) = 5.0 \times 10^{-2}$ mole/liter

$$\frac{dx}{dt} = 4.6 \times 10^{-3} \times 5.0 \times 10^{-2}$$

$$\frac{dx}{dt} = 2.3 \times 10^{-4} \text{ mole liter}^{-1} \text{ sec}^{-1}$$

Thus, as expected, the instantaneous rate decreases with time, just as did the crudely defined ratio $\Delta[B]/\Delta t$ whose values are given in Table 6-1.

* Because of the inherent error in determining k_1 from a measure of the slope of the tangent drawn to the rate curve, the use of more than two significant figures is not justified. The k_1 value calculated above as 4.606×10^{-3} is based on further treatment of the system, and on the basis of the treatment thus far it should be written with only two significant figures, or perhaps more realistically simply as 5×10^{-3} sec^{-1}.

The problem must now be faced that although these calculations are perfectly valid, they are nevertheless based on a somewhat uncertain value of k_1. It is difficult to draw the tangent with any degree of real accuracy, and the possibility of significant error in k_1 and subsequent calculations is considerable. In addition, the rate equation in its present form does not provide for the second kind of calculation discussed earlier—that of determining the amount of B formed in time t.

A second type of rate equation therefore must be developed both for this latter determination and to avoid the problems of mechanical slope determination. This derivation must make use of calculus, in a way analogous to the evaluation of PV work under variable pressure conditions, as discussed in Sec. 1-1-2.

If the differential rate law Eq. (6-8) is rearranged to the form

$$dx = k_1(a - x)\, dt \qquad (6\text{-}10)$$

we can interpret this relation as giving the change in the concentration of product dx when the reaction proceeds for an infinitesimal time period dt. The total change in x from time zero to t can be determined by taking a sum of expressions, $k(a - x)\, dt$. Each term in the sum is the product of the rate constant k, the concentration of reactant $(a - x)$, and the infinitesimal reaction duration dt. The graphical presentation of this summation is shown in Fig. 6-5. The shaded area under the curve represents the amount of product

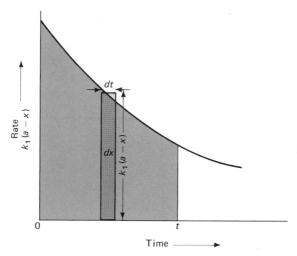

FIG. 6-5 *Variation of the rate of a first order reaction with time. The rectangle is an enlarged view of the area representing the infinitely small increase in concentration of product dx during the time interval dt. The area under the curve from t = 0 to t = t is the amount of product formed.*

formed from the beginning of the reaction up to time t and is the sum of the dx terms defined by Eq. (6-10). Each dx is the area of an infinitely small rectangle, shown enlarged in the figure, with sides dt and $k_1(a - x)$.

As seen previously in similar situations, the area under the curve can be determined by direct mechanical measurement if the exact shape of the curve is known, or more conveniently by using calculus. Integration of Eq. (6-10) yields the following information: The integral of dx from 0 to x gives a Δx—the amount of product formed to time t; integration of the right-hand side from 0 to t gives the sum of the $k_1(a - x)\, dt$ terms to be included in the area.

Equation (6-10) cannot be integrated in its present form, since before the calculus rules can be invoked terms containing x must be collected as follows:

$$\frac{dx}{(a - x)} = k_1\, dt$$

The left-hand side is to be integrated between the limits 0 and x and the $k_1\, dt$ term between 0 and t. Taking the rate *constant* out of the integration gives

$$\int_0^x \frac{dx}{(a - x)} = k_1 \int_0^t dt$$

By the rules of integration the left-hand term becomes

$$-\ln(a - x) + \ln(a - 0) = \ln \frac{a}{(a - x)}$$

while the right-hand integration gives

$$k_1(t - 0) = k_1 t$$

Therefore, the final equation is

$$\ln \frac{a}{(a - x)} = k_1 t$$

Conversion to common logarithms yields the more convenient working relation

$$2.303 \log \frac{a}{(a - x)} = k_1 t \tag{6-11}$$

This is the *integrated rate law* for first order reactions. Note that although this equation does not give the desired value of x at time t *directly* (because of the algebraic manipulations required for the integration) this quantity can be calculated from the equation in a straightforward manner.

Equation (6-11) is of the general form of a straight line equation and, therefore, a graph of $2.303 \log[a/(a - x)]$ will be a linear function of reaction

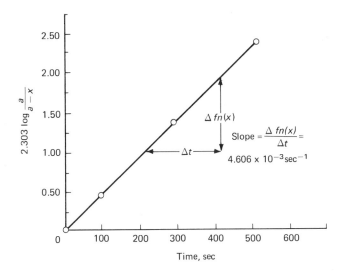

FIG. 6-6 Plot of the integrated rate law for the A → B *example reaction, determining the value of the first order rate constant* k_1.

time. The slope gives k_1 directly, and the intercept is zero. This is an important point. The implication is that not only does the integrated rate law permit calculation of the amount of reaction, it also provides a means of graphical analysis. The rate constant value is obtained with appreciably greater accuracy than can be obtained via the tangent method.

For the example reaction values of x at time t can be obtained from Fig. 6-1, and the numerical values of the left-hand side of Eq. (6-11) can be calculated. The graph of these data plotted against time is shown in Fig. 6-6. The resulting straight line has a slope equal to $4.606 \times 10^{-3} \, \text{sec}^{-1}$, the value of the first order rate constant.

Although the basis of the kinetic treatment is the differential rate law, Eq. (6-8), in practice the integrated form is considerably more useful. For a first order reaction where the initial reactant concentration a is known, the integrated rate law can be used to evaluate

x if k_1 and t are known: the extent of reaction after time t has elapsed

k_1 if x and t are known: an accurate value of the rate constant from a measurement of product formed in given time periods

t if k_1 and x are known: the time required for a particular amount of reactant to be consumed or product formed

One important application of the third type of calculation is the determination of the time required for one-half of the starting material to be consumed—the period known as the *half-life of the reaction* $t_{\frac{1}{2}}$. For a first order process this quantity turns out to be independent of the starting concentration of reactant, being a function of the rate constant only. The relation between $t_{\frac{1}{2}}$ and k_1 is obtained from the integrated rate law in the following way:

At $t_{\frac{1}{2}}$, $x = a/2$ and, therefore, $a/(a - x) = a/[a - (a/2)] = 2$. Substitution into Eq. (6-11) gives

$$2.303 \log 2 = 0.693 = k_1 t_{\frac{1}{2}} \qquad t_{\frac{1}{2}} = \frac{0.693}{k_1} \qquad (6\text{-}12)$$

For example, the $A \rightarrow B$ reaction has $t_{\frac{1}{2}} = 0.693/(4.606 \times 10^{-3}) = 150$ sec. Reference to Fig. 6-1 shows that, as predicted by the calculation, the concentration of product B has risen to a value of 0.50 mole/liter after 150 sec, that is, one-half of the amount of A initially present has been consumed.

In many practical kinetic investigations it is convenient to measure only the concentration of products formed and to determine the exact initial concentration of reactant by carrying the reaction to completion. Strictly speaking, since the rate diminishes as the reactant is consumed, it would take an infinitely long time to obtain complete conversion to product, even if the process were completely chemically irreversible. Practically, however, it is possible to allow the reaction to proceed for a finite time interval until no further changes in concentration can be detected. The concept of experimental determination of the extent of reaction at what is sometimes referred to as *infinite time* can be rationalized by considering an alternative form of Eq. (6-7):

$$\frac{d[A]}{[A]} = -k_1 \, dt \qquad (6\text{-}13)$$

If this differential form of the rate law is integrated, using the same approach as in the derivation of Eq. (6-11), the result is

$$\ln \frac{[A]}{[A]_0} = -k_1 t \qquad (6\text{-}14)$$

Taking antinatural logarithms of both sides gives

$$\frac{[A]}{[A]_0} = e^{-k_1 t}$$

or

$$[A] = [A]_0 e^{-k_1 t} \qquad (6\text{-}15)$$

Thus, as can be inferred from Fig. 6-1, as the concentration of product increases, the value of [A] decreases exponentially but will be exactly zero only at t_{infinity}. When the reaction has proceeded to an extent where further changes cannot be detected physically, we can assume with negligible error that all of the starting material has been consumed. Thus the initial concentration of reactant can be calculated from the stoichiometry. In accurate work infinite time might appropriately be taken as $10t_{\frac{1}{2}}$; thus for the example reaction, $t_{\text{infinity}} \approx 1{,}500$ sec. Substitution of this value into the integrated rate law gives $[A]_{t=1500 \text{ sec}} = [A]_0/1{,}024$, that is, less than 0.1 percent of the initial concentration of 1.0 mole/liter.

Equation (6-14) provides, in addition, an alternative method of handling experimental determinations of reactant concentrations at various extents of reaction. The relation rearranges directly to

$$\ln[A] = -k_1 t + \ln[A]_0$$

or, by conversion to common logarithms,

$$\log[A] = -\frac{k_1 t}{2.303} + \log[A]_0 \tag{6-16}$$

A plot of the log of reactant concentration as a function of time gives a straight line with slope $-k_1/2.303$ and intercept $+\log[A]_0$. Thus we can begin to follow a first order reaction at any particular convenient time after it has begun. By plotting the data in the form of Eq. (6-16), the initial concentration can be found without having to go to long reaction times. Note that this relation is not a new equation but rather an alternative approach that can be taken when it is more convenient to handle experimental data in terms of reactant concentrations rather than the transformation variable x.

6-1-4 Second and Zero Order Reactions

The necessary techniques for dealing with reactions of any particular order have now been developed. The differential rate law is written on the basis of the experimentally observed concentration dependence, the transformation variable is introduced to give equations that can be used directly with either reactant or product concentrations, and the integrated rate law is obtained.

To illustrate how this procedure, developed in some detail for first order processes, is applied to reactions of other order, consider a second order reaction of the type $C \rightarrow D$. The reaction equation as written indicates just a 1:1 mole stoichiometric ratio, while the statement that the process is second order means that the rate is dependent on the second power of the concentration of C.

The differential rate expression is

$$-\frac{d[C]}{dt} = \frac{d[D]}{dt} = k_2[C]^2 \tag{6-17}$$

From the stoichiometry, the transformation variable is $x = [C]_{reacted} = [D]_{formed}$, and therefore if the initial concentration of C is represented by a, Eq. (6-17) becomes

$$\frac{dx}{dt} = k_2(a - x)^2 \qquad (6\text{-}18)$$

Rearrangement of this equation and integration from time zero to t, as for first order reactions, results in the integrated rate law, which takes the form

$$\frac{x}{a(a - x)} = k_2 t \qquad (6\text{-}19)$$

The expression for the half-life is easily obtained by substituting $x = a/2$ at $t_{\frac{1}{2}}$, whence Eq. (6-19) gives

$$t_{\frac{1}{2}} = \frac{1}{k_2 a} \qquad (6\text{-}20)$$

The equations incorporating x are applicable only to cases characterized by the $A \rightarrow B$-type stoichiometry. For a second order process such as $A + B \rightarrow C$, x again can be used to represent the amount of A or B consumed, or the amount of C formed, but the differential rate law is then $dx/dt = k_2(a - x)(b - x)$, where a and b are the initial concentrations of A and B. On the other hand, if the stoichiometry were $2A \rightarrow B$ or $A + B \rightarrow 2C$, the concentration represented by x would depend on whether the definition is in terms of reactant or product concentrations. Here there is no particular merit in using the transformation variable. The necessary rate equations are better developed from the differential rate law written directly in terms of reactant or product concentrations, for example, $d[A]/dt = -k_2[A][B]$ for the reaction $A + B \rightarrow 2C$.

There is another order of reaction worthy of brief but separate comment. There are a number of reactions that proceed at a constant rate, independent of the concentration of reactant—reactions of zero order. In addition, some systems that normally exhibit first or second order kinetics may become zero order under some conditions. Mathematically the coefficients m and n in Eq. (6-4) are zero, and the differential rate law is

$$dx/dt = k_0 \qquad (6\text{-}21)$$

The integrated rate law is easily obtained as

$$x = k_0 t \qquad (6\text{-}22)$$

Equation (6-22) indicates simply that since the rate of reaction is constant, the amount of decomposition in a given time is the product of time and the amount of decomposition per unit time—the only circumstances under which the reaction rate does not vary continuously with time. Of course, the overall rate will change as the concentration of reactant approaches zero.

TABLE 6-2.　RATE EQUATIONS FOR PROCESSES OF THE TYPE $A \rightarrow B$

Rate equation

Order	Differential form	Integrated form	$t_{\frac{1}{2}}$ Equation	Units of k
0	$\dfrac{dx}{dt} = k$	$k = x/t$	$t_{\frac{1}{2}} = a/2k$	mole liter^{-1} sec^{-1}
$\frac{1}{2}$	$\dfrac{dx}{dt} = k(a-x)^{\frac{1}{2}}$	$k = \dfrac{2}{t}[a^{\frac{1}{2}} - (a-x)^{\frac{1}{2}}]$	—	mole$^{\frac{1}{2}}$ liter$^{-\frac{1}{2}}$ sec^{-1}
1	$\dfrac{dx}{dt} = k(a-x)$	$k = \dfrac{2.303}{t} \log \dfrac{a}{(a-x)}$	$t_{\frac{1}{2}} = 0.693/k$	sec^{-1}
2	$\dfrac{dx}{dt} = k(a-x)^2$	$k = \dfrac{1}{t} \dfrac{x}{a(a-x)}$	$t_{\frac{1}{2}} = 1/ak$	liter mole^{-1} sec^{-1}
2^a	$\dfrac{dx}{dt} = k(a-x)(b-x)$	$k = \dfrac{2.303}{(a-b)t} \log \dfrac{b(a-x)}{a(b-x)}$	—	liter mole^{-1} sec^{-1}
3	$\dfrac{dx}{dt} = k(a-x)^3$	$k = \dfrac{1}{2t} \dfrac{2ax - x^2}{a^2(a-x)^2}$	—	liter^2 mole^{-2} sec^{-1}

^{a} For the process $A + B \rightarrow C$.

The differential and integrated rate laws, the half-life expression, and units of the rate constant are summarized for various order reactions in Table 6-2.

6-2 DETERMINATION OF REACTION ORDER

It is apparent from the preceding discussion that a necessary prerequisite for any meaningful and quantitative treatment of a given reaction is the knowledge of its order.

There are a number of methods available for handling experimental kinetic data to determine reaction order. It is useful to examine these techniques. They yield information about a fundamental characteristic of the reaction and provide a different and useful approach to understanding the physical significance of rate equations.

The practical limitations of these methods of analysis must be appreciated. Perhaps one of the least appreciated aspects of order determination is that any method that emphasizes values obtained at small extents of reaction is by nature a superior one. As the reaction proceeds, the concentration of products increases, and these species may have the effect of altering, perhaps quite significantly, the course of the subsequent reaction. There are very few reactions that proceed without at least some complicating factor that becomes more important at larger percent conversions.

The essence of all the methods is one of trial and error, attempting to determine which rate law best represents the course of the reaction. The amount of work required is reduced by using as a starting point an intelligent guess as to the reaction order. Thus, for example, if it were thought that a reaction is likely first order, the procedure would be to first determine whether its behavior is adequately described by the rate laws for first order processes.

6-2-1 Mechanical Slope Determination

This procedure is based on the determination of the instantaneous rate by taking a tangent to the curve giving the rate of formation of product or rate of reactant consumption. This procedure has been discussed previously in connection with the example $A \rightarrow B$ reaction in Fig. 6-3. The tangent is to be taken at several points along the curve. This gives dx/dt at various values of x. The value of the rate constant then can be calculated by means of differential rate laws and the known value of a, using dx/dt and x values from the graph. The calculated values of k will be the same for all dx/dt values only when the appropriate equation—that for the order of the reaction in question —is used. For a first order reaction, then, k values from Eq. (6-8) in the form $k = (dx/dt)/(a - x)$ would be the same. Those determined on the basis of Eq. (6-18) written as $k = (dx/dt)/(a - x)^2$, on the other hand, would be different for each rate since this equation is valid only for second order processes.

Previous comments on the inherent error in obtaining rate values from tangents are equally applicable here. In some systems the scatter in rate values might be sufficiently large so that variations in k values calculated on the basis of, for example, 3/2 and second order might make it impossible to decide which of the two orders is applicable. The method therefore is of limited practical usefulness, but it does illustrate the most fundamental approach to the problem.

6-2-2 Integrated Rate Law Method

The data required here are values of x at various times and the initial concentration a. These sets of values are substituted into the integrated rate law for each order. The applicable equation will yield the same k value for each set of x and t data. This method has the advantage over slope determination in that the accuracy of the k value calculated is dependent only on the error in the original data.

Thus, for example, from Fig. 6-1 the following data can be extracted:

$$x = 0.425 \text{ at } t = 120 \text{ sec} \qquad x = 0.684 \text{ at } t = 250 \text{ sec}$$

Substitution of these values, along with $a = 1.0$ mole/liter, into Eq. (6-11), the integrated first order rate law, gives $k_1 = 4.606 \times 10^{-3}$ sec^{-1} for both points. On the other hand, k_2 values are determined from these data by use of Eq. (6-19)

$$\frac{0.425}{1(1 - 0.425)} = k_2 \, 120 \qquad k_2 = 6.15 \times 10^{-3} \, \text{mole}^{-1} \, \text{liter sec}^{-1}$$

$$\frac{0.684}{1(1 - 0.684)} = k_2 \, 250 \qquad k_2 = 8.66 \times 10^{-3} \, \text{mole}^{-1} \, \text{liter sec}^{-1}$$

indicating that the equation does not apply and that the reaction is not second order.

6-2-3 Empirical Fit Method

All of the integrated rate equations listed in Table 6-2 are of the form $f(a, x) = kt$ and, therefore, as pointed out specifically in the discussion of first order and second order reactions, a plot of values of the function of a and x vs. time gives a straight line with slope k. The linear graph will be obtained, however, only with the equation applicable to the reaction in question.

When this method is applied to the A $\rightarrow$ B example reaction, the results shown in Fig. 6-7 are obtained. The graphs of $f(a, x)$ vs. t appropriate to zero and second order reaction are curves, while that valid for a first order process is the expected straight line. The k_1 value can be obtained from the slope as follows:

Taking the extremes of the plot in the figure we have

$$k_1 = \frac{\Delta Y}{\Delta X} = \frac{y_2 - y_1}{x_2 - x_1} = \frac{2.303 - 0}{500 - 0} = 4.606 \times 10^{-3} \, \text{sec}^{-1}$$

This is perhaps the most useful method since in addition to giving a k value directly, it allows immediate detection of any peculiarities in the reaction. From a comparison of how well the linear relation is followed for similar orders, there also is an indication of the certainty of the order assignment. A reaction could exhibit first order kinetics in the early stages of the process but become zero order at higher percent decompositions. In such circumstances the plot of $2.303 \log[a/(a - x)]$ would be a straight line at small values of x but would begin to deviate significantly from linearity as t increased.

6-2-4 Half-Life Method

The equations presented in Table 6-2 show that for each order of reaction the half-life has a different dependence on rate constant and initial

x (m/l) from Fig. 6–1	t sec	$(a - x)$		Zero order x	$f(a, x)$ for first order $2.303 \log \dfrac{a}{a - x}$	Second order $\dfrac{x}{a(a - x)}$
0	0	1.00		0	0	0
0.37	100	0.63		0.37	0.460	0.595
0.75	300	0.25		0.75	1.38	3.00
0.90	500	0.10		0.91	2.303	9.10

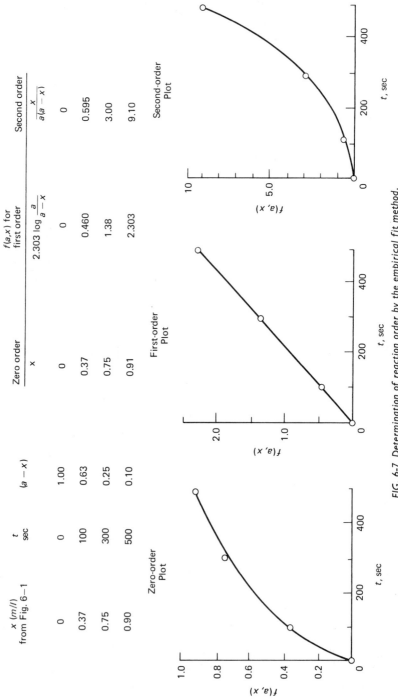

FIG. 6-7 *Determination of reaction order by the empirical fit method.*

concentration. In terms of application, the half-life method is very similar to that based on the integrated rate equation. For a number of different initial concentrations, the time required for consumption of one-half a is determined. Values of $t_{\frac{1}{2}}$, and where necessary a, are substituted into the half-life equations for each order. The applicable equation yields the same k value for all starting concentrations. In a first order process, where $t_{\frac{1}{2}} = 0.693/k_1$, the time required for consumption of one-half the starting material will be the same in each of the experiments performed for this method.

6-3 COMPLEX REACTIONS

In the processes that have been examined thus far, it has been assumed that the reaction proceeds in some simple way. Even if the stoichiometric statement of the chemical event did not directly give the experimental order, it was still possible to obtain a simple correlation between overall rate and the concentration of one or two reactant species.

There are many reactions, however, in which this kind of approach is inadequate, and the rate exhibits a more complex concentration dependence than suggested by any of the equations listed in Table 6-2. It is useful to examine several types of complex reactions that exhibit this type of behavior. In some circumstances they can be treated by the methods developed previously, or at least by an uncomplicated extension of these techniques.

Any complex reaction can be characterized as a series of single step processes, the reaction mechanism. The route from reactants to products can be described by some number of individual processes. When these steps occur in part consecutively or simultaneously, the overall effect is that represented by the stoichiometric equation.

6-3-1 Types of Complex Reaction

One of the least troublesome, in principle at least, of the kinds of process that are not single step events are those where the reaction proceeds through one or more intermediates in sequence to yield the final products. In processes of this nature, known as *consecutive reactions*, there is sometimes one step in the reaction mechanism, that will be significantly slower than all others. In this situation the kinetic behavior of the system is rather simple. Consider a consecutive reaction that could be written as

$$A \xrightarrow{k_1} B \quad (1)$$

$$B \xrightarrow{k_2} C \quad (2)$$

$$C \xrightarrow{k_3} D \quad (3)$$

The subscripts on the rate constants here conventionally refer to the reaction numbering sequence and not the reaction order, and A, B, C, and D can represent one or more species. If k_1 is very much smaller than either k_2 or k_3, then the rate of production of D will be determined by the rate of reaction (1), thus known as the *rate determining step*. No matter how rapidly the intermediates B and C react in sequence to form D, the overall process will depend in a simple way on the concentration of species A. Suppose Reaction (1) follows second order kinetics and has a relatively small rate constant. Once an intermediate species B is formed it "immediately" gives the final product D, whose rate of production therefore is directly dependent on the rate of Reaction (1). The latter step is a single step process and therefore can be treated by the methods developed above to give an expression for this process and consequently for the overall reaction.

Many reactions are *chemically reversible* to an appreciable extent. If the mechanism includes one or more such processes there are some additional complicating features that can be represented as follows:

$$A + B \underset{k_{-1}}{\overset{k_1}{\rightleftharpoons}} C \rightarrow \text{further reactions}$$

The reaction will be controlled not only by the rate at which A and B form the intermediate C but also by the rate at which C either re-forms the starting materials or participates in subsequent reactions. Depending on the overall complexity of the mechanism, such systems can be treated occasionally in terms of a rate-determining step but more often require other methods.

Another type of complicating factor is the occurrence of *side reactions*. There is always the possibility that either the starting materials or some intermediate may participate in two possible reactions that can lead to a final product or another intermediate. Thus we might find, for example, a scheme

$$A \underset{\overset{k_2}{\longrightarrow} C}{\overset{\overset{k_1}{\longrightarrow} B}{\Big|}}$$

Reactions of this type usually show rather complex kinetic behavior and cannot be treated by any simple method.

Another class of complex process is the *chain reaction*. The set of single-step processes comprising the reaction mechanism is such that each sequence has the effect of forming a product or products and a reactive species capable of further reaction with the starting material to initiate the same sequence. Thus with a relatively small energy input to the system a very large amount of reactant decomposition can occur.

The classic example of a chain reaction is the formation of HCl from H_2 and Cl_2. When these latter gases are mixed at room temperature in the dark there is no reaction. If ultraviolet light is allowed to fall on the reaction vessel, a reaction occurs with explosive violence. The mechanism is well established and can be used to illustrate the main features of chain reactions.

The process that begins the sequence of reactions is the light-induced dissociation of molecular chlorine into Cl atoms,

$$Initiation: \quad Cl_2 + h\nu \rightarrow 2\ Cl \quad\quad (1)$$

The chlorine atom can then attack hydrogen and the H atom formed in this step reacts with Cl_2 to regenerate a Cl atom.

$$Propagation: \quad Cl + H_2 \rightarrow HCl + H \quad\quad (2)$$

$$H + Cl_2 \rightarrow HCl + Cl \quad\quad (3)$$

In these two processes, the chain propagating steps, a Cl atom produces two product HCl molecules and is itself regenerated in Process (3). The chlorine atom so formed then restarts the sequence in Reaction (2). The series of reactions continues in this way until either all of the H_2 and Cl_2 present initially is consumed or hydrogen and chlorine atoms are removed from the system through some other reaction. The latter species are known as the *chain carriers*, and the reactions that remove them from the chain sequence are the *chain terminating steps*. In this system there are three possibilities,

$$Termination: \quad H + H \rightarrow H_2 \quad\quad (4)$$

$$Cl + Cl \rightarrow Cl_2 \quad\quad (5)$$

$$H + Cl \rightarrow HCl \quad\quad (6)$$

The reaction of either H or Cl atoms with the reagents H_2 and Cl_2 are quite rapid and, therefore, at least in the initial stages before significant reactant depletion occurs, the concentration of atoms in the system is extremely small. This results in a small rate for Reactions (4)–(6). It is possible, however, to vary the *chain length*, the number of times the sequence of Reactions (2) and (3) occurs from a single initiating Cl atom, by adding to the system varying amounts of some other gas known as an *inhibitor*. This will react rapidly with the Cl and H atoms. The presence of such an addend requires the inclusion of an additional step:

$$H \ (or\ Cl) + inhibitor \rightarrow stable\ product \quad\quad (7)$$

As the concentration of inhibitor is increased, a larger fraction of atomic hydrogen and chlorine will be removed via Reaction (7). Therefore, the chain length will be reduced since H and Cl can no longer participate in chain propagation. The kinetic treatment of such systems is not simple. Other factors, such as the effect of light intensity and total pressure in the system on the rate of the chain-terminating steps, must be considered.

6-3-2 Reaction Mechanisms

Having summarized the various types of kinetic complications that may arise in chemical reactions, we can examine the chemical aspects of reaction mechanisms and indicate how it is sometimes possible to determine the exact nature of the component processes in a mechanism.

It is possible in general to distinguish between different types of chemical reactions by basing the classification primarily on the nature of the intermediate species involved. Gas phase reactions are most often characterized by *free radical* intermediates. In a reaction initiated by heat or light, the reactant molecule is fragmented into particles, free radicals, which have an odd electron and therefore are quite reactive. In the thermal decomposition of ethane, for example, the major portion of the decomposition proceeds initially via the scission of the carbon–carbon bond as

$$H_3C:CH_3 + heat \rightarrow 2 \cdot CH_3$$

On breakage of the covalent bond the two electrons, shown as : in the reaction equation, are split between the two fragments. The methyl radicals, $\cdot CH_3$, each have an odd electron. The radical participates in subsequent reactions such as *abstraction* of a hydrogen atom from another C_2H_6 molecule

$$\cdot CH_3 + C_2H_6 \rightarrow CH_4 + H_3C—\dot{C}H_2$$

to form methane and an ethyl radical.

In solution, reactive intermediates are more often ionic or polar species. In the elimination of HBr from an alkyl bromide, for example, the mechanism involves the intermediacy of a *carbonium ion*,

$$
\begin{array}{c}
Br—CH—CH_2 \\
\quad | \quad\;\; | \\
\quad R \quad\;\; R'
\end{array}
\underset{slow}{\rightleftharpoons}
Br^- +
\begin{array}{c}
\overset{+}{C}H—CH_2 \\
\;| \quad\;\; | \\
R \quad\;\; R'
\end{array}
\tag{1}
$$

$$
Br^- +
\begin{array}{c}
H—\overset{+}{C}—CH_2 \\
\quad | \quad\;\; | \\
\quad R \quad\;\; R'
\end{array}
\underset{fast}{\rightleftharpoons}
HBr +
\begin{array}{c}
CH\!\!=\!\!CH \\
| \quad\;\; | \\
R \quad\;\; R'
\end{array}
\tag{2}
$$

The second step is rapid, and thus the slow formation of the carbonium ion is the rate-determining step.

Free radical intermediates also can occur in the liquid phase. For example, the light-induced dimerization of the amino acid cysteine to cystine in solution occurs through the formation of a *thiyl* radical, $RS\cdot$

$$HSCH_2CH(NH_2)COOH + h\nu \rightarrow H + \dot{S}CH_2CH(NH_2)COOH \tag{1}$$

$$2\,\dot{S}CH_2CH(NH_2)COOH \rightarrow HOOC(NH_2)CHCH_2SSCH_2CH(NH_2)COOH \tag{2}$$

In addition, there are reactions that apparently proceed through a concerted process. An intermediate is formed in this process, but the species gives rise to products directly rather than through some number of subsequent reactions. An example of this kind of mechanism is the addition to olefinic double bonds, which proceeds through a *four-center* intermediate:

$$
\begin{array}{c}
\begin{array}{cc}
\text{H} & \text{R} \\
\diagdown & \diagup \\
\text{X} & \text{C} \\
| \; + & \parallel \\
\text{Y} & \text{C} \\
\diagup & \diagdown \\
\text{H} & \text{R}'
\end{array}
\rightarrow
\left[
\begin{array}{c}
\text{H} \\
| \\
\text{X} \text{---} \text{C} \text{---} \text{R} \\
\vdots \quad \vdots \\
\text{Y} \text{---} \text{C} \text{---} \text{R}' \\
| \\
\text{H}
\end{array}
\right]
\rightarrow
\begin{array}{c}
\text{H} \\
| \\
\text{X} \text{---} \text{C} \text{---} \text{R} \\
| \\
\text{Y} \text{---} \text{C} \text{---} \text{R}' \\
| \\
\text{H}
\end{array}
\end{array}
$$

Finally, there are numerous rearrangement reactions wherein the chemical events are essentially confined to processes within the reactant molecule itself in which changes occur leading to different product molecules. In recent years numerous examples of this type of mechanism have been discovered in the area of organic photochemistry, where the effect of irradiation with ultraviolet light often can be a profound and quite specific rearrangement. An example of such a reaction is the light-induced formation of a bicyclic ketone from the photolysis of α-tropolone methyl ether:

The task of determining the nature of the mechanism by which a particular reaction proceeds is very often a formidable one. Evidence for particular intermediate steps frequently must be obtained in an indirect manner. Furthermore, many of the intermediate species exist for a very short time. Elaborate methods may be required to detect their presence and establish their identity.

The elucidation of reaction mechanisms usually begins with a determination of the nature of the final products and how their rates of formation varies with a number of experimental parameters such as concentration and temperature.

To illustrate the approach taken at this stage of a kinetic study some general techniques can be discussed in terms of their application to biochemical systems. Many biological compounds absorb light in a convenient region of the visible or near ultraviolet portions of the electromagnetic spectrum. In a particular system light absorption at some specific wavelength may result from only one reactant or product. If this is the case the variation in concentration of this species during the reaction could be followed by measuring the change in absorbed intensity of that wavelength in the reaction vessel. Another method of following the reaction involves the measurement of gas pressure. A biological process may involve consumption of oxygen or production of CO_2. Thus the amount of reaction is proportional to the gas pressure. Pressure measurements can be carried out as the reaction process in a closed vessel. An indirect method of employing CO_2 pressure measurements is to perform a reaction involving hydronium ions in the carbonic acid system buffer discussed in Chapter 4. As the pH changes the carbonic acid $\rightleftharpoons$ carbon dioxide and water equilibrium is shifted, resulting in a change in the pressure above the reaction solution. Alternatively, the pH change could be monitored directly using the methods described in Chapter 4 or those outlined in Chapter 7. The most straightforward method of all, but one that cannot always be used, is simply to stop the reaction at various times and carry out chemical qualitative and quantitative analysis.

The information obtained from investigations of this type is usually sufficient to permit the proposal of at least a tentative mechanism that can be subjected to further tests. For example, a study of the thermodynamics of the component processes may indicate that certain of the steps are unfavorable from an energetic point of view.

Perhaps the most difficult task is to establish unequivocally the nature of intermediates. Through spectroscopic methods it often is possible to detect and measure their concentration, even though they do not appear as final products. Alternatively, various reagents can be added to the system to allow trapping of the intermediates as analyzable products. The final mechanism that emerges can be treated in terms of the rate equations for individual steps. These relations then can be combined algebraically into an overall kinetic equation, which may be shown to be consistent with the experimental data, usually by methods of graphical analysis.

6-3-3 Kinetic Treatment of Complex Reactions

It is important to realize that the most complex reaction mechanisms are no more than an assembly of single-step processes to which we can apply individually the rate laws for various order reactions.

Although the presence of highly reactive intermediates makes the determination of the nature of these steps more difficult, their participation does have a simplifying effect on the kinetic analysis of the reaction.

The method of kinetics that can be applied most effectively in such circumstances is known as the *steady-state treatment*. The basic assumption is that if the intermediate is quite reactive, its participation in various subsequent reactions will be very rapid. Therefore, its concentration in the system never will rise above some rather small value. Furthermore, soon after the beginning of the reaction the concentration should become essentially constant with the intermediate being consumed as rapidly as it is being formed. The essential elements of this method can be illustrated in the following simple schematic reaction mechanism:

$$A \underset{k_{-1}}{\overset{k_1}{\rightleftharpoons}} B + C \tag{1}$$

$$A + B \xrightarrow{k_2} D \tag{2}$$

where B represents the reactive intermediate. The change in concentration with time of this species is the difference between the rate of the reaction in which it is formed and that of the two reactions responsible for its consumption.

$$\frac{d[B]}{dt} = k_1[A] - k_{-1}[B][C] - k_2[B][A]$$

Assuming that [B] attains a steady-state value, $d[B]/dt$ will be zero under these conditions and, hence,

$$k_1[A] = k_{-1}[B]_{ss}[C] + k_2[B]_{ss}[A]$$

The subscript ss indicates explicitly that this equation is applicable only in the steady state. Solving this equation for $[B]_{ss}$ gives

$$[B]_{ss} = \frac{k_1[A]}{k_{-1}[C] + k_2[A]}$$

The rate of production of the final product D is given by

$$\frac{d[D]}{dt} = k_2[A][B]_{ss}$$

Elimination of the $[B]_{ss}$ term between these two equations gives

$$\frac{d[D]}{dt} = \frac{k_1 k_2 [A]^2}{k_{-1}[C] + k_2[A]} \tag{6-23}$$

When an equation of this type is required to describe a particular process, the concept of reaction order becomes meaningless. Often, however, the relative magnitudes of the rate constants are such that a considerable simplification of the rate equation can be made. For the system under discussion there are two such possibilities.

If $k_{-1}[C]$ is very much smaller than $k_2[A]$, then $k_{-1}[C] + k_2[A] \approx k_2[A]$ and Eq. (6-23) reduces to

$$\frac{d[D]}{dt} = k_1[A]$$

indicating that the reaction exhibits simple first order kinetics. The physical reality of this situation is that the rate of conversion of the intermediate B in the final process is appreciably faster than its reaction to re-form A. Thus Process (1) becomes the rate-determining step.

The alternative situation is when $k_{-1}[C] \gg k_2[A]$, that is, when B re-forms the reactant very much faster than it reacts with A in Reaction (2). In this case the denominator in Eq. (6-23) is approximately equal to $k_{-1}[C]$ and

$$\frac{d[D]}{dt} = \frac{k_2 k_1 [A]^2}{k_{-1}[C]}$$

and thus the reaction is second order in A.

In more complex mechanisms, particularly those involving free radicals, there may be a number of reactive intermediates. By assuming that each of these species attains a steady-state concentration, an expression for this concentration can be obtained by setting $d[X]/dt$ equal to zero. The relations then can be substituted into the rate equations for the step giving the final products. For mechanisms comprising a large number of steps, the ultimate equation obtained in this way will be considerably more complex than Eq. (6-23).

Note finally that the concept of the steady state is an approximation in that, just as for first order reactions, the concentration of the intermediate will never become truly constant. The approximation is a good one, however, and the approach is one of the most useful ways of dealing with complex mechanisms.

6-4 TEMPERATURE DEPENDENCE OF REACTION RATES

The systems discussed up to this point all have been at a single temperature. The rate "constant" defined by the general differential rate law

$$\frac{dx}{dt} = k_y (a - x)^y$$

is a function of temperature. If a reaction is investigated at a number of different temperatures, the k values found usually show a quite dramatic increase, often several orders of magnitude, over a relatively small temperature range. For many reactions there is approximately a two- to three-fold increase in rate for every 10 °C rise in temperature.

To obtain the desired quantitative expression for the relation between k and T, use is made initially of what is essentially an empirical approach. In the subsequent discussion of kinetic theories the origin of this expression is examined.

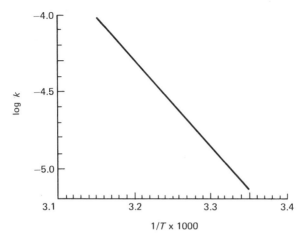

FIG. 6-8 Arrhenius plot for the temperature dependence of the rate constant for the reaction discussed in Example 6-1.

If rate constant values for a given reaction at different temperatures are analyzed graphically, it is found that there is a linear dependence, taking the form shown in Fig. 6-8, of the logarithm of the rate constant on the reciprocal of reaction temperature in °K. This behavior is indicative of the applicable equation, being of the form

$$\log k = m\left(\frac{1}{T}\right) + C$$

The slope m indicates essentially how rapidly k will change with temperature. The exact form of this equation, using the natural logarithm of k, $\ln k$, was suggested originally by Arrhenius. On the basis of thermodynamic arguments, he equated the slope of the $\ln k$ vs. $1/T$ plot with the factor $-E_a/R$, where R is the gas constant and E_a is the *activation energy*. As will be seen presently, E_a represents the temperature independent amount of energy required of the reactants before reaction can occur. The constant was represented as $\ln A$, where A is the *frequency factor* or *pre-exponential factor*. A is also temperature independent to a first approximation and has a unique value for a given reaction. (The physical significance of A is discussed later.) Thus the *Arrhenius equation*,

$$\ln k = -\frac{E_a}{RT} + \ln A \qquad (6\text{-}24)$$

represents a specific form of the empirical relation that can be used to describe the observed variation in the rate constant with temperature. From the slope of the straight line obtained by plotting Eq. (6-24), the value of the activation energy can be determined. As with previously encountered equations involving natural logarithms, it is convenient to convert to common logs for actual calculations. Thus Eq. (6-24) would be written in the form

$$\log k = -\frac{E_a}{2.303\ RT} + \log A \qquad (6\text{-}25)$$

Another frequently used form of Eq. (6-24) is that obtained by taking antinatural logarithms of both sides, giving

$$k = Ae^{-E_a/RT} \qquad (6\text{-}26)$$

This form is particularly useful in discussing the qualitative dependence of k on activation energy and A.

Usually it is convenient to obtain only E_a from the graphical analysis using Eq. (6-25), the pre-exponential factor A being evaluated by calculation from the equation when E_a has been determined from the graph. While the graphical method has the advantage of conveniently handling experimental errors, a value of k at only two temperatures is sufficient to write two equations of the form of Eq. (6-25) and solve simultaneously for A and E_a. If the temperatures and the applicable k values are designated by the subscripts 1 and 2, Eq. (6-25) can be written specifically for each temperature. Subtraction of the two relations gives

$$\log k_2 - \log k_1 = -\frac{E_a}{2.303\ R}\left(\frac{1}{T_2} - \frac{1}{T_1}\right)$$

or more conveniently for activation energy calculations,

$$E_a = \log \frac{k_2}{k_1} \times 4.576 \times \frac{T_2 T_1}{T_2 - T_1} \qquad (6\text{-}27)$$

EXAMPLE 6-1.   A first order reaction has a half-life of 955 sec at 25 °C and 126 sec at 40 °C. Calculate the activation energy and pre-exponential factor for this process.

at T_1 (298 °K) $t_{\frac{1}{2}} = 0.693/k_{T_1}$ $k_1 = 0.693/955 = 7.25 \times 10^{-4}\ \text{sec}^{-1}$

at T_2 (313 °K) $t_{\frac{1}{2}} = 0.693/k_{T_2}$ $k_2 = 0.693/126 = 5.50 \times 10^{-3}\ \text{sec}^{-1}$

$$? \ E_a = \log \frac{7.25 \times 10^{-4}}{5.50 \times 10^{-3}} \times 4.576 \times \frac{298 \times 313}{313 - 298} = \underline{24.5 \times 10^3\ \text{cal}}$$

$$\log k_1 = -\frac{E_a}{2.303\ RT_1} + \log A$$

$$-3.14 = -24,500/1,364 + \log A$$

$$A = \underline{1.26 \times 10^{15}\ \text{sec}^{-1}}$$

6-5 REACTION RATE THEORIES

It now should be apparent how it is possible to deal adequately with the rates of chemical reactions in terms of the rate constant and reactant concentrations. In this section attention is focused on a more detailed examination of the nature of the rate constant itself and the two principal theories that can be used to rationalize the observed k values. The possibility of using these theoretical ideas to predict rate constants not yet determined experimentally also can be examined. These discussions beome involved in looking at what transpires in a chemical reaction on an atomic and molecular basis— the events in which individual species participate as they are converted from reactant to product. The insight gained here is invaluable in understanding the important type of reaction—those in which the rate is significantly increased by a catalyst—that is discussed later.

6-5-1 Collision Theory

The theory is based primarily on consideration of reactions in the gas phase, although the same type of general approach is applicable to processes occurring in solution.

The fundamental principle of the collision theory is that the *reactant species must collide before reaction can occur.* Therefore, the rate of the reaction will be proportional to the number of collisions per unit time. Consider as a working example a gas phase bimolecular and second order process $A + B \rightarrow$ products. The number of collisions between A and B per second at unit concentration, Z_{AB}°, can be calculated from an equation obtained from the kinetic theory of gases:

$$Z_{AB}^{\circ} = \text{constant} \sqrt{T} \tag{6-28}$$

The explicit form of the constant term is not relevant to the present level of examination. Since the number of collisions per unit time depends on the concentration, the reaction rate will be given by a product, proportionality factor $\times Z_{AB}^{\circ} \times$ actual concentration of A and B in moles per liter. The essential step in the development, therefore, is the elucidation of this proportionality factor.

Only a very small fraction of the total number of collisions actually result in a reaction. If the only requirement for a reaction were a collision, then the rate of *all* gas phase reactions would be very high, if they occurred at all. For example, in the case of pure oxygen at 500 °K and 1 atm pressure, substitution of the appropriate factors in Eq. (6-28) indicates that each O_2 molecule suffers about 10^9 collisions/sec. It is almost self-evident that the additional requirement for reaction is an energetic one. When two species collide there must be associated with the collision sufficient energy to cause whatever reaction actually may occur. This is the basic postulate of the Arrhenius theory—*reaction-producing collisions must involve at least some minimum energy.*

Another characteristic of reaction rates that points to the influence of some factor over and above the collision frequency is the doubling or tripling of rate in a relatively small temperature interval. This increase is very much larger than the slightly increased Z_{AB}° value. For example, using Eq. (6-28) and canceling the temperature independent constant terms, shows us that the ratio of collision frequencies at 310 and 300 °K is only

$$Z_{310}^{\circ}/Z_{300}^{\circ} = \frac{(310)^{\frac{1}{2}}}{(300)^{\frac{1}{2}}} = 1.016$$

That is, by increasing the temperature through 10°K the collision frequency is increased by only about 1.6 percent while the rate typically is at least doubled.

In an assembly of gas molecules at a particular temperature, say T_1, there is a distribution of energies that takes the form of the curve in Fig. 6-9. At some higher temperature T_2 the distribution is shifted to higher energies as shown in the figure, but the average energy is moved only slightly. The shape of the curve is such, however, that if we examine at each temperature the number of molecules having energies equal to or greater than some specific E value significantly larger than the average energy, it is apparent that this number is greatly increased by even a slight shift in the distribution. From

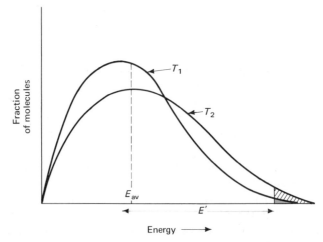

FIG. 6-9 Maxwell-Boltzmann distribution of energies at two different temperatures $T_1 < T_2$.

the treatment of the Maxwell–Boltzmann distribution, the fraction of all molecules in a particular assembly having energy *greater than the average*, by an amount equal to or greater than E', is given by

$$\text{Fraction molecules with } E \geq (E_{av} + E') = e^{-E'/RT} \qquad (6\text{-}29)$$

The identity of this expression and the Arrhenius equation is apparent. Considering a chemical reaction, E' can be interpreted as the activation energy E_a, the minimum energy requirement for reaction. Thus Eq. (6-29) gives the fraction of all collisions involving enough energy to produce reaction. This fraction therefore should be the proportionality factor expressing the dependence of the reaction rate on the actual number of collisions occurring per unit time and volume.

These arguments indicate that in the general relation for the rate.

Rate = number of collisions per unit concentration per unit time

$$\times \text{ fraction with } E \geq (E_{av} + E_a) \times \text{concentrations}$$

the explicit form for each of these terms can be substituted as follows:

$$\text{Rate} = Z_{AB}^{\circ} e^{-E_a/RT}[A][B] \qquad (6\text{-}30)$$

Comparing this to the general differential rate law for a second order process, Rate = $k[A][B]$, shows that the rate constant must correspond to

$$k = Z_{AB}^{\circ} e^{-E_a/RT} \qquad (6\text{-}31)$$

For a given E_a and collision frequency, the k values predicted by this equation are correct for only a very few reactions. One approach has been to include an additional pre-exponential term p, the *steric factor*. This term allows for the fact that only a fraction of the collisions with sufficient energy will involve molecules colliding with the specific spatial orientation that may be required for a particular process. The value of p normally would be less than unity, or exactly 1.0 if there were no steric requirements. Even this refinement, however, cannot adequately handle some reactions that appear to have steric factors greater than unity. In addition to this inadequacy, there is a more fundamental lack of predictive character in collision theory. There is no way to deduce the activation energy except by direct experimental measurement.

Comparison of the collision theory equation, modified with the steric factor,

$$k = p Z_{AB}^{\circ} e^{-E_a/RT} \qquad (6\text{-}32)$$

and the Arrhenius equation, Eq. (6-26), shows that we can identify the frequency factor A with the $p Z_{AB}^{\circ}$ term. Although the effect normally cannot be detected in graphs of Eq. (6-25), the frequency factor should vary as the square root of temperature, since according to Eq. (6-28), $Z_{AB}^{\circ} \propto T^{\frac{1}{2}}$.

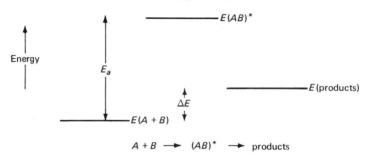

FIG. 6-10 *Energy diagram for the reaction* A + B → *products.*

The physical events that occur in a single step process, as visualized by the collision theory, can be described in terms of the formation of an intermediate species (AB)*. The activation energy is considered to be the energy required for its formation. The energetics of the system can be illustrated by the schematic energy diagram in Fig. 6-10. The reaction A + B → products can be represented as

$$A + B \rightarrow (AB)^* \rightarrow products$$

Those collisions involving a total energy equal to or greater than the activation energy result in the formation of the intermediate that decomposes to give the observed products.

The energy representation in Fig. 6-10 allows for consideration of the thermochemistry of the reaction. If it is assumed that the difference between ΔH and ΔE can be neglected (although the difference may be numerically significant, it would normally not be sufficient to alter the sign of ΔH compared to ΔE), then the example reaction is endothermic since the energy of the products is greater than that of the reactants. For an exothermic process the level representing $E_{products}$ would be below that of the reactants.

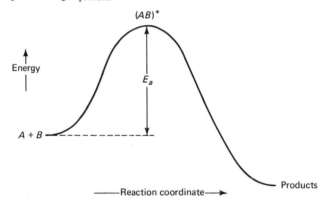

FIG. 6-11 *Energy vs. reaction coordinate for an exothermic reaction of the type* A + B → *products.*

These energy considerations also emphasize the important point made previously that thermodynamics cannot predict the rate of a reaction. Thus a process may be strongly exothermic and, therefore, may (but not necessarily) have a large negative free energy change. It nevertheless may proceed very slowly indeed because of a large activation energy, even though when thermodynamic equilibrium is reached the system will be very predominantly in the product form.

An alternative method of presentation of data of this type is through a plot of energy vs. *reaction coordinate*—a kind of time-distance scale—which gives a graph indicating how the energy varies as the single-step process proceeds from reactant species, through the intermediate, to the products. Figure 6-11 gives a plot of this type for an exothermic process. This visualization is quite useful in that it is often instructive to think of the activation energy requirement as an energy barrier over which the reactants must move in order for reaction to occur.

6-5-2　Transition State Theory

In some ways collision theory is a rationalization of experimental facts, particularly with respect to the activation energy. The equations allow evaluation of the activation energy from measured k values but provide no means whereby E_a can be predicted.

The main distinguishing feature of the transition state theory is that it does permit at least the possibility of making this kind of prediction if sufficient information about atomic and molecular properties is available. The essential idea is that the reaction proceeds through a transition state that is in thermodynamic equilibrium with the reactants. The overall rate of the process is then determined by the rate at which the transition state undergoes "decomposition" into products.

The concept of an intermediate such as (AB)* has not been abandoned; rather, this species now specified as a transition state $AB^\ddagger$ for the process $A + B \rightarrow$ products. In this theory the intermediate is considered formally to be involved in a thermodynamic equilibrium. All functions associated with the transition state are designated by the superscript $\ddagger$. The advantage of assuming the existence of an equilibrium is that profitable use can be made of the various thermodynamic relations involving the equilibrium constant.

The participation of the transition state can be represented as follows:

$$A + B \underset{k_2}{\overset{K^\ddagger}{\rightleftarrows}} AB^\ddagger \xrightarrow{k_d} \text{products}$$

The overall process is comprised of the equilibrium between the reactants and

$AB^\ddagger$ and the subsequent decomposition of the transition state. The latter step has its rate determined by the equilibrium concentration of the transition state species and the magnitude of the rate constant for its unimolecular decomposition k_d; that is,

$$\frac{dx}{dt} = k_d[AB^\ddagger] \tag{6-33}$$

The overall rate can be represented by the ordinary differential rate law,

$$\frac{dx}{dt} = k_2[A][B] \tag{6-34}$$

Combining these two equations gives

$$k_2[A][B] = k_d[AB^\ddagger]$$

$$k_2 = k_d \frac{[AB^\ddagger]}{[A][B]} \tag{6-35}$$

Since it has been assumed that a bonafide thermodynamic equilibrium exists between $AB^\ddagger$ and the reactants, techniques already developed can be employed. Specifically, the equilibrium constant $K^\ddagger$ is defined as

$$K^\ddagger = \frac{[AB^\ddagger]}{[A][B]} \tag{6-36}$$

The concentration terms in Eq. (6-35) therefore can be replaced by $K^\ddagger$,

$$k_2 = k_d K^\ddagger \tag{6-37}$$

The rate constant for the decomposition of the transition state is directly related to the frequency of vibration of the atoms in the intermediate pseudo-molecule $AB^\ddagger$. Making use of classical vibrational theory, k_d can be written as

$$k_d \equiv \nu_D = \frac{RT}{Nh} \tag{6-38}$$

where ν_D is the frequency of vibrations that result in decomposition of $AB^\ddagger$, h is Planck's constant, R is the gas constant, and N is Avogadro's number. (The term R/N thus amounts to the gas constant per molecule.) Putting in this expression for k_d in Eq. (6-37), we get

$$k_2 = \frac{RTK^\ddagger}{Nh} \tag{6-39}$$

The final step in the derivation is to make use of the relations between the equilibrium constant and thermodynamic functions. For the activation equilibrium we have from Eq. (3-5),

$$\Delta G^{\circ\ddagger} = -2.303 \, RT \log K^\ddagger \tag{6-40}$$

Frequently the superscript zero is dropped here, and $\Delta G^{\ddagger}$ is referred to as the *free energy of activation*, although strictly speaking it is a standard free energy change. The distinction may be ignored, however, in this context.

To permit direct substitution in Eq. (6-39) for $K^{\ddagger}$, the $\Delta G^{\ddagger}$ equation is written using natural logarithms and rearranged to the form

$$\ln K^{\ddagger} = -\frac{\Delta G^{\ddagger}}{RT} \qquad (6\text{-}41)$$

Equation (2-23) written for the transition state is

$$\Delta G^{\ddagger} = \Delta H^{\ddagger} - T\,\Delta S^{\ddagger} \qquad (6\text{-}42)$$

where $\Delta H^{\ddagger}$ and $\Delta S^{\ddagger}$ are the *enthalpy and entropy of activation*, respectively. Replacing $\Delta G^{\ddagger}$ in Eq. (6-41) from this relation gives

$$\ln K^{\ddagger} = \frac{\Delta S^{\ddagger}}{R} - \frac{\Delta H^{\ddagger}}{RT}$$

or

$$K^{\ddagger} = e^{\Delta S^{\ddagger}/R} e^{-\Delta H^{\ddagger}/RT}$$

and, therefore, Eq. (6-39) assumes the final desired form

$$k_2 = \frac{RT}{Nh} e^{\Delta S^{\ddagger}/R} e^{-\Delta H^{\ddagger}/RT} \qquad (6\text{-}43)$$

Using this equation we find it possible to make some predictions about the magnitude of k_2 by initially predicting likely values for the enthalpy and entropy of activation by considering the probable nature of the transition state. This is by no means a simple task, but the transition state approach does have at least an inherent capability for prediction that is not a characteristic of collision theory.

It is useful to compare the three rate constant expressions that have been discussed:

$$\text{Arrhenius:} \quad k = Ae^{-E_a/RT}$$

$$\text{Collision:} \quad k = pZ^{\circ}e^{-E_a/RT}$$

$$\text{Transition state theory:} \quad k = \frac{RT}{Nh} e^{\Delta S^{\ddagger}/R} e^{-\Delta H^{\ddagger}/RT}$$

Some of the interrelations are obvious. If differences between ΔH and ΔE (which are unimportant except in gas phase systems) are neglected, both the form of these relations and the graphical presentations in Figs. 6-10 and 6-11 show that the enthalpy of activation and activation energy are equivalent and represent the increase in "energy" when the reactants form the intermediate.

It is also interesting to note that there is an implicit relation between p and $\Delta S^{\ddagger}$. Both are indicative of the reaction-producing efficiency of an encounter between reactant species. A requirement for a specific orientation in this encounter is manifested in collision theory by a small p value and in transition state theory by a relatively large and negative $\Delta S^{\ddagger}$ and hence a value of less than unity for the $e^{\Delta S^{\ddagger}/R}$ term.

6-6 CATALYSIS: ENZYME CATALYZED REACTIONS

Many chemical reactions can exhibit a very marked increase in rate in the presence of a catalyst. Since the catalyst itself is not consumed in the reaction, very small amounts of material are capable of producing profound acceleration of the reaction.

A catalyst usually operates by participating in the formation of the activated complex, or some intermediate in the rate-determining step of a stepwise mechanism, and <u>decreases the activation energy</u>. Since the rate constant is exponentially dependent on E_a, relatively small changes in the activation energy produce large alterations in k. Consider, for example, a typical second order reaction with $E_a = 20$ kcal and $A = 5 \times 10^{11}$ liter mole^{-1} sec^{-1}. From Eq. (6-25) it is simple to calculate that at $T = 298 \,^{\circ}$K, $k = 1.0 \times 10^{-3}$ liter mole^{-1} sec^{-1}. If we add to this system a catalyst that participates in the formation of the activated complex and thereby lowers the activation energy by 2.5 to 17.5 kcal, the same k calculation indicates that the rate constant is increased to a value of 80×10^{-3} liter mole^{-1} sec^{-1}, an increase of almost two orders of magnitude.

There are a number of different mechanisms of catalytic action. In *homogeneous catalysis* the reactants and catalyst are together in a single phase. An example of this type of process is the acid-catalyzed hydrolysis of organic esters in aqueous solution. The stoichiometry of the process is

$$RCOOR' + H_2O \rightarrow RCOOH + R'OH$$

The mechanism consists of a number of steps but can be treated by conveniently utilizing the rate-determining step approach. The overall rate is determined by the rate at which the solvent molecules attack the carbon atom of the carbonyl (C=O) group. In neutral solution this process is rather slow, but on the addition of acid the rate-determining step is preceded by the rapidly established equilibrium

$$\begin{matrix} R-C=O \\ | \\ O-R \end{matrix} + H^+ \rightleftharpoons \begin{matrix} R-\overset{+}{C}-OH \\ | \\ O-R \end{matrix}$$

The rate of attack of water molecules on the carbonium ion formed here has a lower activation energy, and therefore a larger rate constant, than the analogous process involving the neutral ester itself. Since the process remains the rate-determining step, the overall rate is significantly higher when the acid catalyst participates in forming the key intermediate. The species formed by the attack of H_2O on the carbonium ion undergoes rapid electron redistribution and proton transfer to yield the final products, alcohol and ester.

In biological systems by far the most important catalytic processes are those involving *enzymes*.

Enzymes of various types can catalyze almost all known types of chemical reaction. These catalysts exhibit some quite unique properties in the sense that their behavior is somewhat different from that of "ordinary" chemical catalysts for nonbiological processes. The catalytic action of most enzymes is specific—either absolutely, in that only one particular reaction involving a specific reactant is catalyzed, or relatively, in that an enzyme may catalyze only reactions involving a definite class of compound or only a particular type of reaction. The effect of temperature also is remarkable in that the rate initially increases with temperature in the normal way, but an optimum temperature is usually reached, after which the rate declines. This phenomenon results from inactivation of the enzyme at the higher temperatures. The important implication of this phenomenon is that activation energy studies must be carried out over relatively small temperature ranges where inactivation effects legitimately can be neglected. Most enzymes also show maximum activity at a particular pH. It has been seen in previous discussions of acid-base equilibria involving protein molecules that the ionic form in which the molecule exists in aqueous solution is dependent on solution pH. If the catalytic action is a characteristic of just one ionic form of the protein portion of the enzyme, the maximum activity is observed at the pH where this particular form predominates.

It is not possible here to give an account of the fascinating developments in the elucidation of enzyme-catalyzed reactions that have taken place in the last several years. The discussion here attempts to outline in a general way how the basic principles of chemical kinetics developed in this chapter can be applied to relatively straightforward enzymatic reactions. On this foundation can be built more advanced discussions of biological systems of current interest.

Most discussions of the kinetics of enzyme-catalyzed reactions follow a conventional system of nomenclature outlined in the following chart. The essential nature of the mechanism is the formation of the intermediate by a combination of the enzyme and the reactant, followed by decomposition of the intermediate or regeneration of the starting material.

$$\text{Enzyme} \quad + \quad \text{Substrate} \quad \rightleftharpoons \quad \text{Complex} \qquad\qquad \text{I}$$

| Enzyme | Substrate | Complex |
| *protein* | *reactant* | *enzyme-substrate* |

or *protein–prosthetic group* *reactant* *enzyme–substrate*

or *protein · · ·coenzyme*

$$\text{Complex} \rightarrow \text{products} + \text{enzyme} \qquad\qquad \text{II}$$

The group that may be associated with the parent protein is termed a *prosthetic group* or *coenzyme* on the basis of a strong or weak attachment to the protein. Most enzymes are named by adding the suffix *-ase* to the name of the reactant(s) or type of reaction for which they are specific. Thus, for example, the enzyme that catalyzes the hydrolysis of the phosphate from phosphoprotein

$$\text{Phosphoprotein} + H_2O \xrightarrow{\text{enzyme}} \text{protein} + H_2PO_4^-$$

is known as *phosphoprotein phosphotase.*

The kinetics of enzyme-catalyzed reactions have been investigated for many years, and a great deal of information is available about numerous systems. The typical variation in the rate of reaction, with substrate concentration conventionally designated as v, is shown in Fig. 6-12. It is apparent that at high substrate concentrations the reaction is zero order in [S]. For a given enzyme concentration and temperature, this level portion of the curve corresponds to the maximum possible rate v_{max}. Application of any of the methods discussed in Sec. 6-2 to initial rate data obtained at very low substrate concentrations shows that the reaction, under those conditions, is first order in substrate concentration. Any proposed kinetic treatment therefore must explain this alteration in reaction order and the attainment of a limiting rate.

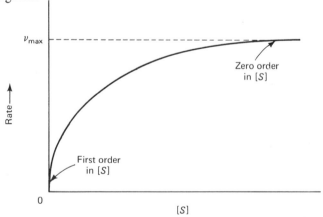

FIG. 6-12 The rate of a typical enzyme catalyzed reaction as a function of substrate concentration for a specific concentration of enzyme at constant temperature.

The fundamental techniques used in handling kinetic data from these systems are based on the classic work of Michaelis and Menten.* This approach represents a very useful application of the *general* methods of kinetics discussed in this chapter. The essential features of the Michaelis–Menten mechanism in fact have been summarized in the nomenclature scheme written above. The scheme now can be rewritten in the more usual reaction equation form:

$$E + S \underset{k_{-1}}{\overset{k_1}{\rightleftharpoons}} ES \qquad\qquad I$$

$$ES \overset{k_2}{\longrightarrow} E + P \qquad\qquad II$$

The most satisfactory approach to the treatment of this mechanism is to assume that the complex ES attains a steady-state concentration. This is a modification of the original Michaelis–Menten treatment, which utilized the rate-determining step method.

For the steady state in [ES],

$$\frac{d[ES]}{dt} = 0 = k_1[E][S] - k_{-1}[ES] - k_2[ES] \qquad (6\text{-}44)$$

The enzyme is present in both the free and *bound* (with the substrate in the complex) forms and therefore its total concentration $[E]_t$ is given by

$$[E]_t = [E] + [ES]$$

or

$$[E] = [E]_t - [ES] \qquad (6\text{-}45)$$

It now is possible to eliminate the concentration of free enzyme, a quantity not susceptible to direct experimental observation determination, between Eqs. (6-44) and (6-45) to obtain

$$k_1([E]_t - [ES])[S] - k_{-1}[ES] - k_2[ES] = 0$$

which can be rearranged to the form

$$[ES] = \frac{k_1[E]_t[S]}{k_{-1} + k_2 + k_1[S]} \qquad (6\text{-}46)$$

The overall rate of product formation in process II will be given by the differential rate law,

$$v = k_2[ES] \qquad (6\text{-}47)$$

Substitution of the steady-state concentration term from Eq. (6-46) into the differential rate equation gives

$$v = \frac{k_2 k_1[E]_t[S]}{k_{-1} + k_2 + k_1[S]}$$

* L. Michaelis and M. L. Menten, *Biochem. Z.*, **49**, 333 (1913).

or

$$v = \frac{k_2[E]_t[S]}{[(k_{-1} + k_2)/k_1] + [S]} \tag{6-48}$$

The ratio of rate constants appearing in the denominator is a fundamental property of the enzyme–substrate system and is referred to as the *Michaelis constant* K_m.

$$K_m = \frac{k_{-1} + k_2}{k_1} \tag{6-49}$$

Substitution of this term in Eq. (6-48) gives the simpler and more convenient expression

$$v = \frac{k_2[E]_t[S]}{K_m + [S]} \tag{6-50}$$

The dissimilar kinetics that characterize enzyme-catalyzed reactions at low and at high substrate concentrations can be rationalized in terms of this equation in the following manner: At small [S], $K_m + [S] \approx K_m$, and Eq. (6-50) becomes

$$v = \frac{k_2}{K_m}[E]_t[S] \tag{6-51}$$

For a particular total enzyme concentration, then, the rate is given by $v = $ constant [S]—a first order reaction with respect to the substrate. This corresponds to the initial portion of the curve in Fig. 6-12.

When the substrate concentration is large, $K_m + [S] \approx [S]$, and Eq. (6-50) then would assume the very simple form

$$v = k_2[E]_t$$

This is consistent with the form of the rate plot indicating v independent of [S] under these conditions. This relation also gives the maximum rate for a particular value of $[E]_t$. It is more useful to indicate this explicitly as

$$v_{max} = k_2[E]_t \tag{6-52}$$

Finally, v_{max} can be substituted into Eq. (6-50). This gives the *Michaelis–Menten equation:*

$$v = \frac{v_{max}[S]}{K_m + [S]} \tag{6-53}$$

Both v_{max} and K_m are fundamental parameters that describe a particular enzyme system. Two important points can be made in connection with K_m. In Eq. (6-49) k_{-1} and k_2 are first order rate constants (in $\sec^{-1}$). k_1 refers to the initial second order reaction and will have units of liter mole^{-1} sec^{-1} if concentrations are expressed in molarities. K_m therefore has the units of a concentration term, moles per liter. A useful expression involving K_m also can be extracted from Eq. (6-50). For this purpose the equation is rearranged to the form

$$K_m = \left(\frac{v_{max}}{v} - 1\right)[S]$$

from which it is apparent that

$$K_m = [S] \quad \text{at} \quad v = 0.5v_{max} \tag{6-54}$$

The Michaelis constant thus is numerically equal to the concentration of substrate when the rate of the process is one-half the maximum rate attainable at the relevant total enzyme concentration and temperature. K_m itself, however, is independent of the concentration of substrate and enzyme.

The form of Eq. (6-53) does not lead to facile evaluation of system parameters from an analysis of rate data. A very convenient modification, due to H. Lineweaver and D. Burk, consists in expressing this relation in reciprocal form:

$$\frac{1}{v} = \frac{K_m}{v_{max}[S]} + \frac{[S]}{v_{max}[S]}$$

and therefore

$$\frac{1}{v} = \frac{K_m}{v_{max}} \cdot \frac{1}{[S]} + \frac{1}{v_{max}} \tag{6-55}$$

A plot of experimentally determined rates at various substrate concentrations, in the form $1/v$ vs. $1/[S]$, yields a straight line of slope K_m/v_{max} and with intercept on the $1/v$ axis of $1/v_{max}$. It also is sometimes useful to utilize the fact that Eq. (6-55) reduces to the form $1/[S] = -1/K_m$ when $1/v = 0$. Thus, by extending the $1/[S]$ axis to the left of the origin (hypothetical negative [S] values), the intercept of the straight line plot with the $1/[S]$ axis gives the negative reciprocal of the Michaelis constant. A Lineweaver–Burk plot, or *double reciprocal* plot as graphs of Eq. (6-55) are usually called, is shown for a typical enzyme-catalyzed reaction in Fig. 6-13. In utilizing this equation it must be remembered that the treatment rests on the simple mechanism comprised of processes I and II indicated previously. If other possible reactions become important, the treatment is no longer adequate and the double reciprocal plots will not be linear. In an essentially straightforward system the most probable cause of such complications is the interfering reactions by products. Thus it is important that the rate values used in the analysis be those determined experimentally at the smallest extents of reaction consistent with accurate analysis.

The catalytic effectiveness of enzymes can be reduced by the action of *inhibitors*, which in some instances may be products themselves, leading to the complications just mentioned. The mode of operation of the inhibitor

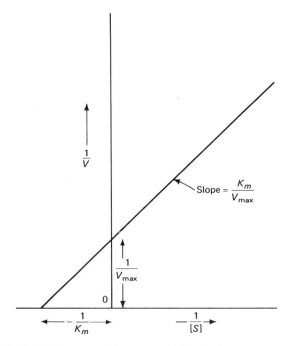

FIG. 6-13 A double-reciprocal (Lineweaver-Burk) plot for a typical enzyme-catalyzed reaction.

may be to combine directly with the enzyme to reduce its effective concentration, to combine with the complex ES, or both. Combination with the complex leads to an inactive species.

The simplest case is that of *competitive inhibition*. Here the inhibitor can attack only the enzyme itself. Thus there is a competition for enzyme between substrate and inhibitor. The extent of inhibition can be reduced by increasing the concentration of substrate relative to that of the inhibitor. The mechanism now must include an additional equilibrium involving the inhibitor as follows:

$$E + S \underset{k_{-1}}{\overset{k_1}{\rightleftharpoons}} ES \overset{k_2}{\longrightarrow} E + P$$

$$E + I \overset{K_I}{\rightleftharpoons} EI$$

Since the complex EI is formed in a chemically reversible process, this sequence of reaction can be handled kinetically by assuming a steady state in both complexes ES and EI. The resulting modified Michaelis–Menten equation is

$$v = \frac{v_{\max}[S]}{[S] + K_m \{1 + ([I]/K_I)\}} \tag{6-56}$$

Again, conversion to the reciprocal straight line form allows for graphical presentation in the double reciprocal form. The relevant equation is

$$\frac{1}{v} = \frac{K_m}{v_{max}}\left(1 + \frac{[I]}{K_I}\right)\frac{1}{[S]} + \frac{1}{v_{max}} \tag{6-57}$$

A plot of $1/v$ vs. $1/[S]$ will have an intercept on the vertical axis of $1/v_{max}$, and on the horizontal $1/[S]$ axis the intercept will be $-1/K_m(1 + [I]/K_I)$.

The second case, so-called *noncompetitive inhibition*, has a second additional equilibrium to be considered:

$$ES + I \overset{K_I{}^n}{\rightleftharpoons} ESI$$

The physical significance of this situation is that the inhibitor can combine with the enzyme completely independently of whether the enzyme has already attached itself to a substrate molecule. Under these conditions $K_I^n \equiv K_I$, and a steady state treatment gives an expression for the rate that can be recast to the double reciprocal form

$$\frac{1}{v} = \frac{K_m}{v_{max}}\left(1 + \frac{[I]}{K_I}\right)\frac{1}{[S]} + \frac{1}{v_{max}}\left(1 + \frac{[I]}{K_I}\right) \tag{6-58}$$

On the $1/v$ vs. $1/[S]$ graph the straight line arising from this relation has a $1/v$ axis intercept of $(1/v_{max})(1 + [I]/K_I)$ and a $1/[S]$ axis intercept of $-1/K_m$.

The forms of all three equations—(6-55), (6-57), and (6-58)—are summarized in a comparison graph given in Fig. 6-14, where the relation

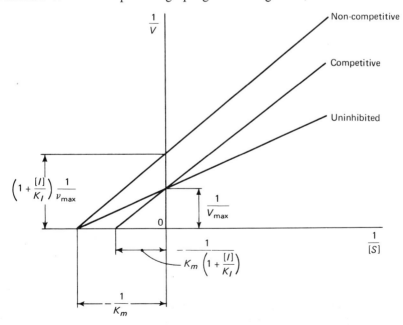

FIG. 6-14 Plots of Eqs. (6-55), (6-57), and (6-58) for enzyme-catalyzed reactions.

of the various intercepts to the terms in the three equations for the different types of enzyme-catalyzed behavior are indicated.

The detailed kinetic treatment of many enzyme-catalyzed reactions is considerably more complex than the rather straightforward cases outlined here. Particularly in the case of product inhibition, wherein the concentration of inhibitor will vary significantly during the course of the reaction, the detailed considerations required can lead to complex algebra. The basis indicated in this discussion, however, indicates the approach that can be taken for less simple systems. It also demonstrates how we can apply to a particular class of reaction of very great practical importance, by considering simple, single-step processes, the concepts of the steady-state and elementary rate laws that have been developed.

PROBLEMS

6-1. A second order reaction in the gas phase has a rate constant equal to 7.36×10^8 liter mole^{-1} sec^{-1} at 25 °C. Calculate the numerical value of this constant when the concentrations are expressed in terms of (a) torr and (b) atmospheres.

6-2. Derive an expression for the half-life of a reaction whose differential rate equation is $dx/dt = k[A]^{\frac{3}{2}}$.

6-3. A reaction was found to go 50 percent to completion in 30 min and 66 percent to completion in 60 min, when the initial concentration of reactant is 1.00 mole/liter. On the basis of the half-life equations for first and second order reactions, calculate possible k values for each order. For both cases calculate the amount of reaction in 60 min and, hence, find the order of the reaction.

6-4. Using the steady-state treatment, derive an expression for the rate of formation of product E in the sequence

$$A + B \underset{k_{-1}}{\overset{k_1}{\rightleftharpoons}} C$$

$$C + A \xrightarrow{k_2} D$$

$$C + B \xrightarrow{k_3} E$$

Assume that C is a highly reactive intermediate and that D and E undergo no further reactions.

6-5. Rate constant determinations for a certain reaction showed that k at 45.3 °C was 2.35 times that at 34.7 °C. Calculate the energy of activation for the reaction.

6-6. A solution decomposition reaction gives 1 mole of gaseous product per mole of reactant consumed. If the following pressures were observed, show that the reaction is first order and calculate k_1 and $t_{\frac{1}{2}}$ (take $t = 1,000$ min as infinite time and, hence, take $P_{t=1,000}$ as a):

t (min)	0	25	50	100	300	1,000
P (torr)	0	95	168	290	461	500

6-7. A second order reaction with $[A]_0 = 1$ mole/liter $2A \rightarrow$ products is observed to go 15 percent to completion in 250 sec. Calculate the value of k_2 and the time required for 60 percent of the starting material to be consumed.

6-8. From the following information for the rate constant of a second order reaction, find k_2 at 385 °K by a graphical method that does *not require* the determination of E_a:

k (liter mole^{-1} sec^{-1})	3.02×10^{-4}	6.31×10^{-3}	3.80×10^{-2}
T (°K)	308	335	360

6-9. For the first order gas phase reaction $D \rightarrow C$, if the initial pressure is 10 atm and after 2.30 sec 4 atm of C are formed, calculate the rate of formation of C, in the appropriate units, when 8, 4, 2, 1, and 0.5 atm of A have been consumed.

6-10. An enzyme-catalyzed reaction is found to have $K_m = 3.8 \times 10^{-3}$ mole/liter and $k_2 = 6.93 \times 10^{-4}$ sec^{-1}. If the total enzyme concentration is 1×10^{-5} mole/liter, determine v_{max}. Calculate the rate v when $[S] \times 10^3$ (moles/liter) is 0.5, 1.0, 1.5, 2.0, 2.5, 3.0, 3.5, and 4.0. Represent the data in a double reciprocal plot and show that the graphical parameters are consistent with the K_m and k_2 values given.

Electrochemistry

7

7-I INTRODUCTION

The treatment of thermodynamics in Chapter 2 showed that the free energy change is a measure of the useful work obtainable in a particular process. In a thermodynamically reversible reaction, the amount of useful work obtained, w'_{max} is identical with the decrease in free energy.

$$\Delta G_{TP} = -w'_{max} \tag{7-1}$$

where the term w'_{max} is the total work w_{max} less the pressure–volume work $P\Delta V$. In an actual irreversible process the amount of work involved will be less than w_{max}, but ΔG remains unaltered provided the same initial and same final state are used.

In chemical reactions characterized by a transfer of electrons, oxidation-reduction reactions, a negative free energy change can be envisaged as a driving force for electron transfer. In these processes w'_{max} takes the form of electrical work. If the electrochemical reaction occurs reversibly, ΔG gives directly the amount of such work to be obtained, In this type of system the physical reality of the concept of *useful work* is most easily appreciated.

The designation oxidation-reduction is applicable to a variety of processes. A convenient classification may be based on the kind of species involved. Three types of reaction that are of interest in subsequent discussions can be distinguished:

222

Ionic:

$$Fe + 2 Ag^+ \rightarrow Fe^{2+} + 2 Ag$$

Organic:

$$5 C_2H_5OH + 4 MnO_4^- + 12 H^+ \rightarrow 5 CH_3COOH + 4 Mn^{2+} + 11 H_2O$$

Biological:

$$2 \text{ cytochrome } c\text{-}Fe^{2+} + 2 H^+ + \tfrac{1}{2} O_2 \xrightarrow{\text{enzyme}} 2 \text{ cytochrome } c\text{-}Fe^{3+} + H_2O$$

The hydronium ion is frequently involved in all three types of system and, therefore, the course of the reaction is dependent on solution pH.

The ease with which w'_{max} can be determined in an electrochemical system is quite variable. It is in ionic systems that measurements of this type are most easily made. In all redox reactions it nonetheless is possible to calculate the maximum possible electrical work and the *electromotive force* (emf) associated with the process. This latter quantity represents the free energy driving force for electron transfer as the voltage obtainable if the reaction were operated reversibly as a current-producing cell. The fact that such a cell may not be physically possible does not prevent its being utilized to calculate emf values as a fundamental property of oxidation-reduction processes.

The central arguments are developed in terms of reactions involving ions, and later the implications of this approach with respect to other types of redox process are examined. Although a number of new terms and definitions must be introduced here, it is helpful to appreciate that the essence of the treatment is an application of the *general* free energy relations developed earlier to a particular kind of reaction.

Consider the redox reaction

$$Zn + Cu^{2+} \rightarrow Zn^{2+} + Cu$$

The standard free energy change for this process ΔG° can be calculated from

$$\Delta G^\circ = \sum_{\text{products}} n \, \Delta G_f^\circ - \sum_{\text{reactants}} n \, \Delta G_f^\circ \tag{2-24}$$

Since the two pure metals are in their standard state by definition, their free energy of formation is zero and the equation assumes the form

$$\Delta G^\circ = \Delta G_{f,Zn^{2+}}^\circ - \Delta G_{f,Cu^{2+}}^\circ$$

Substituting the values for the standard free energy of formation of the two ions,* we obtain $\Delta G_{298}^\circ = -50.73$ kcal. Although this is a *standard* free

---

* ΔG_f° values for ions are usually based on the arbitrary standard that $\Delta G_{f,H^+}^\circ = 0$. The main difficulty with the assignment of values of this term for ions in a general way is that it is not possible to obtain a solution of a single ion. Having set a value for one ion, however, the data for others can be evaluated using Hess' law-type manipulations.

energy change, it is also ΔG for the commonly encountered conditions of 1 atm pressure and ions present at unit activity. Thus, if the physical system consists of an amount of pure zinc metal immersed in a solution containing zinc ions at unit activity and pure copper in a solution with $a_{Cu^{2+}} = 1.0$, the process would be spontaneous. In mechanistic terms, the decrease in free energy in going from reactants to products indicates the driving force for electron transfer. Suppose the same reaction were carried out under conditions where the loss of electrons by metallic zinc and the acquisition of electrons by copper occurred in physically separated parts of the system. That is, the two processes

$$Zn \rightarrow Zn^{2+} + 2\,e^-$$

$$Cu^{2+} + 2\,e^- \rightarrow Cu$$

occur separately so that transfer of electrons takes place through an external circuit. There then should be a direct relationship between the free energy change and the voltage that could be measured in this circuit.

This possibility points up an aspect of electrochemical reactions of great importance. In all previous systems studied, the technique has been to evaluate ΔG *indirectly* by calculating from observed heat effects (to obtain ΔH values) and absolute entropy measurements (for ΔS). Alternatively, if it were possible to observe the system at equilibrium, $\Delta G°$ could be determined from the equilibrium constant and the (nonstandard) free energy change could be calculated from

$$\Delta G = \Delta G° + 2.303\,RT \log Q \tag{2-37}$$

In electrochemical systems, however, the existence of a quantitative relation between the observed emf and the ΔG for a process provides a relatively simple and direct method for determination of this important thermodynamic quantity. This important relation will be obtained presently.

It is not enough, however, to measure the electrical work associated with a reaction. The free energy decrease and the electrical work are equal only when the process is reversible. The key to the utility of electrochemical measurements of this kind is that thermodynamic reversibility may be approached very closely in practice. Provided the system can be arranged so that electrons move from one site to another, the current flow can be reduced to very low levels and an emf measured under almost reversible conditions. The techniques involved are discussed later.

The amount of electrical work determined under these conditions is virtually identical to the theoretical maximum and, therefore, to the ΔG defined by Eq. (7-1). This capability makes electrochemical reactions one of the most fruitful sources of free energy data.

7-2 BASIC ELECTROCHEMICAL CONCEPTS

The unfortunate confusion arising from lack of uniform usage of terms and symbols in physical chemistry is perhaps most acute in electrochemistry. Although there is international agreement on nomenclature and conventions, the student will find a number of different conventions in actual use. Care must be exercised, particularly in using tabulated data, to ensure that the precise definition of the quantities is understood. Often the difference is "only" that of sign but, of course, the use of an incorrect sign invariably leads to an incorrect answer.

7-2-1 Units and Definitions

The electron transfer reactions that have been discussed are of the general category of oxidation-reduction, or *redox*, reactions. A process can be so classified if the conversion from reactants to products is accompanied by a change in *oxidation state*. The latter term may be defined as follows:

OXIDATION STATE: THE NUMBER OF PROTONS LESS THE NUMBER OF ELECTRONS ASSOCIATED WITH A GIVEN CHEMICAL SPECIES.

The oxidation state of free elements is taken as zero. For ionic species the oxidation state is simply the charge on the ion, for example, Na^+ is in the $+1$ oxidation state. The same convention is applied to species formed through covalent bonds as well; and although it is convenient to deal with such compounds by treating them as ionic species, the assignment is quite artificial. For example, in ammonia (NH_3), which is a covalent compound, the nitrogen is considered to be in the -3 oxidation state and hydrogen in the $+1$. The total *oxidation number* of the three hydrogen atoms is $+3$ and, therefore, the total oxidation number of the compound is zero, as required for a neutral species. The process of oxidation and reduction are defined in terms of the direction of change of the oxidation state.

OXIDATION: THE ELEMENT INVOLVED UNDERGOES A CHANGE TO A HIGHER (MORE POSITIVE, LESS NEGATIVE) OXIDATION STATE. THE PROCESS IS ACCOMPANIED BY A LOSS OF ELECTRONS.

REDUCTION: THE ELEMENT INVOLVED UNDERGOES A CHANGE TO A LOWER OXIDATION STATE. THE PROCESS IS ACCOMPANIED BY A GAIN OF ELECTRONS.

In the broad sense oxidation includes the loss of hydrogen or gain of oxygen. In strictly ionic systems, however, it is necessary to become directly involved only with the loss or gain of electrons.

Since electrons are transferred, oxidation and reduction must occur simultaneously—the electrons lost by one species being taken up by the other. The reactant oxidized is considered to "cause" the reduction of the other species and therefore is known as the *reducing agent*. The element reduced brings about the oxidation of the other and is the *oxidizing agent*. These terms are applied, as convenient, to either the element whose oxidation state changes or to the compound containing that element.

These conventions are summarized in the following reaction:

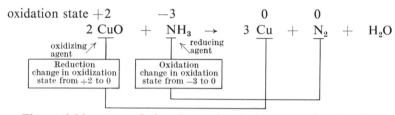

The stoichiometry of the electrochemical process is related to the electrical quantities involved by Faraday's laws, which for present purposes can be condensed into the single statement

FARADAY'S LAW: ONE GRAM-EQUIVALENT OF CHEMICAL CHANGE PRODUCES 96,487 COULOMBS OF ELECTRICITY.

For a compound involved in a redox reaction, the *gram-equivalent weight* (GEW) is defined as the *formula weight divided by the total change in oxidation number*. One (gram) equivalent weight is a weight in grams equal to the gram equivalent weight (GEW). In the example reaction, the GEW of CuO is one-half its formula weight. The *coulomb* is defined as that quantity of electricity equivalent to a current of 1 amp flowing for 1 sec.

By definition of the GEW, 1 eq. of a substance undergoing oxidation produces Avogadro's number, a mole, of electrons as

$$X \rightarrow X^+ + e^-$$

N atoms of X N X$^+$-ions N electrons

According to Faraday's law, the passage of 96,487 coulombs of electricity corresponds to the passage of N, 6.023×10^{23}, electrons. This fundamentally important quantity is known as the *Faraday $\mathscr{F}$*:

$$\mathscr{F} = 96,487 \text{ coulombs/eq.} \equiv 6.023 \times 10^{23} \text{ electrons}$$

More often it will be convenient to use the alternative units for $\mathscr{F}$:

$$\mathscr{F} = 23,060 \text{ cal/volt-eq.} \tag{7-2}$$

EXAMPLE 7-1. Calculate the number of electrons released and the coulombs of electricity generated when 1.00 gm of magnesium is oxidized.

Reaction: $Mg \rightarrow Mg^{2+} + 2e^-$

$$GEW = gm\ at.\ wt./2 = \frac{24.312}{2} = 12.15\ gm$$

$$Number\ of\ equivalents\ reacting = \frac{1.00}{12.15} = 8.23 \times 10^{-2}$$

Number of electrons $= 8.23 \times 10^{-2} \times 6.023 \times 10^{23} = \underline{4.96 \times 10^{22}}$

Number of coulombs $= 8.23 \times 10^{-2} \times 96,487 = \underline{7,940}$

7-2-2 Cells and Electrodes

The *galvanic cell* is a device that permits a spontaneous electrochemical reaction to produce a detectable electric current. If a redox reaction is to be utilized in this way, the reaction must be capable of being separated physically so that the electrons are lost in one part of the system and gained in another. In the conventional nomenclature, oxidation occurs at an *electrode* in one *half-cell* and reduction occurs at an electrode in another half-cell. The electrode at which oxidation occurs is referred to as the *anode*, while the site of reduction is the *cathode*. Since electrons are lost in the oxidation process they leave the cell at the anode, traverse the external circuit, and reenter at the cathode. Basing the system on the designation that electrons leave the cell at the anode avoids confusion with designations of $+$ and $-$ electrodes and "conventional currents." To prevent physical mixing of the solutions containing the electrolyte and other complicating effects to be discussed below, while allowing for migration of ions in the cell, the two half-cells are connected by a *salt bridge*. This is usually a saturated solution of potassium bromide or chloride suspended in a gel. The physical arrangement is shown in Fig. 7-1 for the schematic general redox process

$$A + B^+ \rightarrow A^+ + B$$

A and B are pure metals, and A^+ and B^+ ions are present in aqueous solution along with a negatively charged ion X^-. As the latter species does not participate in the reaction, it is usually excluded from the reaction equation. For this system the half-cells, that is, the separated oxidation and reduction processes, are

Oxidation: $A \rightarrow A^+ + e^-$

Reduction: $B^+ + e^- \rightarrow B$

Simple algebraic summation of these two equations gives the overall cell reaction.

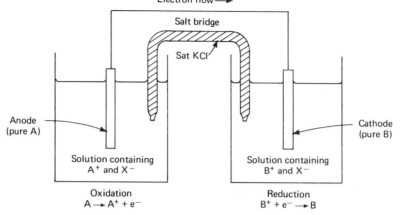

FIG. 7-1 *Schematic diagram of a galvanic cell with potassium chloride salt bridge. The reaction taking place in the cell is* A + B⁺ → A⁺ + B.

The conventional representation of an individual half-cell is as follows:

$$\textit{Oxidation:} \quad \text{electrode} \mid \text{electrolyte}$$

$$\textit{Reduction:} \quad \text{electrolyte} \mid \text{electrode}$$

The vertical bar indicates a phase separation. For materials in their pure state the physical state is indicated as necessary. The activities (or molarities, pressure, etc. if the treatment is approximate) of components of variable concentration also are given. In the example half-cells shown in Fig. 7-1, if pure metal A is in an electrolyte containing A^+ ions at unit molarity, the designation would be $A \mid A^+$ (1 mole/liter). In a general discussion, where no particular electrolyte concentration is involved, the half-cell or *couple*, as it is sometimes called, can be referred to simply as $A \mid A^+$. In the second half-cell, where reduction occurs, the cell similarly would be $B^+ \mid B$, with the value of $[B^+]$ again being specified if necessary.

There are a variety of types of electrodes involved in redox half-cells. A few of the most common types are listed in Table 7-1 along with an example

TABLE 7-1. ELECTRODE TYPES

Electrode	*Example half-cell reaction*	*Reaction type*	*Conventional half-cell (electrode) designation*
Metal–metal ion	$Zn \rightarrow Zn^{2+} + 2\,e^-$	oxidation	$Zn \mid Zn^{2+}(x \text{ moles/liter})$
	$Ag^+ + e^- \rightarrow Ag$	reduction	$Ag^+(x \text{ moles/liter}) \mid Ag$
Gas–ion	$2\,Cl^- \rightarrow Cl_2 + 2\,e^-$	oxidation	$Pt \mid Cl_2(y \text{ atm}) \mid Cl^-(x \text{ moles/liter})$
	$2\,H^+ + 2\,e^- \rightarrow H_2$	reduction	$H^+(x \text{ moles/liter}) \mid H_2(y \text{ atm}) \mid Pt$
Metal–insoluble metal salt	$AgCl(s) + e^- \rightarrow$ $Ag(s) + Cl^-(aq)$	reduction	$Cl^-(x \text{ moles/liter}) \mid AgCl(s) \mid Ag(s)$
Redox	$Fe^{3+}(aq) + e^- \rightarrow$ $Fe^{2+}(aq)$	reduction	$Fe^{3+}(x \text{ moles/liter}),$ $Fe^{2+}(x \text{ moles/liter}) \mid Pt$

of each type. The writing of a particular electrode reaction as oxidation or reduction and the resulting conventional half-cell representation is purely arbitrary. Depending on the nature of the particular electrode comprising the second half-cell, the electrodes listed could be involved in *either* oxidation *or* reduction. The table also indicates some further conventions of cell designations. In the gas-ion and redox electrode, the species actually involved in the reaction cannot be used as electrodes and, therefore, platinum metal is used to carry the electrons to and from the reaction site. Although the Pt electrode itself can be involved in a redox reaction, for reasons that will become apparent such a reaction normally will not occur. Caution must be exercised, however, in referring to Pt as an "inert electrode" since it can react but normally does not do so. For a gas the pressure in atmospheres is specified. When the reactants are in the same phase, the vertical bar is replaced by a comma; thus Fe^{3+} and Fe^{2+} are both in solution and the designation is Fe^{3+}, Fe^{2+}. Electrodes of the type involved in the metal–insoluble salt redox reaction usually consist physically of the salt carefully precipitated out on the metal electrode.

To examine the conventions used in representing the overall cell, which is obtained by combination of any two half-cells, consider a specific case, the *Daniell cell*, in which the Zn–Cu reaction discussed previously occurs. The half-cell reactions and conventions for the half-cells individually have been listed in Table 7-1. For the overall cell the representation is just a combination of the two electrode notations $Zn \mid Zn^{2+} \parallel Cu^{2+} \mid Cu$. The double vertical bar* indicates that between the two half-cells there is a junction that does not contribute to the cell. Cells considered are those where a salt bridge that eliminates the junction potential through a mechanism not yet completely understood is utilized. We also can have no liquid junction, a cell where the same electrolyte can supply the ions for both half-cells. The voltage developed by the working cell is designated as $\mathscr{V}$.

There is more or less universal adherence to the conventions associated with the overall cell representation. The variations mentioned previously arise in dealing with the emf generated by individual half-cells, which data can be used to calculate the total cell emf. For the moment the discussion will deal with the complete cell.

The essential feature of the method is to write the cell in the same manner regardless of whether the actual process will be spontaneous in that direction. If we follow the convention the sign of the observed or calculated voltage will indicate the direction of the spontaneous reaction. This technique has an advantage in that it is not necessary to anticipate the direction in which

---

* A single bar in this position indicates a liquid junction that develops its own emf because of differences in mobilities of ions and thus introduces further complications. This type of situation is not dealt with here.

electron transfer will proceed. The approach is analogous to that discussed in the introductory treatment of thermodynamics. Difficulty is avoided by setting up a perfectly general equation or convention and allowing the signs to "take care of themselves." The conventions for cell representations are summarized below:

> CONVENTION I. IN REPRESENTING THE OVERALL CELL, THE HALF-CELL IN WHICH OXIDATION OCCURS, IN THE REACTION AS WRITTEN, IS PLACED ON THE LEFT IN THE NOTATION OXIDATION ‖ REDUCTION.

> CONVENTION II: A POSITIVE VALUE OF $\mathscr{V}$ INDICATES THAT THE PROCESS OCCURS SPONTANEOUSLY AS WRITTEN, WITH ELECTRONS FLOWING FROM LEFT TO RIGHT.

> CONVENTION III: A NEGATIVE VALUE OF $\mathscr{V}$ INDICATES THAT THE SYSTEM AS WRITTEN IS AN ELECTROLYTIC CELL RE-QUIRING THE APPLICATION OF AN EXTERNAL VOLTAGE GREATER THAN v VOLTS TO CAUSE THE REACTION TO PRO-CEED.

The result of convention I is that the left-hand cell in the notation $A \mid A^+ \parallel B^+ \mid B$ is the anode, the site at which electrons leave the cell. Strict adherence to these conventions is of primary importance in dealing with half-cell voltages individually.

For the Daniell cell the process is spontaneous as written. For both ions present at 1 mole/liter, the theoretical maximum voltage is $+1.10$ volt. It would be equally correct to write the reaction as

$$Zn^{2+} + Cu \to Zn + Cu^{2+}$$

in which case the cell is $Cu \mid Cu^{2+} \parallel Zn^{2+} \mid Zn$. The voltage calculated for this cell, however, would be -1.10 volt, indicating that the reaction is not spontaneous in the direction indicated but rather from right to left.

Before examining the details of half-cell voltages and how they may be used to evaluate emf for various galvanic cells of interest, the application of thermodynamic principles to electrochemical processes will be examined.

7-3 THERMODYNAMICS OF GALVANIC CELLS

7-3-1 Measurement of Reversible emf

It was noted in the introduction to this chapter that one of the important features of electrochemical reactions is that it is possible to approach reversibility quite closely. In thermodynamic systems in general, reversible process are characterized by maximum efficiency. Thus in a galvanic cell that operates reversibly the *reversible emf* $\mathscr{E}$ is the maximum possible voltage.

$\mathscr{E}$ invariably will be greater than the observed voltage $\mathscr{V}$ of a current-producing cell. In general the electromotive force of a cell operating under reversible conditions is referred to as an emf, while that observed when conditions are irreversible is termed a voltage.

The conditions for electrochemical reversibility can be stated by analogy with the case of gas expansion. In Chapter 1 the general condition for reversibility that the driving force and opposing force be only infinitesimally different was stated in equation form for a gas volume change as

$$P_{internal} = P_{external} + dP \qquad (1\text{-}20)$$

A galvanic cell can be set up in conjunction with an appropriate external circuit so that the emf produced is opposed by an almost identical, but opposite in direction, voltage from an external source. The cell then will operate more or less reversibly according to

$$\mathscr{E}_{cell} = \mathscr{V}_{external} + d\mathscr{V} \qquad (7\text{-}3)$$

In other words, the direction of cell reaction—charge or discharge—could be reversed by an infinitesimal alteration in the external applied voltage. It also is apparent that the reaction occurring in each half-cell must be chemically reversible. Thus, for example, the $Zn \mid Zn^{2+}$ half-cell would qualify since the zinc can go into solution as zinc ion or the latter can deposit as metallic zinc.

The device that permits cell operation in this way is known as a *potentiometer*. Figure 7-2 shows a very simple apparatus of this kind. For actual determination of reversible emf values in the laboratory the instrumentation is somewhat more elaborate than the simple system shown in the figure;

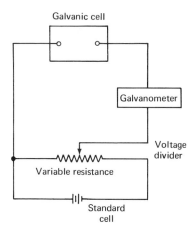

FIG. 7-2 A simple potentiometer circuit for the determination of the reversible emf of a galvanic cell.

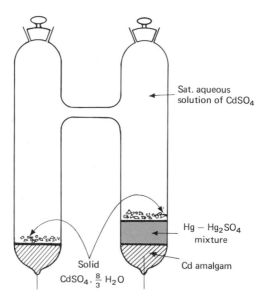

Sat. aqueous
solution of CdSO$_4$

Hg — Hg$_2$SO$_4$
mixture

Solid
CdSO$_4 \cdot \frac{8}{3}$ H$_2$O

Cd amalgam

FIG. 7-3 Illustration of the standard Weston cell.

the principle of operation, however, is the same. The potentiometer allows balancing of the output of the standard cell against that of the cell whose $\mathscr{E}$ value is unknown. The full voltage of the former is reduced by the calibrated variable resistance of the voltage divider until the galvanometer shows no deflection, indicating the absence of current flow. The exact input voltage to the unknown cell at this point is the value of $\mathscr{E}$.

The Weston cell, shown in Fig. 7-3, is the universal reference cell for measurements of this type. The voltage of the cell is 1.01463 volts at 298 °K. The advantage of using saturated solutions (CdSO$_4$ and Hg$_2$SO$_4$) in the two half-cells is that ion concentrations remain constant while the cell is in operation. The overall cell reaction for this reference galvanic cell is

$$Cd + Hg_2SO_4 \rightarrow Cd^{2+} + 2\,Hg + SO_4^{2-}$$

If very great accuracy is not required, quite good results can be obtained using the vacuum tube voltmeter of the pH meter. This instrument is discussed in detail in Sec. 7-4-2. For the moment it is necessary only note to that the electronics are specifically designed to measure small voltages while drawing minimal currents. Thus the voltage of a cell determined in this manner approaches potentiometric measurement. Except in very dilute solutions where the chemical reaction at the electrodes may cause significant changes in electrolyte concentration, the values obtained are close to those observed using considerably more elaborate potentiometers.

7-3-2   Relation of $\mathcal{E}$ to ΔG

Having seen the way in which the reversible emf may be determined experimentally, it is now necessary to obtain the fundamental relation between the emf and the free energy change in the process. This will permit utilization of a variety of thermodynamic equations based on ΔG.

Again it is useful to draw an analogy with other systems. In gas expansion the PV work is a product of an intensity and capacity factor:

$$w = P\,\Delta V \tag{1-22}$$

For electrical work there is a similar product, with the electrical charge q being the capacity factor and the voltage $\mathcal{V}$ being the intensity factor:

$$w_{el} = q\mathcal{V}$$

The electrical charge is a product of the number of coulombs per equivalent and the number of equivalents of reactant undergoing oxidation or reduction n. The former term, as indicated earlier, is the Faraday $\mathcal{F}$, and therefore

$$w_{el} = n\mathcal{F}\mathcal{V}$$

If the voltage is a reversible emf, this is the maximum value. Therefore, the work obtained must be a maximum, that is,

$$w'_{max} = w_{el,max} = n\mathcal{F}\mathcal{E} \tag{7-4}$$

Substitution into Eq. (7-1) yields the fundamentally important equation

$$\Delta G = -n\mathcal{F}\mathcal{E} \tag{7-5}$$

To obtain ΔG in calories, the numerical value of $\mathcal{F}$ given by Eq. (7-2) must be used. It should be appreciated that Eq. (7-5) is essentially the only "new" equation required to handle the thermodynamics of electrochemical processes. Although a variety of additional equations are developed here, they all originate in a substitution for ΔG from Eq. (7-5) in previously derived general thermodynamic relations.

The electromotive force of a cell is an *intensive* property. Like temperature and pressure, it is independent of the size of the system. Thus if we double the physical size of the electrochemical cell the emf is unchanged, whereas an *extensive* property such as volume would be doubled.

The use of Eq. (7-5) in calculations and the distinction between the extensive free energy change and intensive emf can be illustrated by reconsidering the Daniell cell reaction. If both the zinc and copper ions are present at concentrations of 1 mole/liter, then $\Delta G = \Delta G°$. Assuming the equivalence of molarity and activity is valid for this case, the activity coefficients will cancel if both half-cells have the same electrolyte concentration. (The mechanism for this algebraic simplification will be apparent when the equation for the concentration dependence of $\mathscr{E}$ is discussed in Sec. 7-3-3.) The ΔG or $\Delta G°$ value for the system therefore is -50.73 kcal (Sec. 7-1). For the reaction as written, $n = 2$. We can consider that on an atomic and ionic basis two electrons are transferred or on a mole basis that two equivalents are reduced and oxidized. From Eq. (7-5), therefore,

$$\mathscr{E} = -\frac{\Delta G}{n\mathscr{F}} = \frac{+50,730 \text{ cal}}{2 \text{ eq.} \times 23,060 \text{ (cal/volt eq.)}} = +1.10 \text{ volts}$$

If the reaction is written as

$$2 \text{ Zn} + 2 \text{ Cu}^{2+} \rightarrow 2 \text{ Zn}^{2+} + 2 \text{ Cu}$$

and standard state conditions are again assumed, the value of ΔG is $-2 \times 50,730$ cal and $n = 4$. Reapplication of Eq. (7-5) gives

$$\mathscr{E} = \frac{+(2 \times 50,730)}{4 \times 23,060} = +1.10 \text{ volts}$$

The positive value of $\mathscr{E}$ has the same significance as a negative value for ΔG, the reaction as written is spontaneous. In fact, the voltage generated by a galvanic cell can be thought of as a potential electric difference. This kind of interpretation will prove useful in discussions of the emf of individual half-cells.

EXAMPLE 7-2. Calculate the free energy change in the redox reaction $\text{Cd} + 2 \text{ Ag}^+ \rightarrow \text{Cd}^{2+} + 2 \text{ Ag}$ under conditions where the emf is 1.202 volts.

For the reaction as written $n = 2$, and Eq. (7-5) gives

$$\Delta G = -n\mathscr{F}\mathscr{E} = -2 \times 23,060 \times 1.202 = -55.43 \text{ kcal}$$

Subsequent discussion is concerned exclusively with electrochemical systems at 298 °K. It is useful to see, however, how Eq. (7-5) can be used to evaluate the change in $\mathscr{E}$ with temperature by making use of some previously obtained thermodynamic relations. Although there are other equations somewhat more convenient for this purpose, the calculation can be carried out by employing only equations given here. At the two temperatures of interest, ΔG and $\mathscr{E}$ can be calculated one from the other via Eq. (7-5) and the difference in ΔG between the two temperatures can be obtained from

$$\Delta G_2 = \Delta H + \frac{T_2}{T_1}(\Delta G_1 - \Delta H) \qquad (2\text{-}38)$$

Thus if $\mathscr{E}$ is known at two values of T, ΔG_2 and ΔG_1 are calculated from Eq. (7-5), ΔH is obtained from Eq. (2-38), and finally ΔS is derived from the defining equation for the free energy change, $\Delta G = \Delta H - T\Delta S$. Emf measurements thus provide an additional valuable source of thermodynamic data.

7-3-3 Concentration Dependence of $\mathscr{E}$

For convenience the examples discussed in the last section dealt with conditions under which ΔG and $\Delta G°$ were identical. Very frequently, however, this is not the case, and a distinction must be made between the (reversible) emf $\mathscr{E}$ and the *standard* (reversible) emf $\mathscr{E}°$, which is the voltage obtained from the reversibly operated cell when all components are present in their standard state. For this case Eq. (7-5) is written

$$\Delta G° = -n\mathscr{F}\mathscr{E}° \tag{7-6}$$

The general chemical reaction

$$b\text{B} + d\text{D} \rightleftharpoons m\text{M} + n\text{N}$$

for which

$$\Delta G = \Delta G° + 2.303\, RT \log Q \tag{2-37}$$

can be used to obtain an expression for the concentration dependence of the emf. If the general reaction involves electron transfer, then we can substitute for ΔG and $\Delta G°$ from Eqs. (7-5) and (7-6), respectively, to obtain

$$-n\mathscr{F}\mathscr{E} = -n\mathscr{F}\mathscr{E}° + 2.303\, RT \log Q \tag{7-7}$$

which can be rearranged to the more convenient form

$$\mathscr{E} = \mathscr{E}° - 2.303\, \frac{RT}{n\mathscr{F}} \log Q \tag{7-8}$$

When all components are present at unit activity, $Q = 1$ and the second term in this equation vanishes. The emf $\mathscr{E}$ then is identical to $\mathscr{E}°$. At 298 °K, putting in the numerical values of the constants gives the working equation

$$\mathscr{E}_{298} = \mathscr{E}°_{298} - \frac{0.05915}{n} \log Q \tag{7-9}$$

Q has the usual significance of the ratio of activities of products to activities of reactants.

Equation (7-9) is known as the *Nernst equation*. There are a number of important applications of this relation that will be encountered. In the context of the present discussion, primary interest is in utilizing it to obtain the emf of a galvanic cell when the products and reactants are present in other than their standard states. The use of the equation involves mainly evaluation of the activity ratio Q. Values for the standard emf are available from either $\Delta G_f°$ data for the ions or from emf values for the half-cells. The techniques involved in using the latter data are discussed in Sec. 7-3-5.

Almost invariably, ionic species are involved. Thus the assumption that activity coefficients are unity may introduce considerable error. In the electrochemical processes, however, the algebraic form of Q is often such that γ terms may cancel. Alternatively, the concentration of the ions in the two half-cells may not be appreciably different, and the *difference* in $\gamma_{\pm}$ values can be neglected and the cancellation still effected without error. Each half-cell will contain an ion of opposite charge, in addition to the species participating in the redox process, and this must be included in calculating the ionic strength. Furthermore, as will be seen presently, even if the activity coefficients make a significant difference in the value of Q, the effect on the calculated $\mathscr{E}$ value often is not particularly large.

For the Daniell cell reaction, the activity ratio takes the form

$$Q_a = \frac{a_{Zn^{2+}}}{a_{Cu^{2+}}} = \frac{[Zn^{2+}]\gamma_{Zn^{2+}}}{[Cu^{2+}]\gamma_{Cu^{2+}}}$$

In this case, if the concentrations of zinc and copper ions and the ionic strength in both half-cells are the same, the activity coefficients will cancel, but so do the concentrations, so that $Q_a = 1$. If, on the other hand, the concentrations are not identical, $Q_a \neq 1$ and the activity coefficients must be determined for each half-cell.

The cell $Zn \mid Zn^{2+} \parallel Fe^{3+}, Fe^{2+} \mid Pt$ represents another type of situation. The reaction equation consistent with this designation is

$$\tfrac{1}{2} Zn + Fe^{3+} \rightarrow Fe^{2+} + \tfrac{1}{2} Zn^{2+}$$

For this system $n = 1$. Substituting the product of concentration and activity coefficient for each species of variable concentration, we get a Nernst equation of the form

$$\mathscr{E} = \mathscr{E}^\circ - 0.05915 \log \frac{[Fe^{2+}]\gamma_{Fe^{2+}}([Zn^{2+}]\gamma_{Zn^{2+}})^{\frac{1}{2}}}{[Fe^{3+}]\gamma_{Fe^{3+}}}$$

Although Fe^{2+} and Fe^{3+} are present in the same solution, their activity coefficients will be different, and therefore all three $\gamma_{\pm}$ terms must be retained.

In cells involving a gas electrode, the activity of the gas can be expressed with negligible error by the pressure in atmospheres. Thus for the reaction

$$Zn + Cl_2 \rightarrow Zn^{2+} + 2 Cl^-$$

the expression for the cell emf is

$$\mathscr{E} = \mathscr{E}^\circ - 0.0296 \log \frac{[Zn^{2+}]\gamma_{Zn^{2+}}([Cl^-]\gamma_{Cl^-})^2}{P_{Cl_2}}$$

When the reaction in one of the half-cells involves hydronium ions (almost invariably abbreviated as H^+ in electrochemical relations), the Nernst equation is written in the usual way. If desired, the log a_{H^+} term can be separated out so that the equation indicates explicitly the dependence of emf on pH. This technique can be illustrated for the following reaction involving the reduction of quinone to hydroquinone:

$$C_6H_4O_2 + 2\,H^+ + Cu \rightarrow C_6H_4(OH)_2 + Cu^{2+}$$

In the reaction two equivalents are oxidized and reduced, $n = 2$, and since both quinone and hydroquinone are molecular species their activity is adequately represented by molar concentration. The Nernst equation for the system is

$$\mathscr{E} = \mathscr{E}^\circ - 0.0296 \log \frac{[HQ][Cu^{2+}]\gamma_{Cu^{2+}}}{[Q](a_{H^+})^2}$$

where $C_6H_4O_2$ and $C_6H_4(OH)_2$ are abbreviated as Q and HQ, respectively. If the log $1/a_{(H^+)^2}$ term is taken out and expressed in terms of pH, the result is

$$\mathscr{E} = \mathscr{E}^\circ - 0.0296 \log \frac{[HQ][Cu^{2+}]\gamma_{Cu}^{+2}}{[Q]} - 0.05915 \text{ pH}$$

It is apparent from the form of this expression that if all concentrations in the system, except $[H^+]$, are held constant, the observed emf will be a linear function of pH. The important implication is that this or any other cell involving hydronium ions can be used, in theory at least, to determine the pH of a solution. This application of emf measurements is discussed in some detail later.

In calculations using the Nernst equation it must be kept in mind that the second term generally will not make a very large contribution to the cell emf value. In addition, variations in the value of Q_a are attenuated by the logarithmic nature of the term. Unless the value of $\mathscr{E}$ happens to be particularly small, the percentage error in the caculated values of $\mathscr{E}$ owing to errors in the activity ratio will be quite minor. Except when high accuracy is required, then, activity coefficients frequently can be assumed to be unity. This is permissible not because the assumption is particularly valid but because the error so introduced may be partially compensated for by similarities in product and reactant γ values and because of the form of the equation.

On the other hand, if the emf can be measured experimentally with great accuracy, the observed and calculated values can be compared to evaluate activity coefficients.

The general techniques involved in the use of the Nernst equation, as well as the effects of neglecting activity coefficient considerations, are illustrated in the following example:

EXAMPLE 7-3. An electrochemical cell is set up in the manner illustrated in Fig. 7-1 and utilizes the redox reaction

$$Mn + 2\,Ag^+ \rightarrow Mn^{2+} + 2\,Ag$$

for which $\mathscr{E}^\circ = +1.979$ volts. Half-cell I contains an aqueous solution of $AgNO_3$ at 5.0×10^{-2} mole/liter, and half-cell II contains 1.0×10^{-4} mole/liter of $Mn(NO_3)_2$. Calculate the reversible emf using (1) activity coefficients and (2) assuming activity coefficients of unity throughout.

In half-cell I $\qquad\qquad \mu = c = 5.0 \times 10^{-2}$

$$Z = +1 \qquad \gamma_{Ag^+} = 0.81 \text{ (Fig. 5-3)}$$

In half-cell II $\qquad \mu = \frac{1}{2}[C_{Mn^{2+}}(2)^2 + C_{NO_3^-}(-1)^2] = 3 \times 10^{-4}$

$$Z = +2 \qquad \gamma_{Mn^{2+}} = 0.91 \text{ (Fig. 5-3)}$$

$$Q_a = \frac{[Mn^{2+}]\gamma_{Mn^{2+}}}{([Ag^+]\gamma_{Ag^+})^2} = \frac{1.0 \times 10^{-4} \times 0.91}{(5.0 \times 10^{-2} \times 0.81)^2} = 5.67 \times 10^{-2}$$

$$Q_c = \frac{[Mn^{2+}]}{[Ag^+]^2} = \frac{1.0 \times 10^{-4}}{(5.0 \times 10^{-2})^2} = 4.00 \times 10^{-2}$$

For the reaction as written $n = 2$, and the Nernst equation is

$$\mathscr{E} = \mathscr{E}^\circ - 0.0296 \log Q$$

Including activity coefficients (Q_a),

$$\mathscr{E} = 1.979 - 0.0296 \log(5.67 \times 10^{-2}) = 1.979 + 0.037 = \underline{2.016 \text{ volts}}$$

Assuming activity coefficients of unity (Q_c),

$$\mathscr{E} = 1.979 - 0.0296 \log(4.00 \times 10^{-2}) = 1.979 + 0.041 = \underline{2.020 \text{ volts}}$$

Thus the assumption that $\gamma_\pm = 1$, which is in error by 7 and 19 percent for the Mn^{2+} and Ag^+ ions individually, introduces an error of only 0.2 percent in the calculated reversible emf.

The Nernst equation also indicates the possibility of another type of galvanic cell that utilizes the same species in each half-cell but at different concentrations. This type of system is known as a *concentration* cell.

Consider a galvanic cell in which the half-cells consist of a metallic zinc electrode in $ZnSO_4$, solutions at concentrations of 0.010 and 0.005 mole/liter. The conventional notation for the cell is $Zn \mid Zn^{2+}$ (0.005 mole/liter) $\parallel Zn^{2+}$(0.010 mole/liter) $\mid Zn$. In the left-hand cell the reaction is oxidation of zinc metal to Zn^{2+}, while in the right-hand cell Zn^{2+} ions are reduced to Zn. The overall cell reaction is

$$Zn' + Zn^{2+}(0.010 \text{ mole/liter}) \rightarrow Zn^{2+\prime}(0.005 \text{ mole/liter}) + Zn$$

where the Zn' and $Zn^{2+\prime}$ are in the same half-cell.

The standard emf for this or any other half-cell is zero since, when all species are present in their standard state, both half-cells are identical and ΔG° is zero. The Nernst equation for the zinc concentration cell $n = 2$ therefore is

$$\mathscr{E} = -0.0296 \log \frac{a'_{Zn^{2+}}}{a_{Zn^{2+}}}$$

The actual activity values for this system are $a'_{Zn^{2+}}$ (at $[Zn^{2+}] = 0.005) = 0.0032$ and $a_{Zn^{2+}}$ (at $[Zn^2] = 0.010) = 0.0047$ and, therefore,

$$\mathscr{E} = -0.0296 \log \frac{0.0032}{0.0047} = -0.0296(-0.168) = \underline{+0.00496 \text{ volt.}}$$

The required equation can be written for a general concentration cell as follows:

$$\mathscr{E} = -\frac{0.05915}{n} \log \frac{(\text{activity of ion in Cell II})}{(\text{activity of ion in Cell I})} \tag{7-10}$$

The final use that can be made of Eq. (7-6) is to obtain a value for the value of the thermodynamic equilibrium constant. Usually these values are extremely large, or extremely small, depending on the direction of spontaneity of the reaction as written. The standard emf and the equilibrium constant can be related most simply by the combination of Eq. (7-6) and the previously obtained expression,

$$\Delta G^\circ = -2.303 \, RT \log K \tag{3-5}$$

These two equations give

$$-n\mathscr{F}\mathscr{E}^\circ = -2.303 \, RT \log K$$

$$\log K = \frac{n\mathscr{F}\mathscr{E}^\circ}{2.303 \, RT}$$

which at 298 °K takes the form

$$\log K = \frac{n\mathscr{E}^\circ}{0.05915} \tag{7-11}$$

For the Daniell cell reaction K can be evaluated as follows by use of Eq. (7-11):

$$\log K = \frac{n\mathscr{E}^\circ}{0.05915} = \frac{2 \times 1.10}{0.05915} = 37.2$$

$$K_{298} = 1.6 \times 10^{37}$$

This very large value is another manifestation of the large negative ΔG° value and indicates that the equilibrium for practical purposes is 100 percent in favor of the products Zn^{2+} and Cu.

Equilibrium constants are not of very great practical importance for process of this type. In certain redox reactions, however, the net reaction corresponds to the solution of a slightly soluble salt, and the K value obtained is the thermodynamic solubility product K_{sp}°. This kind of determination is illustrated in the following example

EXAMPLE 7-4. The standard emf for the cell Ag | AgCl(s) | Ag$^+$, Cl$^-$ (sat) | Ag(s) is $+0.577$ volt. Calculate the thermodynamic solubility product of silver chloride. The two half-cell reactions in this system are

$$Ag + Cl^- \rightarrow AgCl(s) + e^- \qquad Ag^+ + e^- \rightarrow Ag$$

The overall reaction is $Ag^+ + Cl^- \rightarrow AgCl(s)$

Note that this galvanic cell does not require a salt bridge since the single electrolyte solution supplies the ions for both half-cell reactions.

For the overall reaction, Q_a in Eq. (7-11) takes the form $1/(a_{Ag^+})(a_{Cl^-})$. The equilibrium constant calculated from the Nernst equation therefore is $1/K_{sp}$.

$$\log K_{298} = \frac{n\mathscr{E}^\circ}{0.05915} = \frac{0.577}{0.05915} = 9.76$$

$$K = 5.6 \times 10^9 = \frac{1}{K_{sp}^\circ} \qquad K_{sp}^\circ = 1.8 \times 10^{-10}$$

7-3-4 Calculation of $\mathscr{E}^\circ$ from Half-cell emf Values

It would be entirely possible to evaluate the standard emf data for all possible cells either directly under standard state conditions or indirectly by measurements when the species are in nonstandard states and calculations using the Nernst equation. A complete set of such values, however, would be inordinately long and quite inconvenient. A much more satisfactory approach is to consider known half-cells individually, to determine the potential developed within each, and to use such values to obtain $\mathscr{E}^\circ$ for any combination of half-cells of interest.

It is not possible to measure absolutely the electrical potential developed in an isolated half-cell. To determine experimentally, for example, the electrical potential between a metal and its ions in aqueous solution necessitates setting up an electrochemical cell. This requires a second electrode that will have its characteristic potential. The electromotive force measured will be the *sum* of the potentials of *both* half-cells. This fundamental difficulty can be circumvented by using the same reference cell at all times. We compare all half-cells against the same standard and then simplify the calculations by arbitrarily assigning a half-cell potential of zero to the reference half-cell. In such a galvanic cell with two half-cells—one of unknown potential and the other the reference half-cell—the overall emf is the voltage assigned to the unknown cell, since the reference potential is set at zero. The reference potential assigned could be any value—the use of zero volts is entirely a matter of convenience and can have no effect on the calculated $\mathscr{E}$ value. This can be demonstrated in the following way:

In effect, the calculations involve setting up the hypothetical double galvanic cell

(half-cell I)(reference half-cell)(reference half-cell)(half-cell II)

$\underbrace{\qquad\qquad}_{\text{emf}_I} \qquad \underbrace{\qquad\qquad}_{\text{emf}_{II}} \qquad \underbrace{\qquad\qquad}_{\text{emf}_{III}}$

The observed total emf is the sum of the indicated values; emf_I is assigned entirely to half-cell I. Similarly, emf_{III} is attributed to half-cell II. In the arrangement shown there is no potential difference between the two reference cells, that is, $\text{emf}_{II} = 0$, and therefore the total emf is effectively $\text{emf}_I + \text{emf}_{III}$. The particular value assigned to the reference electrode therefore is immaterial. The same total emf would be observed, or calculated, regardless of the particular half-cell chosen as reference, or its assigned emf, although the absolute numerical values of emf_I and emf_{III} depend on the reference electrode value.

The absolute reference standard for half-cell measurements is the *standard hydrogen electrode*. Figure 7-4 shows a typical arrangement of apparatus for this electrode. The choice of a gas-ion electrode as standard may appear to be undesirable, but the hydrogen electrode is quite reproducible and this, along with the availability of very pure H_2 gas, outweights any mechanical inconvenience associated with a gas electrode. When the cell is connected to another half-cell via a salt bridge, oxidation or reduction occurs depending on the potential of the other half-cell. The two possible reactions are

Oxidation: $\text{Pt} \mid H_2(1 \text{ atm}) \mid H^+(a = 1)$ $\frac{1}{2} H_2 \rightarrow H^+ + e^-$

Reduction: $H^+(a = 1) \mid H_2(1 \text{ atm}) \mid \text{Pt}$ $H^+ + e^- \rightarrow \frac{1}{2} H_2$

The half-cell emf of both reactions (the electrical potential between H_2

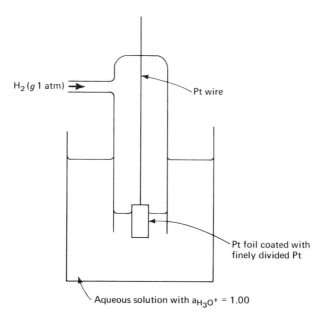

$H_2 (g \ 1 \text{ atm}) \rightarrow$

Pt wire

Pt foil coated with finely divided Pt

Aqueous solution with $a_{H_3O^+} = 1.00$

FIG. 7-4 The standard hydrogen electrode.

and its ions) is assigned a value of zero volts under the conditions of 1 atm pressure of H_2 gas and unit activity of hydronium ions. Thus, when a galvanic cell is constructed from the standard hydrogen electrode (SHE) as one half-cell and an electrode whose emf we wish to determine (EUP, electrode of unknown potential) as the other, the observed emf is taken as resulting entirely from the EUP. There are two methods of presenting half-cell emf values determined in this way: The basis of the comparison is the tendency of the EUP, relative to the standard hydrogen electrode, of the EUP to undergo either oxidation or reduction. Although it would be preferable to discuss just one convention here and to use this throughout, to avoid confusion later it is necessary that both be detailed since the two conventions are in current use.*

Oxidation potentials. In this convention the half-cell of interest is considered in terms of its oxidation reaction $X \rightarrow X^{+n} + ne^-$. The resulting electromotive force is termed the *oxidation potential E;* if both species are present in their standard state, it is the *standard oxidation potential $E°$.* This quantity measures the tendency of the electrode to undergo oxidation—to drive electrons from the half-cell out into the external circuit. In the galvanic cell set up to measure $E°$, we consider that both the EUP and the SHE have some characteristic driving force for the oxidation process and that the half-cell with the larger *E actually* will be oxidized while the other is reduced. If the standard oxidation potential of the EUP is greater than that of the SHE, the half-cell reactions as they actually occur would be

$$X \rightarrow X^{+n}(a = 1) + ne^- \quad \text{and} \quad H^+(a = 1) + e^- \rightarrow \tfrac{1}{2} H_2(P = 1 \text{ atm})$$

and electrons flow from the EUP to the SHE. Since $E°$ for the reference electrode is zero, the cell emf is the standard oxidation potential of the $X \mid X^+$ half-cell, and this will be a positive quantity, that is, $E°_{X|X^+} = +A$ volts.

Conversely, if the SHE has a greater oxidation potential, it undergoes oxidation and the EUP is reduced. Electron flow then is SHE $\rightarrow$ EUP, with the half-cell reactions being

$$Y^+(a = 1) + e^- \rightarrow Y \quad \text{and} \quad \tfrac{1}{2} H_2(P = 1 \text{ atm}) \rightarrow H^+(a = 1) + e^-$$

$E°_{Y^+|Y}$ will be a negative quantity.

By combining $E°$ values for any particular pair of half-cells, the value of the overall cell emf for all species in their standard state $\mathscr{E}°$ can be determined. The cell is written in the invariable general convention (the anode

---

* In making use of electrochemical data obtained from various sources, it is imperative that the user be certain of the conventions employed. Unfortunately it is not enough to examine the half-cell reaction as presented in the table giving the emf values since the sign of the potential does not always correspond to the indicated reaction. Perhaps the best procedure is to keep in mind the numerical value and, even more important, the sign for a particular half-cell in *one* of the conventions. Comparing this with the value given in a particular table then will indicate which convention is being used.

as the left cell) without reference to either the direction in which the cell will actually operate or the type of half-cell potential, oxidation or reduction, to be employed.

Consider a cell utilizing the two half-cells

$$Cr \rightarrow Cr^{2+} + 2\,e^-,\ E^\circ = 0.76\ \text{volt} \quad \text{and} \quad Pb \rightarrow Pb^{2+} + 2\,e^-,\ E^\circ = 0.13\ \text{volt}$$

In the real galvanic cell only one of these reactions will proceed in the indicated direction. In setting up the cell representation either can be chosen as the anode. Suppose the cell is written as $Cr \mid Cr^{2+} \parallel Pb^{2+} \mid Pb$. The standard cell emf will be

$\mathscr{E}^\circ =$ standard emf resulting from reaction in left half-cell
  $+$ standard emf due to reaction in right half-cell

Since the reaction in the left half-cell is oxidation (in the cell as written) the first term in this equation is E_L°, the standard *oxidation* potential of the $Cr \mid Cr^{2+}$ couple. At the right electrode the process is reduction, $Pb^{2+} + 2\,e^- \rightarrow Pb$, and the emf is $-E_R^\circ$.

E_R° refers to an oxidation process, and for the reverse reaction the emf will be of opposite sign. Substitution of these terms into the equation gives

$$\mathscr{E}^\circ = E_L^\circ - E_R^\circ \tag{7-12}$$

For this particular system, $\mathscr{E}^\circ = 0.76 - 0.13 = +0.63$ volt, where the positive sign indicates that the cell reaction proceeds as written with the electrons flowing from left to right.

It would have been equally correct to represent the cell as $Pb \mid Pb^{2+} \parallel$, $Cr^{2+} \mid Cr$. Equation (7-12) then will give $\mathscr{E}^\circ = 0.13 - 0.76 = -0.63$ volt showing that the cell reaction as written in the indicated direction is not spontaneous.

Reduction potentials. Using this approach the EUP is considered from the point of view of its tendency to undergo reduction, $Y^+ + e^- \rightarrow Y$. The emf of this process is the *reduction potential V* or the *standard reduction potential* V°, when both species are in their standard state. V° also is frequently referred to as the *electrode potential*, but since this term is also sometimes applied to values that really are oxidation potentials, it will not be employed here.

When two half-cells are assembled to determine a V° value, we consider the relative tendency of the EUP and SHE to be reduced—to withdraw electrons from the external circuit. If the EUP has the greater tendency to be reduced, its V° will be positive. In the measurement cell electrons will flow from the SHE to the EUP. Conversely, if the half-cell of interest has a smaller tendency to be reduced than does the hydrogen electrode, reduction will occur in the latter and V° is negative.

To determine $\mathscr{E}^\circ$ values from half-cell reduction potentials, the cell is represented in the conventional manner, $A \mid A^+ \parallel B^+ \mid B$. The left cell is shown in this notation as an oxidation process, so that the potential for the reaction as it occurs in that cell is $-V_L^\circ$. Taking the emf of the galvanic cell as the sum of the half-cell emf values, we have, therefore,

$$\mathscr{E}^\circ = V_R^\circ - V_L^\circ \tag{7-13}$$

Reconsidering the chromium–lead cell, $\mathrm{Cr} \mid \mathrm{Cr}^{2+} \parallel \mathrm{Pb}^{2+} \mid \mathrm{Pb}$, we find that the half-cell reactions are

$$\mathrm{Cr}^{2+} + 2\,e^- \rightarrow \mathrm{Cr}, \; V^\circ = -0.76 \text{ volt} \quad \text{and} \quad \mathrm{Pb}^{2+} + 2\,e^- \rightarrow \mathrm{Pb}, \quad V^\circ = -0.13 \text{ volt}$$

Application of Eq. (7-13) gives

$$\mathscr{E}^\circ = V_R^\circ - V_L^\circ = -0.13 - (-0.76) = +0.63 \text{ volt.}$$

The standard (reversible) emf of any galvanic cell therefore can be calculated from tabulated values of the standard oxidation or reduction potentials.

Table 7-2 lists the standard potentials of a number of half-cells. Note that the reaction equation given in the table is an oxidation process that has the indicated emf, E°. The corresponding reduction potentials also are listed; these have the same numerical value but are of opposite sign. The standard convention is to present the data as reduction potentials. However, since both E° and V° values continue to be used, it is necessary to be familiar with both types of notation.

The general utility of the Nernst equation, the central relation in electrochemistry, is summarized in the following chart. The arrows indicate the directions in which calculations can be made from experimentally determined values or experimental parameters can be determined from known information. The equation permits calculation of several important quantities from a relatively small number of experimental determinations or from tabulated thermodynamic data.

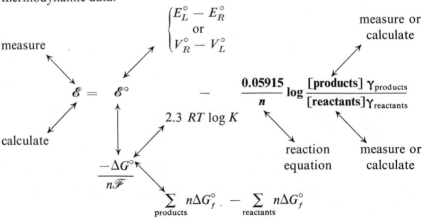

TABLE 7-2. STANDARD OXIDATION AND REDUCTION POTENTIALS AT 298 $^\circ K^a$

Half-cell	Oxidation reaction	E° (volts) Oxidation potential	V° (volts) Reduction potential
Li $\mid$ Li$^+$	Li $\rightarrow$ Li$^+$ + e^-	+3.045	−3.045
Ba $\mid$ Ba^{2+}	Ba $\rightarrow$ Ba^{2+} + 2 e^-	+2.90	−2.90
Sr $\mid$ Sr^{2+}	Sr $\rightarrow$ Sr^{2+} + 2 e^-	+2.89	−2.89
Ca $\mid$ Ca^{2+}	Ca $\rightarrow$ Ca^{2+} + 2 e^-	+2.87	−2.87
Mg $\mid$ OH$^-$ $\mid$ Mg(OH)$_2$	Mg + 2 OH$^-$ $\rightarrow$ Mg(OH)$_2$ + 2 e^-	+2.69	−2.69
Mg $\mid$ Mg^{2+}	Mg $\rightarrow$ Mg^{2+} + 2 e^-	+2.37	−2.37
Mn $\mid$ CO$_3^{2-}$ $\mid$ MnCO$_3$	Mn + CO$_3^{2-}$ $\rightarrow$ MnCO$_3$ + 2 e^-	+1.48	−1.48
Mn $\mid$ Mn^{2+}	Mn $\rightarrow$ Mn^{2+} + 2 e^-	+1.18	−1.18
Fe $\mid$ S^{2-} $\mid$ FeS	Fe + S^{2-} $\rightarrow$ FeS + 2 e^-	+1.01	−1.01
Zn $\mid$ Zn^{2+}	Zn $\rightarrow$ Zn^{2+} + 2 e^-	+0.763	−0.763
Cr $\mid$ Cr^{3+}	Cr $\rightarrow$ Cr^{3+} + 3 e^-	+0.74	−0.74
S^{2-} $\mid$ S	S^{2-} $\rightarrow$ S + 2 e^-	+0.48	−0.48
Cd $\mid$ Cd^{2+}	Cd $\rightarrow$ Cd^{2+} + 2 e^-	+0.403	−0.403
Co $\mid$ Co^{2+}	Co $\rightarrow$ Co^{2+} + 2 e^-	+0.277	−0.277
Ni $\mid$ Ni^{2+}	Ni $\rightarrow$ Ni^{2+} + 2 e^-	+0.250	−0.250
Sn $\mid$ Sn^{2+}	Sn $\rightarrow$ Sn^{2+} + 2 e^-	+0.136	−0.136
Pb $\mid$ Pb^{2+}	Pb $\rightarrow$ Pb^{2+} + 2 e^-	+0.126	−0.126
Cu $\mid$ NH$_3$ $\mid$ Cu(NH$_3$)$_2^+$	Cu + 2 NH$_3$ $\rightarrow$ Cu(NH$_3$)$_2^+$ + e^-	+0.12	−0.12
Ag $\mid$ CN$^-$ $\mid$ AgCN	Ag + CN$^-$ $\rightarrow$ AgCN + e^-	+0.017	−0.017
H$_2$ $\mid$ H$^+$	H$_2$ $\rightarrow$ 2 H$^+$ + 2 e^-	0.000	0.000
Ag $\mid$ Br$^-$ $\mid$ AgBr	Ag + Br$^-$ $\rightarrow$ AgBr + e^-	−0.095	+0.095
CH$_4$ $\mid$ C $\mid$ H$^+$	CH$_4$ $\rightarrow$ 4 H$^+$ + C + 4 e^-	−0.13	+0.13
H$_2$S $\mid$ S $\mid$ H$^+$	H$_2$S $\rightarrow$ S + 2 H$^+$ + 2 e^-	−0.141	+0.141
Sn^{2+} $\mid$ Sn^{4+}	Sn^{2+} $\rightarrow$ Sn^{4+} + 4 e^-	−0.15	+0.15
Pt $\mid$ OH$^-$ $\mid$ Pt(OH)$_2$	Pt + 2 OH$^-$ $\rightarrow$ Pt(OH)$_2$ + 2 e^-	−0.15	+0.15
Cu$^+$ $\mid$ Cu^{2+}	Cu$^+$ $\rightarrow$ Cu^{2+} + e^-	−0.153	+0.153
CH$_3$OH $\mid$ HCHO(aq)H$^+$	CH$_3$OH$\rightarrow$HCHO(aq)+2 H$^+$+2 e^-	−0.19	+0.19
Ag $\mid$ OH$^-$ $\mid$ Ag$_2$O	2 Ag+2 OH$^-$$\rightarrow$Ag$_2$O+H$_2$O+2 e^-	−0.344	+0.344
Cu $\mid$ Cu^{2+}	Cu $\rightarrow$ Cu^{2+} + 2 e^-	−0.337	+0.337
Cu $\mid$ Cu$^+$	Cu $\rightarrow$ Cu$^+$ + e^-	−0.521	+0.521
C$_2$H$_6$ $\mid$ C$_2$H$_4$ $\mid$ H$^+$	C$_2$H$_6$ $\rightarrow$ C$_2$H$_4$ + 2 H$^+$ + 2 e^-	−0.52	+0.52
I$^-$ $\mid$ I$_2$	2 I$^-$ $\rightarrow$ I$_2$ + 2 e^-	−0.5355	+0.5355
H$_2$O$_2$ $\mid$ O$_2$ $\mid$ H$^+$	H$_2$O$_2$ $\rightarrow$ O$_2$ + 2 H$^+$ + 2 e^-	−0.682	+0.682
Fe^{2+} $\mid$ Fe^{3+}	Fe^{2+} $\rightarrow$ Fe^{3+} + e^-	−0.771	+0.771
Ag $\mid$ Ag$^+$	Ag $\rightarrow$ Ag$^+$ + e^-	−0.7991	+0.7991
NO $\mid$ NO$_3$ $\mid$ H$^+$	NO + 2 H$_2$O $\rightarrow$ NO$_3^-$ + 4 H$^+$ + 4 e^-	−0.96	+0.96
Br$^-$ $\mid$ Br$_2$	2 Br$^-$ $\rightarrow$ Br$_2$(l) + 2 e^-	−1.0652	+1.0652
Cl$^-$ $\mid$ Cl$_2$	2 Cl$^-$ $\rightarrow$ Cl$_2$ + 2 e^-	−1.3595	+1.3595
Au $\mid$ Au^{3+}	Au $\rightarrow$ Au^{3+} + 3 e^-	−1.50	+1.50
Co^{2+} $\mid$ Co^{3+}	Co^{2+} $\rightarrow$ Co^{3+} + e^-	−1.82	+1.82
N$_2$ $\mid$ H$_2$N$_2$O$_2$ $\mid$ H$^+$	N$_2$+2 H$_2$O$\rightarrow$H$_2$N$_2$O$_2$+2 H$^+$+2 e^-	−2.85	+2.85

^{a} Data from W. M. Latimer, *The Oxidation States of the Elements and Their Potentials in Aqueous Solutions*, 2nd ed (Engelwood Cliffs, New Jersey: Prentice-Hall, Inc., 1952, pp. 340–348).

One other application of this versatile relation is to the reaction occurring in an individual half-cell. The approach is quite useful when we wish to focus our attention on just one component of a galvanic cell—usually when the other half-cell is established as a reference electrode with invariable potential. For the calculations utilizing the oxidation potential of a given half-cell, for example, $X \rightarrow X^{+n} + ne^-$, the equation is

$$E = E° - \frac{0.05915}{n} \log \frac{a_{X^{+n}}}{a_X} \tag{7-14}$$

Since X^{+n} represents the oxidized form and X is the reduced form of the species involved, this equation is frequently written, particularly in biological treatments, as

$$E = E° - 0.05915 \log \frac{[\text{oxidized form}]}{[\text{reduced form}]} \tag{7-15}$$

where activities are replaced by concentrations in moles per liter. For the analogous calculation with reduction potentials, the reaction considered is $X^{+n} + ne^- \rightarrow X$, and the relevant equations are

$$V = V° - \frac{0.05915}{n} \log \frac{a_X}{a_{X^{+n}}} \tag{7-16}$$

$$V = V° - \frac{0.05915}{n} \log \frac{[\text{reduced form}]}{[\text{oxidized form}]} \tag{7-17}$$

This last equation is often written with a positive sign before the last term in the form

$$V = V° + \frac{0.05915}{n} \log \frac{[\text{oxidized form}]}{[\text{reduced form}]} \tag{7-18}$$

It is apparent from all of these equations that if [oxidized form] = [reduced form], then $V = V°$ or $E = E°$. Thus, for an individual half-cell, the standard oxidation or reduction potential can be thought of as the potential of the half-cell when reactants and products are present at the same concentration (activity).

These equations also provide an alternative method of computing the overall cell emf. The method discussed previously consisted in determining $\mathscr{E}°$ from Eq. (7-12) or Eq. (7-13) and then $\mathscr{E}$ from the Nernst equation. The equation just obtained, however, can be used to determine E or V for each half-cell as the first step and then $\mathscr{E}$ for the overall cell from the equations analogous to Eqs. (7-12) and (7-13):

$$\mathscr{E} = E_L - E_R \tag{7-12'}$$

$$\mathscr{E} = V_R - V_L \tag{7-13'}$$

7-3-5 Experimental Determination of Standard Oxidation Potentials

The techniques are discussed here in terms of evaluating E° values for given half-cells but, if desired, the corresponding V° value subsequently can be obtained by a simple change of sign. The necessary equations also provide an illustration of the use of the Nernst equation for individual half-cells.

As an example of the general procedure followed, consider the determination of E° for the $\text{Zn} \mid \text{Zn}^{2+}$ half-cell using a common reference half-cell, the saturated calomel electrode shown in Fig. 7-5. The electrode consists of mercury in contact with Hg_2Cl_2 (calomel) and a saturated aqueous KCl solution. The net reaction in the electrode, when it forms the cathode of a galvanic cell, is

$$\tfrac{1}{2}\,\text{Hg}_2\text{Cl}_2 + e^- \rightarrow \text{Hg} + \text{Cl}^-$$

The presence of solid KCl maintains $[\text{Cl}^-]$, and therefore the half-cell emf, constant. Oxidation, $\text{Zn} \rightarrow \text{Zn}^{2+} + 2\,e^-$, occurs in the zinc half-cell when it is combined with the calomel electrode. The oxidation potential of the saturated calomel electrode is -0.244 volt, and thus at any particular concentration of zinc ions the emf of the overall cell will be $\mathscr{E} = E + 0.244$. Thus $E = \mathscr{E} - 0.244$, where E is the oxidation potential of the $\text{Zn} \mid \text{Zn}^{2+}$ couple. The Nernst equation for the latter is

$$E = E^{\circ} - 0.0296\ \log[\text{Zn}^{2+}]\gamma_{\pm}$$

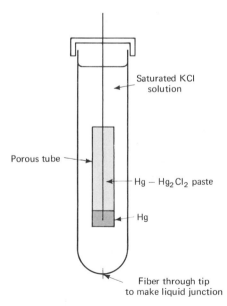

FIG. 7-5 The saturated calomel electrode.

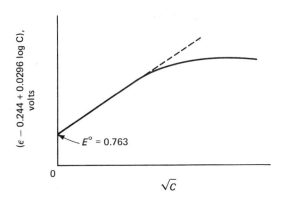

FIG. 7-6 Plot of Eq. (7-20) for the zinc half-cell.

Substitution for E and rearrangement yields

$$\mathscr{E} - 0.244 + 0.0296 \log[\text{Zn}^{2+}] = E° - 0.0296 \log \gamma_\pm \qquad (7\text{-}19)$$

According to the Debye–Huckel equation, Eq. (5-13), $-\log \gamma_\pm$ is proportional to $\sqrt{\mu}$ in dilute solution. If the conversion factor for ionic strength to concentration is included in the proportionality constant, the term $-0.0296 \log \gamma_\pm$ can be replaced by $+\text{constant} \sqrt{C}$, where $C = [\text{Zn}^{2+}]$. With these changes the working equation can be written as

$$\mathscr{E} - 0.244 + 0.0296 \log C = E° + \text{constant} \sqrt{C} \qquad (7\text{-}20)$$

which is of the straight line form. Thus a plot of the left-hand side of Eq. (7-20) is a linear function of $\sqrt{C}$ with the intercept on the y axis being the desired quantity $E°$. A plot of this equation for the zinc calomel reference electrode is shown in Fig. 7-6.

Once the standard oxidation potential has been determined in this manner a value for the mean activity coefficient of the zinc ion can be calculated from Eq. (7-18) at each concentration for which a measured value of C is available. This is just one important application of emf measurements—the determination of activity coefficients. Several other applications are discussed in the next section.

7-4 APPLICATIONS OF EMF MEASUREMENTS

7-4-I Potentiometric Titrations

A given oxidation-reduction process can be carried out by successive additions of a solution containing one reagent to a fixed volume of a solution containing the other. The relative concentrations of the reduced and oxidized forms of the reactant in the latter will change during the titration. If the solution is utilized as one half-cell in conjunction with some convenient reference electrode in a galvanic cell, the course of the reaction can be followed by continuously monitoring the cell emf. The resulting titration curve, in which emf is plotted as a function of the amount of oxidizing or reducing agent added, has the same general form as an acid-base titration curve. This similarity is indicated by a comparison of the Nernst equation for the redox half-cell, Eq. (7-18), with the Henderson-Hasselbalch equation,

$$pH = pK_a + \log \frac{[\text{base form}]}{[\text{acid form}]} \qquad (4\text{-}27')$$

The equivalence point can be determined by plotting the titration curve from emf measurements or by use of a redox indicator. The mode of operation of such species also is similar to acid-base indicators. They exist generally in a colorless and colored form in the reduced and oxidized states. The form that predominates in a given solution and thereby imparts the characteristic color will be determined by the relative concentrations of the principal oxidizing and reducing agents present, in effect, by the half-cell potential. By suitable alignment of the standard oxidation potentials of the indicator and the emf-determining reactants, the potential at which visible color change occurs will fall within the range over which the galvanic cell emf changes rapidly at the end point.

7-4-2 Determination of pH

It was mentioned in Sec. 7-3-3 that in principle any galvanic cell utilizing a redox reaction involving hydronium ions can be used to determine pH. The most convenient procedure is to employ a cell comprised of a reference half-cell whose potential remains constant and a half-cell in which H_3O^+ ions participate in the redox reaction. By comparing the observed emf with that obtained when hydronium ions are present at unit activity, $a_{H_3O^+}$ can be calculated from the Nernst equation.

Although other electrodes are used in practice, the techniques involved are most simply discussed in terms of the hydrogen electrode itself. The conventional notation for the galvanic cell that would be used is $H_2 \mid Pt \mid H^+(a = x) \parallel$ reference electrode. The oxidation potential of the reference is usually not a true $E°$ value. The electrode is set up to have an invariable emf resulting from some conveniently maintained nonstandard state conditions such as a solution saturated with respect to some particular electrolyte. The value therefore is designated here as E'. According to Eq. (7-12) the observed emf for the pH measuring cell will be

$$\mathscr{E} = E_{H_2 \mid H^+} - E'_{\text{ref}} \qquad (7\text{-}21)$$

E' remains constant so that $\mathscr{E}$ depends solely on $E_{H_2|H^+}$. To obtain the dependence of $\mathscr{E}$ on pH the half-cell reaction at the hydrogen electrode is written as an oxidation process, $\frac{1}{2}\,H_2 \rightarrow H^+ + e^-$. The applicable form of the Nernst equation is

$$E_{H_2|H^+} = E^\circ_{H_2|H^+} - 0.05915 \log \frac{a_{H^+}}{(P_{H_2})^{\frac{1}{2}}}$$

At $P_{H_2} = 1$ atm, and since $E^\circ_{H_2|H^+} = 0.00$ volt by definition, $E_{H_2|H^+}$ is equal to $-0.05915 \log (a_{H^+})$. Finally utilizing the pH definition in Eq. (4-2), we get

$$E_{H_2|H^+} = 0.05915 \text{ pH} \qquad (7\text{-}22)$$

Substitution of this expression into Eq. (7-21) yields

$$\mathscr{E} = 0.05915 \text{ pH} - E'_{ref}$$

or

$$\text{pH} = \frac{\mathscr{E} + E'_{ref}}{0.05915} \qquad (7\text{-}23)$$

The most frequently employed reference electrode is the saturated calomel half-cell, which as indicated previously has $E(\equiv E'_{ref})$ of -0.244 volt. For pH determinations using the combination of the hydrogen and calomel electrodes, then, we have

$$\text{pH} = \frac{\mathscr{E} - 0.244}{0.05915} \qquad (7\text{-}24)$$

For routine measurements the hydrogen electrode is not particularly convenient. Frequently a commercially available *glass electrode* is employed. In this apparatus an internal reference electrode, usually $Ag \mid AgCl$, is sealed in a glass tube filled with 0.1-N aqueous HCl solution. A portion of the bottom of the tube consists of a specially fabricated thin glass membrane that develops an electrical potential when the solution in contact with the exterior of the electrode is at a different pH from that of the HCl solution. The internal reference electrode serves, in conjunction with the external reference electrode, usually calomel, to measure this potential, which is a direct measure of the pH. The overall galvanic cell consists of one half-cell containing the calomel electrode separated from the $Ag \mid AgCl$ electrode by the glass membrane and has a potential determined by the invariable potential of the two references and the pH dependent potential across the glass membrane. The glass electrode has the considerable advantage that it can be used to measure the pH of a variety of solutions without contamination.

Commercially available electrodes also include types in which both the glass and calomel electrodes are combined in a single probe, thus permitting readings to be taken on very small sample volumes. The observed emf gives the pH via Eq. (7-22), wherein E'_{ref} now includes terms for both reference electrodes and the glass membrane itself, which must be determined for each electrode individually. The current generated by this arrangement, however, is quite small. An instrument known as a *pH meter* usually is used in conjunction with the glass and calomel electrodes. Through suitable electronics the emf of the cell can be read quite accurately on a millivoltmeter calibrated directly in pH units.

7-5 ELECTROCHEMISTRY OF BIOLOGICAL OXIDATION–REDUCTION PROCESSES

The electron transfer processes that occur in biological systems are appreciably more complex than those in the simple systems that have been utilized in this chapter to illustrate fundamental electrochemical methods. Oxidation–reduction reactions constitute an extremely important type of metabolic process. What is involved here is the utilization in oxidation reactions of the molecular oxygen transported to the cell. The overall reactions consist of a complex series of steps in which successive electron transfers, usually accompanied by proton transfer, occur. Molecular oxygen requires a *carrier* before the process can be initiated and, in fact, each step in the mechanism is enzyme catalyzed.

In Chapter 1 the overall oxidation process was considered from the point of view of a combustion reaction and its associated total enthalpy change—the heat energy released. In a similar manner the free energy change in the overall reaction can be determined, and thus a determination of the maximum useful work obtainable can be made. Approaching biological redox reactions from the electrochemical point of view, each step in the mechanism can be treated by using the techniques developed here as a cell reaction. Thus we can profitably utilize the approach of assigning to each half-cell reaction an oxidation or reduction potential in exactly the same manner as an $E°$ is listed for an ionic reaction such as $Zn \rightarrow Zn^{2+} + 2\,e^-$. The potential associated with each reduced form $\rightarrow$ oxidized form process represents an alternative to free energy calculations as a criterion of spontaneity. The two types of data of course, are, readily interchangeable through Eqs. (7-5) or (7-6). Note, however, that the emf of any half-cell can be made positive, in theory at least, by suitable adjustment of the concentrations. Thus in determining the actual potentials involved in a reaction sequence we must consider not only the standard emf value but also the effect of product removal, in subsequent steps, on the concentrations of species in the real system. The overall emf for a sequence of oxidation processes and their concomitant reduction reactions can be determined from the sum of the potentials for the individual half-cells. Care must be exercised, however, to ensure that the same sign convention is observed for all component steps. Since the process occurs in aqueous solution, and since many of the electron transfer processes in biological systems involve simultaneous dehydrogenation, the potential developed will be pH dependent. Consider, for example, the oxidation process

$$\text{D-Glucose} \rightarrow \text{gluconolacetone} + 2\,H^+ + 2\,e^-$$

This is entirely analogous to the hydroquinone reduction discussed previously.

The Nernst equation takes the form

$$\mathscr{E} = \mathscr{E}^\circ - 0.0296 \log \frac{[K]}{[S]} - 0.0592\,pH$$

where [K] and [S] represent gluconolacetone and glucose, respectively. Very often emf values are given for the "half-oxidized" reactant at pH = 7, and such values are designated as $\mathscr{E}^{\circ\prime}$. The implication of the "half-oxidized"

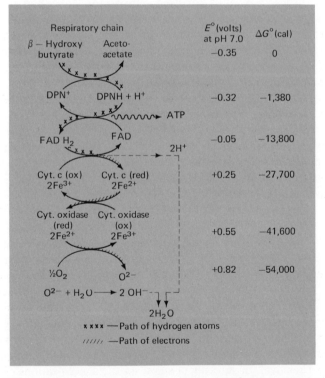

FIG. 7-7 *Schematic representation of the oxidation chain forming a stage in the metabolism of fatty acids. DPN⁻ and DPNH, and FAD and FADH₂, refer to the oxidized and reduced forms of diphosphopyridine neucleotide and flavin adenine dinucleotide respectively. Reprinted by permission from J. H. Linford, An Introduction to Energetics with Applications to Biology, Butterworths, London, 1966, p. 144.*

specification is that [product] $=$ [reactant] and, hence, the activity quotient, from which a_{H^+} has been removed, is zero and the Nernst equation gives $\mathscr{E} = \mathscr{E}^\circ - 0.05915$ pH. $\mathscr{E}^{\circ\prime}$ for this particular example is $+0.36$ volt.

The details of many biological redox mechanisms are not yet known in their entirety. A representative oxidation scheme is shown in Fig. 7-7. For any sequence of steps, provided the intermediate processes are established and their redox potentials are known, the overall emf can be evaluated by a combination of $\mathscr{E}$ values for each process and the free energy change can be calculated as indicated previously. The potential of any individual half-cell can be established by the use of a series of redox indicators, which exhibit the characteristic color changes described in connection with their use in redox titrations, choosing indicators that will bracket the emf of the half-cell of interest. This is the same technique as in the determination of solution pH, and hence K_a values, by the use of acid–base indicators.

PROBLEMS

7-1. Calculate the standard reversible emf $\mathscr{E}^\circ$ at 298 °K for each of the following galvanic cells:
(a) Ba $|Ba^{2+} \| Cr^{3+}|$ Cr
(b) Pb $|Pb^{2+} \| Mn^{2+}|$ Mn
(c) Mg $|OH^- |$ Mg(OH)$_2 \| ClO^- |$ OH$^- |Cl^-|$ Pt
(d) Pt $|I^-| I_2 \| Ni^{2+}|$ Ni

7-2. For each of the cells in Problem 7-1, write the balanced overall cell reaction and calculate ΔG° and K at 298 °K.

7-3. Calculate the weight in grams of copper deposited in a Zn $|Zn^{2+} \| Cu^{2+}|$ Cu galvanic cell that has produced 10,000 coulombs of electricity.

7-4. A galvanic cell utilizes the two electrodes Al $|Al^{3+}$ and $Cd^{2+}|$ Cd.
(a) Write the half-cell reactions (1) to which the V° value given in Table 7-2 applies and (2) as they actually occur in the operating cell.
(b) Write the overall cell reaction, describe the cell in conventional notation, and calculate $\mathscr{E}^\circ$.
(c) Write the Nernst equation and, assuming activity coefficients of unity, calculate $\mathscr{E}$ when the ions are present in their respective half-cells at a concentration of 0.1 moles/liter.

7-5. Calculate $\mathscr{E}^\circ$, ΔG°, and ΔH° for the Cu $|Cu^{2+} \| Ag^+|$ Ag cell from the standard oxidation potentials of the half-cells and the absolute entropy data:

$$S^\circ \text{ (cal/deg mole)} \qquad Ag + 10.206 \qquad Cu + 7.96$$
$$Ag^+ + 17.76 \qquad Cu^{2+} - 23.6$$

7-6. For the cell described in Problem 7-5, use the data calculated to evaluate the emf at 45 °C.

7-7. Calculate $\mathscr{E}$ for the cell $\text{Pt} \mid \text{H}_2(1 \text{ atm}) \mid \text{H}^+(a = x) \parallel \text{Zn}^{2+}(a = 1) \mid \text{Zn}$ when the pH in the hydrogen half-cell is 5.76.

7-8. Calculate the oxidation potential of the Fe^{2+}, Fe^{3+} half-cell when $[\text{Fe}^{2+}] = [\text{Fe}^{3+}] = 0.1, 0.05, 0.01, 0.005, 0.001$, and 0.0001 mole/liter, using the activity coefficient values given in Fig. 5-3. Plot $\mathscr{E}$ vs. $\log(C_{\text{Fe}^{2+}} + C_{\text{Fe}^{3+}})$ and compare the curve obtained with the straight line that would characterize the system if activity coefficient values were constant at 1.00 over this concentration range.

7-9. Calculate $\mathscr{E}^\circ$ for the cell $\text{Pt} \mid \text{H}_2 \mid \text{H}^+ \parallel \text{KCl(sat)} \mid \text{Hg}_2\text{Cl}_2 \mid \text{Hg}$ and write the balanced redox reaction for the cell reaction.

7-10. Given that the mean activity coefficient for a 1-molar aqueous solution of HCl is 0.809, calculate the reversible emf for the cell $\text{Ag} \mid \text{AgCl}(s) \mid \text{HCl}(1 \text{ mole/liter}) \mid \text{H}_2(1 \text{ atm}) \mid \text{Pt}$. Note that this cell does not require a salt bridge since the same electrolyte solution supplies the necessary ions to both electrodes.

Phase Equilibria

8

Chemical processes have dominated the discussion up to this point. In the present and subsequent chapters, attention is focused on processes involving physical changes. Both types of processes are susceptible to the same kind of quantitative treatment in terms of the fundamental thermodynamic principles developed in the first three chapters.

Phase equilibria provide an excellent model to demonstrate the applicability of thermodynamics to physical processes. In addition, the discussion provides the necessary background for an examination of solutions and their properties in the next chapter.

It is necessary initially to describe the three phases individually, along with their properties relevant to presentation of phase equilibria in a single-component system in terms of pressure-temperature and pressure-volume diagrams.

8-1 THE VAPOR PHASE

The familiar ideal gas law

$$PV = nRT \qquad (1\text{-}17)$$

has been utilized previously in several contexts. This law is followed almost exactly by a large number of gases under normal conditions. In Chapter 3, however, it was noted that the product of pressure and volume does not remain constant under extreme pressures and temperatures.

According to Eq. (1-17), a plot of PV vs. P or V at constant temperature should give a straight line of zero slope and intercept equal to nRT. Figure 8-1 illustrates the nature of the deviations from this ideal PV curve for some real gases. In the limit of zero pressure, the curves at both temperatures give the value of nRT predicted by the ideal gas law, that is 30.619 liters atm at 373 °K and 22.414 liters atm at 273 °K.

As is apparent from the figure, deviations are most pronounced at higher pressures and lower temperatures. When a real gas is subjected to high pressures, there are a larger number of molecules per unit volume and the volume of the molecules themselves becomes a significant fraction of the total volume occupied by the gas. The result is that the gas is less compressible than predicted by the ideal gas law, since Eq. (1-17) is based on the assumption that the molecular volume is negligible. At low temperatures a real gas will be more compressible than an ideal gas. Under these conditions the average kinetic energy of the gas molecules is low. Hence, the small but

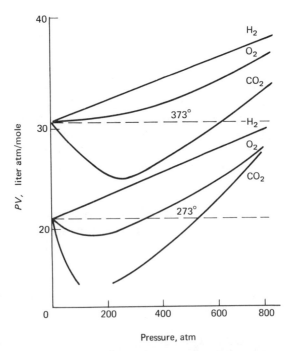

FIG. 8-1 *The PV product per mole of real gases. Adapted from* W. H. Hamill, R. R. Williams, Jr., *and* C. MacKay, *Principles of Physical Chemistry, 2nd ed., Prentice-Hall, Inc., Englewood Cliffs, N.J., 1966, p. 7.*

finite forces of intermolecular attraction that exist between real gas molecules can exert an appreciable influence. This is manifested by the increased facility of compression.

Although in the general discussion of phase equilibria ideal gas behavior is assumed for the vapor phase of all systems, it is instructive to examine several of the equations used to describe the PV product of a real gas.

Perhaps the simplest modification of the ideal gas law is to include an additional term to correct for the increase in the PV product at high pressures. If it is assumed that the magnitude of PV is directly proportional to pressure in this region, the alteration in Eq. (1-17) is simply

$$PV = nRT + nb'P \qquad (8\text{-}1)$$

where b' is a temperature dependent constant characteristic of each gas.

For more accurate treatment of high-pressure deviations, a useful approach is the *virial equation of state*, which has the form

$$PV = nRT + nBP + nCP^2 + \ldots \qquad (8\text{-}2)$$

This type of equation is becoming increasingly important since by the use of computers it is possible to describe a particular gas with virtually any accuracy by using the appropriate number of parameters B, C, etc. in this equation. These constants will be specific for the gas in question.

An equation of state for real gases that is worthwhile examining in some detail is the *van der Waals equation*. The form of this semiempirical relation derives from explicit consideration of the two factors giving rise to deviations —intermolecular forces and molecular volume. The nature of the correction factors can be rationalized in the following manner: The ideal gas law is written in the form

$$P_i V_i = nRT \qquad (1\text{-}17')$$

where P_i represents the pressure that would be exerted and V_i the volume that would be occupied by the gas were it to behave ideally. The observed pressure P, will be less than P_i, so that $P = P_i - x$, where x is a correction for the reduction in pressure owing to intermolecular attractive forces. P_i in Eq. (1-17') therefore can be replaced by the term $(P + x)$. The volume occupied by the real gas will be greater than V_i, and therefore $V = V_i + y$. The correction represented as y arises from the volume occupied by the molecules themselves. Thus V_i in the ideal gas law becomes $V - y$. The explicit forms of the correction factors x and y in the van der Waals equation are n^2a/V^2 and nb, respectively. Making these substitutions gives the conventional form of the equation:

$$\left(P + \frac{n^2a}{V^2} \right)(V - nb) = nRT \qquad (8\text{-}3)$$

TABLE 8-1.   PROPERTIES OF REAL GASES AND VAPORS[a]

Substance	van der Waals constants		Critical constants	
	a (liter^2 atm/mole^2)	b (liter/mole)	$T_c(°K)$	P_c(atm)
CH_3COOH	17.59	0.1068	595	57.2
CH_3COCH_3	13.91	0.0994	508	47
NH_3	4.170	0.03707	405	111.5
C_6H_6	18.00	0.1154	562	47.7
CO_2	3.592	0.04267	303	73
CO	1.485	0.03985	134	35
$CHCl_3$	15.17	0.1022	536	—
C_2H_5OH	12.02	0.08407	516	63.1
$(C_2H_5)_2O$	17.38	0.1344	467	35.5
H_2	0.2444	0.02661	33	12.8
H_2S	4.431	0.04287	173	88.9
CH_4	2.253	0.04278	190	45.8
N_2O	3.782	0.04415	310	71.7
O_2	1.360	0.03183	154	49.7
SO_2	6.714	0.05636	116	77.7
H_2O	5.464	0.03049	647	217.72

[a]Data from *Handbook of Chemistry and Physics*, 42nd ed. and 45th ed. (Cleveland: Chemical Rubber Co., 1960, 1964, p. D90, pp. 2303–5). Reprinted by permission.

where a and b are the *van der Waals constants* for the gas. Values of these constants for some frequently encountered gases, and liquids whose vapor phase is often of interest, are given in Table 8-1, along with some additional data to be discussed presently.

Calculation of the real gas pressure when V and T are known, or evaluation of T given P and V, are straightforward. Evaluation of volume at a known pressure and temperature, however, is difficult since the relation is a cubic equation in V.

EXAMPLE 8-1. Using the van der Waals equation, calculate the pressure exerted by 2.0 moles of acetone confined to a volume of 30 liters at a temperature of 375 °K and compare this with the pressure exerted by an ideal gas under the same conditions. ($R = 0.0821$ liter atm/deg mole.)

$$\left(P + \frac{n^2a}{V^2}\right)(V - nb) = nRT$$

$$\left(P + \frac{(2)^2 13.91}{(30)^2}\right)[30 - (0.0994 \times 2)] = 2 \times 0.0821 \times 375$$

$$(P + 0.0618)(29.8) = 61.6$$

$$P = \underline{2.01 \text{ atm}}$$

For an ideal gas,

$$P = \frac{nRT}{V} = \frac{61.6}{30} = \underline{2.05 \text{ atm}}$$

8-2 EQUILIBRIA BETWEEN TWO PHASES

8-2-I Liquid–Vapor Equilibrium

The property of liquids most directly related to phase equilibria is vapor pressure. The molecules in the liquid state have a distribution of energies represented by the Boltzmann distribution (see Fig. 6-9). Some fraction has sufficient energy to overcome the forces of intermolecular attraction and escape into the vapor phase above the liquid. If the vaporization occurs at constant temperature in a closed container, the concentration of the substance in the vapor phase will increase, with a concomitant increase in the rate at which molecules return to the liquid phase through condensation. After some time equilibrium is attained when the rate of vaporization is identical to that of condensation, and the concentration in the vapor phase becomes constant. The pressure exerted by this equilibrium gas concentration is the *vapor pressure*. It is a characteristic of the liquid at a particular temperature.

In Chapter 1 the process of vaporization was dealt with from the point of view of the First Law. The heat effects involved were discussed in terms of ΔH_{vap}, the amount of heat required to convert 1 mole of a substance from the liquid phase to the vapor phase. Recall that ΔH_{vap} is related to the alteration in energy associated with this change of state by

$$\Delta H_{vap} = \Delta E_{vap} - P\,\Delta V \tag{1-26}$$

The most profitable approach to quantitative description of the vapor pressure of a liquid is through consideration of the free energy. If the vapor pressure is considered as a particular concentration in the gas phase of molecules of substance A, resulting from the equilibrium

$$A(l) \rightleftharpoons A(g)$$

then, according to the general condition for equilibrium, ΔG for the process as written must be zero:

$$\Delta G = G_{A(g)} - G_{A(l)} = 0$$
$$G_{A(g)} = G_{A(l)} \tag{8-4}$$

This equation states the important principle, discussed in Sec. 2-4-2, that the free energy per mole of a substance in two phases in equilibrium must be identical.

The most important implication of this idea for present purposes is that it permits determination of the temperature dependence of vapor pressure. The vapor pressure of liquids increases with temperature as the average kinetic energy increases, and a greater proportion of the molecules have

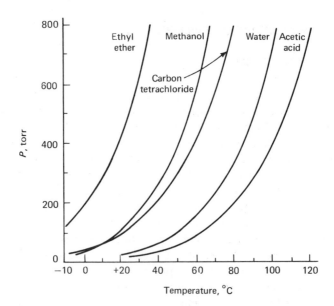

FIG. 8-2 *Temperature dependence of the vapor pressure of some liquids.*

sufficient energy to escape into the vapor phase. Figure 8-2 shows the temperature variation of vapor pressure for some common liquids. It is apparent that the dependence in not a simple proportionality. The *rate* of increase of P with T becomes larger as the temperature rises. Therefore, the equation obtained must deal, initially at least, with infinitesimal temperature changes.

For *any* change in a system comprised of two phases of a substance in equilibrium, the alteration in the free energy in the liquid phase must be accompanied by an identical change in G in the vapor. Considering the free energy change produced by an infinitesimal change in temperature dT, Eq. (8-4) can be written in the form

$$dG_{A(g)} = dG_{A(l)} \qquad (8\text{-}5)$$

To obtain the appropriate expressions for these dG terms, the defining equation for the free energy, $G = H - TS$, is written for an infinitesimal change as follows:

$$dG = dH - T\,dS - S\,dT$$

Substitution of the defining relation for enthalpy, $H = E + PV$, also written for a perfectly general infinitesimal change, gives

$$dG = dE + P\,dV + V\,dP - T\,dS - S\,dT$$

Since the process involves only an infinitesimal change, it is by definition reversible and, therefore, from Eq. (2-6), $T\,dS = q_{rev} = dq$. Considering that we can have PV work only, $P\,dV$ can be represented by dw, and the equation becomes

$$dG = dE + dw + V\,dP - S\,dT - dq$$

According to the first law, $dE + dw - dq = 0$, and thus the equation reduces to the desired form:

$$dG = V\,dP - S\,dT \qquad (8\text{-}6)$$

Equation (8-6) now can be written explicitly for the change in free energy of the substance as a liquid and a vapor. The two expressions can be equated according to the condition expressed by Eq. (8-5) for the maintainence of equilibrium when the system undergoes an infinitesimal change.

$$V_l\,dP - S_l\,dT = V_g\,dP - S_g\,dT$$

$$dP/dT = (S_g - S_l)/(V_g - V_l)$$

$$\frac{dP}{dT} = \frac{\Delta S_{\text{vap}}}{\Delta V_{\text{vap}}} \qquad (8\text{-}7)$$

This is known as the *Clapeyron equation*. The mathematical significance of the equation is analogous to that of differential rate expressions in chemical kinetics. dP/dT represents the slope of the vapor pressure vs. the temperature curve. Its value is given by the ratio of the ΔS and ΔV values for the vaporization process at the temperature where the "tangent is drawn to the curve." This is illustrated in Fig. 8-3.

For purposes of routine calculation Eq. (8-7) may be transformed into an expression that utilizes readily available data somewhat more directly. For a reversible phase change,

$$\Delta S = \frac{\Delta H}{T} \qquad (2\text{-}8)$$

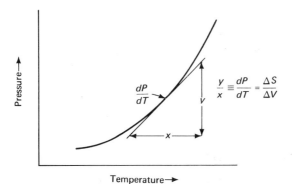

FIG. 8-3 *Physical significance of the Clapeyron equation.*

where the requirement of reversibility here is the general one that the vapor pressure and applied (external) pressure be identical. The ΔV term can be treated as follows: The molar volume of the liquid is assumed to be negligible with respect to that of the vapor, $\Delta V = V_g - V_l \approx V_g$, and thus if the vapor behaves ideally the molar volume will be given by

$$V_g = RT/P \tag{8-8}$$

This expression can be used to replace ΔV in Eq. (8-7). The result is

$$\frac{dP}{dT} = \frac{\Delta H_v P}{RT^2} \tag{8-9}$$

which is an alternative form of the Clapeyron equation.

EXAMPLE 8-2. Ethanol has $\Delta H_v = 9{,}220$ cal/mole. Calculate the rate of change of vapor pressure with the temperature at 62.5 °C where the vapor pressure is 400 torrs.

$$\frac{dP}{dT} = \frac{\Delta H_v P}{RT^2} = \frac{9{,}220 \times 400}{1.987 \times (337)^2}$$

$$\frac{dP}{dT} = \underline{16.3 \text{ torr/deg}}$$

As in the mathematical treatment of reaction rate, it is useful to develop a relationship between vapor pressure and temperature that will avoid difficulties associated with having to deal with an instantaneous rate. This would permit calculation, for example, of heats of vaporization from vapor pressure data without the necessity of taking tangents to the curve.

The required relation can be obtained* by again taking advantage of the fact that the vaporization process is an equilibrium. For $A(l) \rightleftharpoons A(g)$ an equilibrium constant can be written as

$$K = \frac{a_{A(g)}}{a_{A(l)}}$$

The activity of A in the liquid phase will be unity, and in the vapor phase a_A can be represented as the vapor pressure in atmospheres. The equilibrium constant thus is simply the vapor pressure P. The procedure here is the same as in Sec. 3-3-3, where the van't Hoff equation was derived, basing the treatment on

$$2.303 \log K = -\frac{\Delta H^\circ}{RT} + \frac{\Delta S^\circ}{R} \tag{3-7}$$

For the vaporization process, ΔH° and ΔS° are simply ΔH_v and ΔS_v,

---

* The more usual method is definite integration of Eq. (8-9) between two temperatures T_1 and T_2, a procedure similar to the conversion of kinetic rate laws from the differential to integrated form.

respectively, and K can be replaced by the vapor pressure:

$$2.303 \log P = -\frac{\Delta H_v}{RT} + \frac{\Delta S}{R} \qquad (8\text{-}10)$$

Assuming ΔH_v and ΔS_v to be independent of temperature over a moderate range, Eq. (8-10) can be written explicitly for two temperatures T_1 and T_2, and the two equations can be subtracted. The result is

$$\log \frac{P_2}{P_1} = -\frac{\Delta H_v}{2.303\ R}\left(\frac{1}{T_2} - \frac{1}{T_1}\right) \qquad (8\text{-}11)$$

which is known as the *Clausius-Clapeyron equation*. It should be immediately apparent that this procedure and the resulting equation are just the van't Hoff equation written for the specific case of an equilibrium between a liquid and its vapor. It is also evident that a plot of $\log P$ vs. $1/T$ will be a straight line with a slope equal to $-\Delta H_v/2.303\ R$. Thus we have a quite convenient method of representing the temperature dependence of vapor pressure as a linear function. Figure 8-4 shows a plot of this kind for the liquids whose vapor pressure curves were presented in Fig. 8-2.

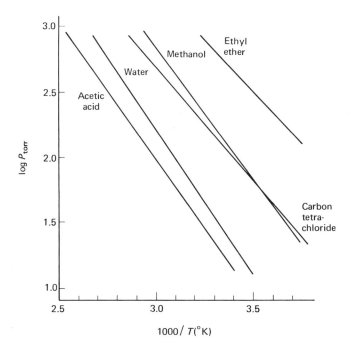

FIG. 8-4 *Clausius-Clapeyron equation type plot for the vapor pressures of liquids in Fig. 8-2.*

TABLE 8-2.   HEATS OF VAPORIZATION OF SOME LIQUIDS[a]

Liquid	ΔH_v, kcal/mole	nbp, °C
CH_3COOH	5.83	118.2
CH_3CHO	5.99	21
$(CH_3)_2CO$	9.19	56.1
C_6H_6	7.35	80.2
$n\text{-}C_4H_9OH$	10.45	117
CS_2	6.40	46.3
CCl_4	7.170	76.8
$CHCl_3$	7.04	61.5
C_2H_5OH	9.40	78.3
C_2H_5Cl	5.96	47
$(C_2H_5)_2O$	6.13	35
$(CH_2)_2O$	6.16	13
$HCOOH$	5.54	101
$(CH_2OH)_2$	11.8	197
$i\text{-}C_5H_{12}$	6.31	13
H_2O	9.717	100

[a] Data from *Handbook of Chemistry and Physics*, 42nd ed. (Cleveland, Chemical Rubber Co., 1960, pp. 2318–2321). Reprinted with permission.

The Clausius-Clapeyron equation can be used to evaluate vapor pressure at any temperature of interest, or the temperature at which a liquid exhibits a particular vapor pressure (for example, the normal boiling point), when the heat of vaporization and the vapor pressure at one temperature is known. Conversely, ΔH_v can be calculated from experimental observations of vapor pressure at a minimum of two different temperatures. This approach is useful in that the data can be presented in the form of a log P vs. $1/T$ plot and the best straight line taken through the points to obtain the slope and hence ΔH_v. Some values of heats of vaporization are given in Table 8-2.

EXAMPLE 8-3.   Calculate the normal boiling point (nbp) of ethanol, using the data given in Example 8-2.

$\Delta H_v = 9{,}220$ cal/mole, $P_1 = 400$ torr, $T_1 = 337\,°K$, and $P_2 = 760$, $T_2 = $ nbp.

$$\log \frac{P_2}{P_1} = \frac{-\Delta H_v}{2.303\,R}\left(\frac{1}{T_2} - \frac{1}{T_1}\right) = \log \frac{760}{400} = -\frac{9{,}220}{2.303 \times 1.987}\left(\frac{1}{T_2} - \frac{1}{337}\right)$$

$$\frac{1}{T_2} = -0.1386 \times 10^{-3} + 2.967 \times 10^{-3}$$

$$T_2 = 353\,°K$$

EXAMPLE 8-4. Methyl acetate has a vapor pressure of 400 torr at 40 °C and 40 torr at -8 °C. Calculate the heat of vaporization of this liquid.

$$\log \frac{400}{40} = -\frac{\Delta H_v}{2.303 \, R} \left(\frac{1}{313} - \frac{1}{265} \right)$$

$$1 = -\frac{\Delta H_v}{4.576} (3.195 - 3.774) \times 10^{-3}$$

$$\Delta H_v = 7,960 \text{ cal/mole}$$

There is an interesting empirical correlation between the normal boiling point and the heat of vaporization, known as *Trouton's rule*, that for nonpolar liquids $\Delta H_v \approx T_b$ (°K) $\times$ 21. Although this relation is quite approximate, heats of vaporization of nonassociated liquids estimated on this basis are useful first approximations in the absence of other data.

8-2-2 Solid–Vapor Equilibrium

The treatment of vapor pressure of solids is entirely analogous to that outlined for liquids. The Clapeyron equation retains the same form but with ΔS and ΔV now referring to the sublimation process. Similarly, the relation analogous to Eq. (8-9) is

$$\frac{dP}{dT} = \frac{\Delta H_{sub} P}{R T^2} \tag{8-12}$$

This is obtained by assuming ideal gas behavior in the vapor phase and that the molar volume of the solid is negligible with respect to that of the vapor. The *heat of sublimation* ΔH_{sub} is the heat required to vaporize 1 mole of solid at constant temperature and pressure. The Clausius-Clapeyron equation also is directly applicable, substituting ΔH_{sub} for the heat of vaporization.

8-2-3 Solid–Liquid Equilibrium

Processes involving the solid and liquid phases are the third type of physical change that must be considered to complete the description of two-phase equilibria. Again the approach is from the point of view of considering an equilibrium,

$$A(s) \rightleftharpoons A(l)$$

The free energy requirement for equilibrium again leads to the Clapeyron equation in the form

$$\frac{dP}{dT} = \frac{\Delta H_{fusion}}{T(V_l - V_s)} \tag{8-13}$$

The physical significance of this relation, however, is quite different from the

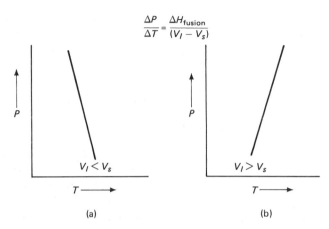

FIG. 8-5 Pressure dependence of the melting point according to the Clapeyron equation for (a) a decrease and (b) an increase, in molar volume on melting.

analogous expressions applicable to equilibria involving the vapor phase. The term dP/dT represents the reciprocal of the rate at which the melting point (the pressure and temperature at which solid and liquid exist in equilibrium) is altered by a change in the applied pressure. The graph of melting point vs. pressure has virtually negligible curvature, and for practical purposes it can be regarded as a straight line. Equation (8-13) thus can be written, using the Δ notation, as

$$\frac{\Delta P}{\Delta T} = \frac{\Delta H_{fusion}}{T\,\Delta V_{fusion}} \qquad (8\text{-}14)$$

Since in general the change in volume attending this phase change is quite small, pressure has only a minor effect on the melting point. For most substances the molar volume of the liquid is greater than that of the solid and a positive slope of the pressure-temperature plot is obtained. For water the opposite is true. The two possibilities are illustrated in Fig. 8-5. It should be appreciated that in this graph, as in most phase equilibrium diagrams covering moderate pressure ranges, the slope of the melting point curve is very much exaggerated for purposes of illustration. In using Eq. (8-14) in calculations, the value of ΔH_{fusion} must be converted from the usual calories per mole to liter atmospheres per mole to obtain a value for $\Delta P/\Delta T$ in useful units, atmospheres per degree.

8-2-4 Properties of Gases and Liquids Near the Critical Point

An interesting feature of equilibria between liquid and vapor phases meriting special attention is the behavior at high temperatures and pressures.

If a system consisting of a substance entirely in the vapor phase is subjected to increasing pressure, a state eventually will be attained where condensation to liquid begins to occur. Such transformations will not invariably occur, however, since for every substance there exists a specific temperature, the *critical temperature* T_c, above which liquefaction will not occur, regardless of the extent to which the pressure is raised. This behavior of the system can be understood best by considering the events taking place when a substance, liquid at room temperature, is heated in a closed container. As the temperature rises the vapor pressure increases in the general manner indicated in Fig. 8-2. This results in an increase in the density of the substance in the vapor phase. At the same time the liquid exhibits the usual thermal expansion with an accompanying decrease in density. The densities of the two phases approach one another, until at T_c they are the same and we no longer can make a meaningful physical distinction between the two phases. The vapor pressure of the substance at this point is the *critical pressure* P_c. This also represents the applied pressure required to liquefy the substance at T_c. Above the critical temperature, where the system exists only as a single phase, the substance frequently is referred to as a *fluid*, since it can no longer validly be considered as either a liquid or a vapor. Values of the critical temperature and pressure for some liquids and gases are given in Table 8-1.

The behavior of substances near the *critical point* can be discussed most conveniently in terms of a pressure-volume diagram such as that given in Fig. 8-6, for isopentane. The curves, known as *isotherms*, represent series of

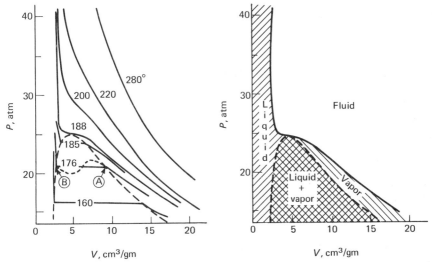

FIG. 8-6 Pressure-volume curves for isopentane. Adapted from W. H. Hamill, R. R. Williams, Jr., and C. MacKay, Principles of Physical Chemistry, 2nd Ed., Prentice-Hall Inc, Englewood Cliffs, N.J., 1966, p. 16.

pressure–volume points for the system at the indicated values of the constant temperature. At temperatures above T_c the fluid is described with varying accuracy by Boyle's law-type curves, such as those drawn for the isopentane system at $T = 280$ and $230\,°C$. As the critical temperature is approached, the curves become increasingly distorted. For the isotherm at $187.8\,°C$, T_c, known as the *critical isotherm* there is an inflection point at the critical pressure. If we consider isopentane as being compressed in the fluid state along this isotherm, with the pressure increasing as shown, then liquefaction occurs when P_c is attained. Along isotherms at lower temperatures the vapor phase follows Boyle's law approximately until condensation begins, for example, at point A on the isotherm at $176\,°C$. Further decreases in volume occur at constant pressure until liquefaction is complete at point B. The isotherm then describes the behavior of the liquid phase, and the incompressibility of the liquid is manifested in the very sharply rising curve at smaller volumes. When the two points on each isotherm, corresponding to the first appearance of liquid and complete condensation, are connected, the result is the dome-shaped curve shown with dashed lines in the figure. PV points within the region represent the system under conditions where the liquid and vapor can coexist. Similar deductions for the physical state of the system when its PVT points fall in other areas of the diagram are summarized in Fig. 8-6b.

The dotted line associated with the isotherm at $176\,°C$ in Fig. 8-6a shows the isotherm predicted by the van der Waals' equation. No relation predicts exactly the plateau region actually observed, so that for an approximation of the behavior of a system that might be expected near the critical point, the van der Waals' relation is quite useful.

8-3 EQUILIBRIA BETWEEN THREE PHASES

For a substance in the solid or liquid phase, as the temperature is increased the vapor pressure is increased. As indicated previously, this variation can be presented as a straight line graph by plotting $\log P$ vs. $1/T$. For each substance there exists a characteristic temperature at which the vapor pressure of both the solid and liquid phases are identical. This is known as the *triple point*, and the temperature and corresponding vapor pressure are designated as T_{tp} and P_{tp}. The value of these constants for a particular system can be obtained by considering simultaneously the temperature dependence of the vapor pressure of both phases. Thus a Clausius-Clapeyron $\log P$ vs. $1/T$ plot gives two straight lines that intersect when $P_{solid} = P_{liquid}$. This is illustrated in Fig. 8-7.

The triple point can be determined by this graphical method or by the simultaneous solution of two Clausius-Clapeyron equations written for the

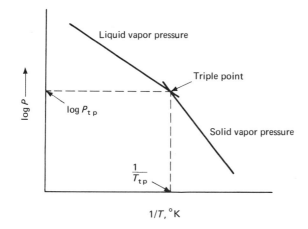

FIG. 8-7 Graphical determination of the triple point from vapor pressure plots for solid and liquid phases.

solid and liquid phase. If ΔH_{vap} and ΔH_{sub} for the substance are known, together with the vapor pressure at one temperature for each phase, the two unknowns that satisfy both equations are T_{tp} and P_{tp}.

The triple point for any substance is an extremely useful reference state that allows correlation of the separate graphs for the various two-phase equilibria discussed previously into single pressure-temperature diagrams, such as the one shown in Fig. 8-8.

In the figure the point B is the triple point for the system. AB is the vapor pressure of the solid with the slope at any point being given by $\Delta H_{\text{sub}} P / RT^2$. Similarly BC, with the slope at any point equal to $\Delta H_v P / RT^2$, is the vapor pressure of the liquid. The upper limit of this curve is determined by the critical temperature. BD represents the vapor pressure of the *supercooled* liquid, where the system is in an unstable state as a liquid below its usual freezing point. The line BE is the melting point curve, at whose pressures and temperatures the solid and liquid can exist in equilibrium.

In addition to regarding the diagram in this way as a method of representing the pressure and temperature characterizing the various phase equilibria for a particular substance, the figure also can be employed to establish the phase of a system under various conditions. Thus, if the point representing the pressure and temperature of the system lies to the left of ABE, the system is in the solid state. Points within the area bounded by EBC are characteristic of the liquid phase, while for those to the right of ABC, the system is vapor.

In terms of this approach a variety of commonly observed phase changes can be interpreted. If the system is in the solid phase at a temperature and pressure represented by the initial point on line 1 in the figure, heating

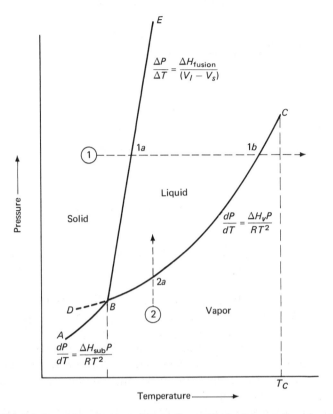

FIG. 8-8 Pressure-temperature diagram for a typical substance. Lines 1 and 2 represent respectively isobaric heating and isothermal compression.

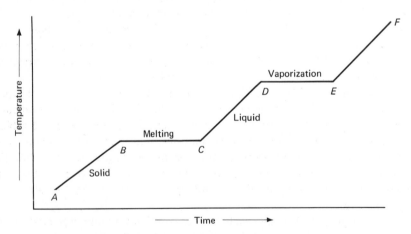

FIG. 8-9 Schematic heating curve for a typical substance. The heat input in cal/min is maintained constant.

at constant pressure is indicated by this horizontal path in the figure. When the line crosses *EB*, point 1a, the solid will melt and the temperature will remain constant until the fusion process is complete. Further heating increases the temperature of the liquid until the curve *CB* is crossed, point 1b, where boiling occurs. Again the temperature is constant during this phase change. Further heating increases the temperature of the vapor phase. If the *PT* point representing the system were at a pressure below P_{t_p}, isobaric heating would result in warming of the solid and sublimation (when the curve for the vapor pressure of the solid is crossed) followed by vapor phase heating.

Vertical line 2 represents a typical isothermal compression. The *PT* points move vertically as pressure is increased until *BC* is crossed where condensation to the liquid phase occurs. This is the process represented in Fig. 8-6a by any of the isotherms below the critical temperature.

The isobaric heating process discussed above also presents an opportunity to indicate how thermochemical techniques may be applied to calculate the heat effects associated with the sequence of phase changes. A graph of temperature vs. time for a typical system, initially in the solid state, being supplied with some number of calories per unit time at constant pressure is shown in Fig. 8-9. For the heating of the solid the number of calories absorbed is calculated from Eq. (1-6),

$$q = C_p(T_2 - T_1) \tag{1-6}$$

providing the heat capacity at constant pressure is constant over the temperature range involved. Heat absorbed in the melting process is simply the product of the heat of fusion and the number of moles present. Equation (1-7), written for the temperature limits appropriate to this system,

$$q = \int_{T_C}^{T_D} C_p \, dT \tag{1-7'}$$

almost invariably will be required for heating of the liquid along *CD*, since the heat capacity will vary appreciably over the range between the melting and boiling points. Vaporization, line *BE*, is treated in the same way as fusion utilizing ΔH_v. Heating of the vapor phase is treated using Eq. (1-6) or (1-7), as appropriate to the behavior of C_p of the gas between the boiling point and terminal temperature.

PROBLEMS

8-1. Assuming ideal behavior, calculate the number of gas molecules in a 3.5-liter vessel at 317 °K in which the pressure is 1.00 μ (1 μ = 1 × 10^{-3} torr).

8-2. Using the van der Waals' equation, determine the temperature at which 1.75 mole of NH_3 would exert a pressure of 9.50 atm when confined to a 5.00-liter flask.

8-3. Calculate the pressure exerted in a 10.0-liter cylinder by 1.00 mole of water vapor when the temperature is 200, 300, 500, and 750 °C, using the van der Waals' equation and the ideal gas law. Plot P vs. $1/T$ for both sets of data on the same graph.

8-4. From the following vapor pressure data for water, determine graphically the heat of vaporization:

T (°C)	20	30	40	50	60	75
P (torr)	17.54	31.83	55.32	92.51	149.4	289.1

8-5. Neglecting the effect of pressure on the melting point, calculate the maximum applied pressure at which naphthalene would sublime. $\Delta H_{fusion} = 35.6$ cal/gm, $\Delta H_v = 75.5$ cal/gm, nfp $= 79.9$ °C, and vapor pressure of solid naphthalene $= 1.00$ torr at 52.6 °C.

8-6. Use the following data to *estimate* the critical pressure of cyclohexane: $\Delta H_{vap} = 7,190$ cal/mole, nbp 80.7 °C, and $T_c = 281$ °C.

8-7. Using the data in Table 8-2, calculate the vapor pressure of ethylene glycol $[(CH_2OH)_2]$ at 125 °C.

8-8. Test the applicability of Trouton's rule to benzene, carbon tetrachloride, ethanol, water, and ethyl chloride. Explain the relative magnitudes of the deviations. What is the implication of Trouton's rule with respect to the entropy of vaporization of nonassociated liquids?

8-9. Calculate the volume change on melting of an organic solid (mp, 125 °C and $\Delta H_{fusion} = 1.50$ kcal/mole), whose melting point is increased by 0.47 °C when the applied pressure is increased from 1 to 175 atm.

8-10. Calculate the triple point pressure and temperature for *t*-butyl alcohol, making use of the following data: $\Delta H_{vap} = 130.5$ cal/gm, nbp $= 98.1$ °C, and vapor pressure of the solid $= 1.00$ torr at -20.4 °C and 40 torr at 24.5 °C.

Properties of Solutions

9

The physical behavior of systems of two or more components is of widespread interest. Such systems can be classified as *solutions*, interpreting this term in its broadest sense, if they consist of a homogeneous mixture of substances.

A large proportion of chemical processes occur in solution. Previously various aspects of solutions, particularly the properties of ions in aqueous media, have been discussed. In this chapter the general principles of thermodynamics are applied to explain qualitatively and describe quantitatively the physical properties of solutions and their behavior. The main emphasis is on solid-liquid and liquid-liquid solutions.

For a two-component (*binary*) system, the conventional nomenclature has the *solute* as the component in the lesser amount and the *solvent* in greater amount. These are the most convenient designations for solid-in-liquid solutions. Homogeneous liquid mixtures, however, frequently are examined over the complete range of concentrations. In such cases it is more suitable to dispense with the solvent-solute nomenclature and designate the components simply as A, B, etc.

The *solubility* is the maximum concentration of solute that can exist in equilibrium with the undissolved solute at a particular temperature. For slightly soluble salts this concentration is determined, as indicated previously, by the value of the solubility product K_{sp}.

Most of the conventional methods of expressing solution concentrations have been mentioned previously, but it is useful to review them at this point. If n_A represents the number of moles of solute (or simply of component A), the concentration can be expressed as

Molarity: n_A per liter of solution

Molality: n_A per 1,000 gm of solvent

Mole fraction: n_A/n_{tot}

Molality and mole fraction have the advantage of being based on the weight of the substances present and therefore are independent of temperature. The other commonly used measure is normality, the number of gram equivalent weights per liter of solution. The use of normalities was discussed in Chapter 4.

9-1 RAOULT'S LAW

Physical processes in solution are most conveniently introduced by considering the situation where the system behaves ideally. In the same way that "ideal gas behavior" is defined by indicating that the gas must obey the ideal gas law, the criterion for an ideal solution can be stated as follows:

IDEAL SOLUTION: A HOMOGENEOUS MIXTURE IN WHICH THE COMPONENTS EXPERIENCE NO MODIFICATION OTHER THAN THAT OF DILUTION. THEY FOLLOW RAOULT'S LAW OVER THE ENTIRE CONCENTRATION RANGE.

Raoult's law is an important relation. It forms the basis for the treatment of binary liquid mixtures as well as for solutions containing nonvolatile solutes and their colligative properties. In its elementary form the law states that the partial vapor pressure of a solution component is directly proportional to its mole fraction. The proportionality factor is the vapor pressure of the pure component:

$$P_A = P_A^\circ X_A \tag{9-1}$$

In an ideal solution the pressure resulting from any component varies linearly from the vapor pressure of the pure material where $X_A = 1$ to zero when $X_A = 0$—in the absence of that component. Nonideality in a real solution is manifested by a deviation of the vapor pressure curve from the linearity predicted by Raoult's law.

Equation (9-1) is a direct result of the requirement that the molar free energy in the vapor and condensed phases be the same at equilibrium. If the expression

$$G = G^\circ + 2.303 \, RT \log a \tag{2-34}$$

is written for a substance at two different activities, the change in free energy associated with the change from a_1 to a_2 is given by

$$\Delta G = 2.303 \, RT \log \frac{a_2}{a_1} \tag{9-2}$$

The activity in the condensed phase can be represented as the mole fraction and that in the vapor phase, as pressure. To maintain equilibrium when the mole fraction of the component changes from X_1 to X_2 in the liquid phase, the ΔG given by Eq. (9-2) must be identical with the change in free energy in the vapor phase:

$$\Delta G_{\text{liquid}} = 2.303 \, RT \log \frac{X_2}{X_1} \equiv \Delta G_{\text{vapor}} = 2.303 \, RT \log \frac{P_2}{P_1}$$

$$\frac{X_2}{X_1} = \frac{P_2}{P_1}$$

If X_1 is unity, $P_1 = P°$, the vapor pressure of the pure substance. Making these substitutions gives

$$X_2/1 = P_2/P°$$

$$P_2 = P° X_2 \tag{9-1'}$$

9-2 BINARY LIQUID MIXTURES

9-2-1 Ideal Solutions

Raoult's law can be applied initially to the case of two completely miscible liquids which behave ideally over the entire concentration range. A great deal of useful information about such systems, and about those showing only minor departures from ideal behavior, can be obtained from this treatment.

Designating the components as A and B, we write the relation between their mole fractions as

$$X_A + X_B = 1 \tag{9-3}$$

From Dalton's law of partial pressure we have that

$$P_{\text{soln}} = P_A + P_B \tag{9-4}$$

and applying Raoult's law to each component, we get

$$P_{\text{soln}} = P_A° X_A + P_B° X_B \tag{9-5}$$

The most convenient graphical representation of the system is a plot of vapor pressure of the solution as a function of mole fraction of one component. An ideal liquid solution whose vapor pressure is described by Eq. (9-5) at some fixed temperature is shown in Fig. 9-1.

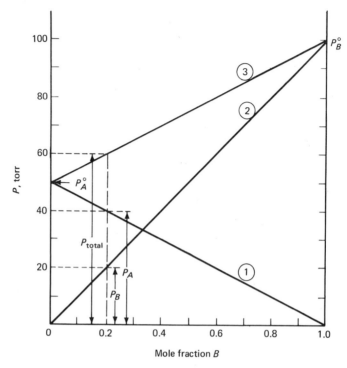

FIG. 9-1 *Total and component vapor pressures for an* ideal binary liquid mixture.

The pressures are plotted as a function of mole fraction of component B. On the left, where $X_B = 0$, the system consists entirely of component A, while the right-hand limit represents pure B. Line 1 is the partial pressure of A, originating at zero at $X_A = 0$ and rising to P_A° at $X_A = 1.0$. Similarly, the partial pressure of B, shown as line 2, begins at P_B° at $X_B = 1$ and decreases to zero. The plot representing the vapor pressure of the solution is obtained through either Eq. (9-5) or by simple addition of the pressure values from lines 1 and 2 at any particular mole fraction. This latter procedure is illustrated in the figure at $X_B = 0.20$.

Line 3 is the *liquid composition curve* and gives the solution vapor pressures as a function of mole fraction of one component in the liquid phase. It also is of interest to examine not only the total pressure but the composition of the vapor as well. To avoid confusion between liquid and vapor composition, the designation X_B will be reserved for the mole fraction in the liquid phase, and Y_B will be employed for the vapor phase mole fraction of component B. The number of moles of each component in the vapor phase can be determined from its pressure, if we assume ideal gas behavior, in the following way:

$$Y_B = \frac{n_B}{n_A + n_B} = \frac{P_B V/RT}{P_A V/RT + P_B V/RT}$$

$$Y_B = \frac{P_B}{P_A + P_B} = \frac{P_B}{P_{soln}} \tag{9-6}$$

It is apparent qualitatively that the mole fraction of either component will not be the same in the liquid and vapor phases because of the differences in volatilities. Y_B and X_B, for example, are related quantitatively by an expression obtained by substitution from Raoult's law in Eq. (9-6) for P_B to obtain

$$Y_B = X_B \frac{P_B^\circ}{P_{soln}} \tag{9-7}$$

This equation reveals a fundamental property of the ideal binary liquid mixture—*the vapor is always richer than the liquid in the more volatile component.* In the system illustrated in Fig. 9-1, since B is the more volatile of the two liquids, P_{soln} is invariably less than P_B° and, thus, Eq. (9-7) shows that Y_B will be greater than X_B. Although Eq. (9-7) is required for numerical calculations, the principle can be demonstrated directly by reference to Fig. 9-1 at $X_B = 0.50$. From the graph, $P_B = 50$ torr and $P_A = 25$ torr and $P_{soln} = 75$ torr. Thus Y_B is $50/75 = 0.67$—as compared to the liquid phase mole fraction of 0.50.

To include the variation in vapor phase composition in a graphical presentation of the same form as Fig. 9-1, the curves for the partial pressures of the individual components can be removed for clarity. For every point giving solution pressure at a particular liquid composition, there is a second point, shifted toward a higher concentration of the more volatile component, representing the vapor composition. A series of such points can be calculated using Eq. (9-7) to construct the second curve, the *vapor composition curve*. This gives the variation in solution vapor pressure with mole fraction in the vapor phase. For the system being used as an example, $P_A^\circ = 50$ torr and $P_B^\circ = 100$ torr, the liquid and vapor composition curves are shown in Fig. 9-2.

To obtain the vapor composition from this graph, the mole fraction of B plotted along the x axis is interpreted as Y_B. From this type of representation we can obtain all the required information about the system. For example, in the binary liquid mixture described in the figure, for an equimolar mixture of A and B, $X_B = 0.5$, we proceed first along line 1 to get the value of the solution vapor pressure, 75 torr. The horizontal line 2 connects the liquid and vapor compositions in equilibrium when $P_{soln} = 75$ torr, and the value of $Y_B = 0.67$ is obtained by following line 3 to the mole fraction axis.

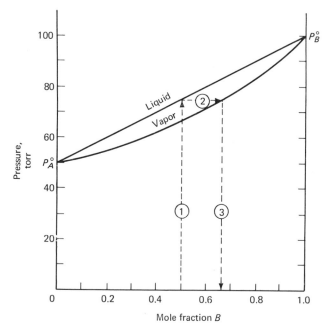

FIG. 9-2 *Vapor pressure diagram for an ideal binary liquid mixture giving the solution vapor pressure as a function of composition of the liquid and vapor phase.*

EXAMPLE 9-1.   At a temperature where the vapor pressure of pure benzene is 100 torr, the vapor pressure above a mixture of liquid benzene and toluene was analyzed as 20 torr pressure of benzene and 16.5 torr pressure of toluene. Assuming ideal behavior, calculate the vapor pressure of pure toluene at this temperature.

$$P_{\text{soln}} = P_{\text{B}} + P_{\text{T}} = 20.0 + 16.5 = 36.5 \text{ torr} \qquad P_{\text{B}} = P_{\text{B}}^{\circ} X_{\text{B}}$$

$$X_{\text{B}} = 20/100 = 0.20 \qquad X_{\text{T}} = 1 - X_{\text{B}} = 1 - 0.20 = 0.80$$

$$P_{\text{T}} = P_{\text{T}}^{\circ} X_{\text{T}} \qquad P_{\text{T}}^{\circ} = P_{\text{T}}/X_{\text{T}} = 16.5/0.80 = \underline{20.6 \text{ torr}}$$

9-2-2 Deviations from Raoult's Law

Raoult's law adequately describes a liquid mixture system in which there are no appreciable interactions between the components that alter their activity coefficients significantly. We have seen that the direct proportionality of vapor pressure to mole fraction is based directly on the free energy change, defined by Eq. (9-2), being identical for the liquid and vapor phases. Although the assumption of ideal gas behavior may be incorrect, and therefore the activity and vapor pressure may not be equivalent, the difficulties arise more often because of the inaccuracy of using mole fraction in the liquid

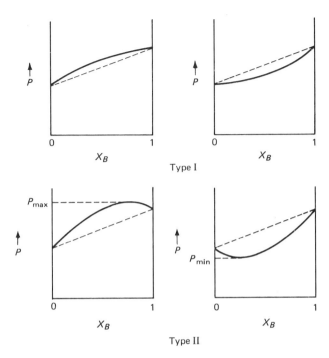

Type I

Type II

FIG. 9-3 *Deviations from Raoult's law exhibited by real binary liquid mixtures. In each case the dashed line represents ideal behavior.*

phase for activity. If intermolecular attractive forces exist between the components, then the arguments cited in the derivation of Raoult's law are no longer valid and the curve of solution vapor pressure vs. mole fraction will deviate at least to some extent from the theoretical straight line. The types of deviation observed are illustrated schematically in Fig. 9-3.

In *Type I* deviations, the solution vapor pressure is greater or less than that predicted by Raoult's law but is intermediate between that of the pure components over the entire concentration range. In *Type II* there is a large deviation from the ideal solution behavior. Over some range of concentrations, the vapor pressure of the solution exceeds that of the more volatile component and exhibits a maximum (P_{max} in the figure) at some characteristic composition. Large deviations also can be in the opposite direction, with the solution exhibiting minimum a in the vapor pressure curve. As indicated in the figure, over some range of X_B values, P_{soln} is lower than $P°$ of the less volatile component.

An important property of systems exhibiting Type II deviations is that at the minimum or maximum in the vapor pressure, the composition of the liquid and vapor phases is the same. Using a Type II case as an example,

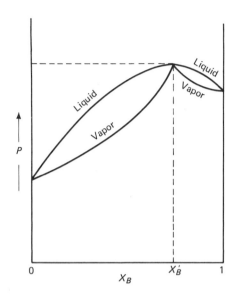

FIG. 9-4 A binary liquid mixture exhibiting a Type II deviation from Raoult's law.

Fig. 9-4 shows both the liquid and vapor composition curves. These latter graphs are, of course, experimental determinations, since Raoult's law is inapplicable to this nonideal case. The liquid composition giving rise to P_{max}, X'_B in the figure, is known as an *azeotrope* or *constant boiling mixture*. This composition represents an invariant system. As will be discussed presently, no fractionation occurs in vaporization since both phases have the same proportion of the two components.

9-2-3 Fractional Distillation

One important application of the treatment of binary liquid mixtures that follow Raoult's law exactly, or exhibit only deviations of Type I, is to the useful technique of fractional distillation. This procedure takes advantage of the increase in concentration of the more volatile component in going from the liquid to the vapor phase. The method essentially involves a series of steps in each of which the vapor is "separated" and condensed to a liquid. In the sequence the vapor becomes progressively more enriched in the more volatile component. In the residual liquid there is a concomitant increase in the proportion of the component of lower volatility. With a sufficient number of successive vaporization-condensation procedures the mixture thus can be separated for practical purposes completely. The boiling point of the various liquid fractions involved will be different because of the changing concentrations.

The pressure-composition diagrams discussed so far all refer to constant temperature. For the discussion of the distillation process it is more convenient to represent the system in terms of a boiling-point composition diagram. This takes the same form as the graphs already presented, except that it is reversed with respect to the two components of different volatility. Thus the more volatile member of the pair will exhibit a lower boiling point. The boiling point curve will rise from this temperature to the boiling point of the less volatile substance. Points representing the vapor composition are shifted toward the more volatile component.

As an example, consider a binary liquid mixture that exhibits a positive (in terms of vapor pressure) Type I deviation from Raoult's law and for which the liquid composition–boiling point curve is *below* the ideal linear plot. The liquid and vapor composition curves are shown for such a system in Fig. 9-5. Here B is the more volatile component. Thus the boiling point of the solution decreases as the mole fraction of B is increased. If the distillation is begun with liquid having composition X_B, as indicated in the figure, the boiling point is T_B and the vapor composition, obtained by following lines 1, 2, and 3, is Y_B. Condensation of this vapor gives a liquid of composition X'_B with the

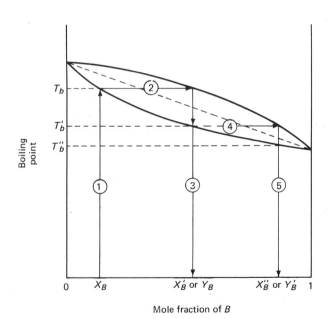

Mole fraction of *B*

FIG. 9-5 *Boiling point–mole fraction diagram for a binary liquid mixture exhibiting Type I positive deviation from Raoult's Law. The dashed line is the b.p. curve for an ideal mixture. The mole fractions associated with a sequence of vaporization-condensation steps in a fractional distillation are indicated by the numbered lines.*

boiling point reduced to T'_B. The vapor that would be in equilibrium with this liquid has composition Y'_B (lines 3, 4, and 5) and gives on condensation a second liquid distillate with composition X''_B and boiling point T''_B. Repetition of this procedure evidently results in a rather rapid increase in the concentration of the more volatile component B. The last distillate has a vapor virtually free from component A, and when condensed it yields pure B. After each vapor "removal" the residual liquid has a higher concentration of A and a higher boiling point. An actual fractional distillation requires very many more steps than indicated in this simple diagram. The procedure outlined involves essentially discarding the liquid residue at each step. In practice these fractions are continually returned to the main bulk of liquid. Thus a whole series of separations, such as the single sequence shown in the figure, are occurring simultaneously during the distillation.

There are a variety of types of distillation column. They have in common the general feature of providing in the vertical column above the main liquid reservoir devices that permit the condensation and vaporization cycles to occur. There is, therefore, more or less continuous redistillation throughout the length of the column. A vertical temperature profile develops as the vapor becomes richer in the component with the lower boiling point. The vapor forms a condensate of reduced boiling point while the residual liquid, now richer in the material of lower volatility, has a higher boiling point and will flow to the lower portions of the column before being vaporized.

The final point to be made here is to reemphasize that mixtures showing Type II deviations cannot be separated completely by this technique. Solutions can be fractionally distilled until they reach the azeotropic composition but not beyond. As this characteristic mole fraction is approached, the compositions of the liquid and vapor phase come closer and closer together until at X'_B in Fig. 9-4 they are identical. No further separation can be obtained since the vapor that is "removed and condensed" at this point has exactly the same proportion of the two components as the parent liquid. The other direct result of the absence of fractionation is that the boiling point of the azeotrope is constant. Changes in composition across this point must be accomplished by chemical means.

able to mix.

9-2-4 Partially Miscible Liquids

The most interesting feature of liquid mixtures where the components have limited mutual solubility is the liquid-liquid equilibrium and its temperature dependence.

The behavior of a pair of partially miscible liquids follows a characteristic pattern resulting from the fact that over some temperature range each component is soluble in the other, but only to a limited extent. It is useful to examine first the effect of concentration that is conventionally given in weight percent for such systems. Consider a liquid system of components A and B at a temperature where the solubilities ate 10 percent by weight of A in B and 20 percent by weight of B in A. The physical significance of these limits becomes apparent on examination of what occurs when increasing amounts of component A are added to pure B. The component continues to dissolve so that the system remains a single phase until its concentration reaches 10 percent by weight, at which point a second phase begins to separate. This is entirely analogous to the separation of a solid solute from a solution in which the solubility has been exceeded. Two phases consisting of A saturated with B and B saturated with A are maintained on further additions of A until the system is 80 percent in A, whereafter further additions of A produce a single phase. At this point the opposite end of the scale has been reached, and the mixture can be considered as a solution of B in A which, as indicated above, has a maximum solubility of 20 percent. Thus, when sufficient A has been added, the concentration of B drops below the solubility limit at that temperature and the system goes over to a single phase.

For many systems, as the temperature is increased the mutual solubilities increase. For the example system, typical solubility limits at 60 °C might be 30 percent A in B and 45 percent B in A. As the temperature is increased further, these values converge until at some characteristic temperature, the *upper consolute temperature* or *critical solution temperature*, they are identical and the liquids are completely miscible. A series of experiments at different temperatures would produce a set of mutual solubility points such as shown in Fig. 9-6a. The regions below the critical solution temperature T' represent temperatures and compositions where the system is a single phase if the point lies outside the dome-shaped curve. Points within that area are temperatures and compositions where the system will exist as two phases. At any temperature above T', the system will be in a single phase

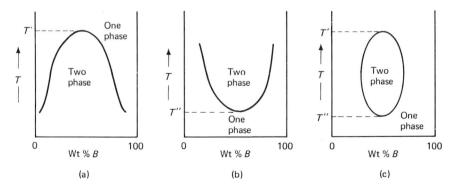

FIG. 9-6 Temperature-composition diagrams for typical partially miscible liquids.

regardless of composition. This type of behavior is exhibited by a number of systems, for example, phenol-water, n-butyl-alcohol–water. The figure also shows two other types of composition temperature behavior that have been observed. If the mutual solubility decreases with increasing temperature (Fig. 9-6b), a lower consolute temperature T'' is found (triethyl amine-water is the classic example), while for a very few systems, for example nicotine-water, both an upper and a lower consolute temperature are observed (Fig. 9-6c). The last type of behavior is usually of importance only at high pressures.

The partially miscible liquid cases discussed here represent the simplest example of two-component, real liquid systems. The phase diagrams and the general treatment become considerably more complex when, for example, boiling of the mixture occurs before the consolute temperature is attained.

9-2-5 Immiscible Liquids

This type of system can be regarded as the limit of partially miscible liquid pairs. Here the mutual solubility is so small that we have two phases always and each liquid exerts its own vapor pressure independent of the presence of the other. Thus a pressure–mole fraction diagram takes the form shown in Fig. 9-7. Raoult's law is totally inapplicable here, and the vapor pressure of the mixture is given simply by

$$P_{\text{mixture}} = P_A^\circ + P_B^\circ \tag{9-8}$$

The important implication of this behavior is that since P_{mixture} is larger than either of the pure components, the boiling point is lower than that of either A or B and will be independent of the composition. The practical application of the reduced boiling point is the technique of *steam distillation*.

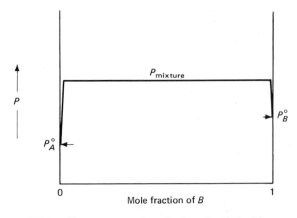

FIG. 9-7 Vapor pressure of a pair of immiscible liquids.

Compounds immiscible with water and unstable at high temperatures can be separated from impurities by the passage of steam through a mixture of the impure compound and water. The additivity of the vapor pressures and consequent depression of the boiling point permits the process to be carried out at a temperature where the material will not suffer thermal decomposition. The boiling point depends on the vapor pressures of the pure components but *not* on the composition. Not only does steam distillation provide this convenient separation method, but it also permits determination of the approximate molecular weight of the material being ·distilled. The calculation involves relating the weights of each component obtained from the distillation to the mole fraction in the vapor phase. Since the vapor pressure and molecular weight of water are known, the mole fraction data can be used to determine the unknown molecular weight. The necessary equation is obtained as follows:

The molecular weight of the compound being distilled, A, is M_A. It has a partial vapor pressure at the boiling point of the mixture of P_A° and a mole fraction in the vapor phase of Y_A. Assuming ideal gas behavior, $Y_A = P_A^\circ/P_{tot}$ and, for water, $Y_{H_2O} = P_{H_2O}^\circ/P_{tot}$. The ratio of the vapor pressures of the two components therefore is

$$\frac{P_{H_2O}^\circ}{P_A^\circ} = \frac{Y_{H_2O}P_{tot}}{Y_A P_{tot}}$$

The P_{tot} terms cancel, as do the terms for the total number of moles that appear when the mole fractions are written as $Y_i = n_i/n_{tot}$, so that

$$\frac{P_{H_2O}^\circ}{P_A^\circ} = \frac{\text{moles } H_2O}{\text{moles A}} = \frac{W_{H_2O}/M_{H_2O}}{W_A/M_A}$$

The mole ratio is that in the vapor phase, and therefore in the condensed distillate. For each component the number of moles is determined from the weight divided by the molecular weight. At the boiling point the total pressure is identical with P_{atm}, and thus from Eq. (9-8) the pressures in the system are related via

$$P_A^\circ = P_{atm} - P_{H_2O}^\circ$$

Substituting this relation for P_A°, along with the numerical value for M_{H_2O}, into the expression for the ratio $P_{H_2O}^\circ/P_A^\circ$, we get

$$\frac{P_{H_2O}^\circ}{P_{atm} - P_{H_2O}^\circ} = \frac{W_{H_2O}/18.02}{W_A/M_A}$$

or, more conveniently,

$$M_A = \frac{W_A}{W_{H_2O}} \frac{P_{H_2O}^\circ}{P_{atm} - P_{H_2O}^\circ} 18.02 \qquad (9\text{-}9)$$

The weight ratio of unknown compound to water in the distillate, the barometric pressure, and the vapor pressure of H_2O at the distillation temperature are the required experimental data. The molecular weight calculated from Eq. (9-9) will be approximate since it is assumed that the compound is quantitatively immiscible with water and that equilibrium conditions are maintained throughout.

EXAMPLE 9-2. Calculate the molecular weight of an organic unknown that steam distills at 90 °C at $P_{atm} = 755$ torr. The ratio of the weight of unknown to that of water in the distillate was 4.87. $P^\circ_{H_2O} = 526$ torr at 90 °C.

$$M_x = 4.87 \times \frac{526}{755 - 526} \times 18.02 = \frac{4.87 \times 526 \times 18.02}{229} = \underline{202}$$

9-3 SOLUTIONS OF GASES IN LIQUIDS

In homogeneous mixtures where one component is in relatively small concentrations, the solute-solvent system of nomenclature is most convenient. For the solvent, Raoult's law takes the form

$$P_{solvent} = P^\circ_{solvent} X_{solvent} \tag{9-10}$$

The law is applicable to real solutions only at low concentrations. This restriction is particularly important when the solution components are initially in different phases, for example, the case of gas-in-liquid solutions. The limitation arises from the fact that as the concentration increases, the mole fraction is an increasingly inaccurate measure of activity. Over the concentration range where Eq. (9-10) is applicable, $P_{solvent}$ is linearly dependent on concentration. The graph of $P_{solvent}$ vs. $X_{solvent}$ extrapolates to the vapor pressure of the pure solvent.

The situation with the solute is somewhat different. Being present in low concentrations, its activity in the condensed phase can be represented by X_{solute}. By the same arguments as outlined in Sec. 9-1, equilibrium is maintained only if $X_{solute,1}/X_{solute,2} = P_1/P_2$ when the mole fraction changes from X_1 to X_2. Unlike the solvent, however, it is not appropriate to take X_1 as unity and, therefore, $P_1 = P^\circ$. This would involve extrapolating solute behavior from a dilute to a more concentrated solution, ultimately to the pure *solute*, and would require the activity–mole fraction equivalence to hold over the entire range. This is true only for an ideal solution. However, over the range from infinite dilution to some concentration where mole fractions begin to deviate significantly from activities, the vapor pressure of the solute is a linear function of X_{solute}, but the proportionality factor evidently is not the pressure of the pure solvent.* The law that expresses this proportionality is known as _Henry's law;_

$$P_1 = KX_1 \qquad (9\text{-}11)$$

The law is another manifestation of the principle that equilibrium is maintained only by equivalent changes in free energy of each component in all parts of the system. Since in dilute solutions all measures of concentration are proportional to the mole fraction, the equation also can be written using molarity or molality. The numerical value of K will depend on the concentration units. Although Eq. (9-11) is applicable to any solute in dilute solution, it is used most often to describe solutions of gases in liquids.

Values of the Henry's law constant K for a few common gases are given in Table 9-1. The process of dissolution of a gas is usually exothermic, and therefore the solubility and hence K decreases with temperature. When the mole fraction is used as the concentration measure, K has the units of P.

TABLE 9-1.   HENRY'S LAW CONSTANTS FOR SOME GASES IN WATER[a]

$K \times 10^{-7}$, torr $(P = KX)$

Gas								
T($^\circ$C):	0	10	20	30	38	40	50	60
CO_2	0.0555	0.0788	0.108	0.139	0.168	0.173	0.217	0.258
H_2	4.42	4.82	5.20	5.51	5.72	5.78	5.82	5.80
N_2	4.07	4.87	5.75	6.68	7.51	7.60	8.20	8.70
O_2	1.91	2.48	2.95	3.52	4.04	4.14	4.50	4.84
N_2O	0.074	0.108	0.155	0.210	0.242	0.246	0.279	—
C_2H_2	0.056	0.072	0.090	0.112	0.131	0.133	—	—
C_2H_4	0.370	0.552	0.753	1.00	1.21	1.23	—	—

[a] Data from _Handbook of Chemistry and Physics_ (Cleveland: 42nd ed. Chemical Rubber Co., 1962, pp. 1708–1709). Reprinted with permission.

Henry's law can be applied in two ways: (1) to determine the mole fraction of a gas that will result when a particular pressure is applied to the solvent or, alternatively, (2) to determine the gas pressure above a solution that contains some known mole fraction of gas. The second interpretation gives rise to the biochemical practice of describing the concentration of dissolved gas in terms of its _tension_—the pressure of gas required to give the observed mole fraction of gas in solution.

The values of K are applicable only to situations where there is no chemical reaction between the gas and the solvent or other dissolved species. Any process of this kind can have a marked effect on the "solubility." For example, carbon dioxide, as discussed in Chapter 4, is involved in the bicarbonate–carbonate buffer system. Thus the total concentration of CO_2 in the

* The proportionality constant is the vapor pressure of the solute in a hypothetical state obtained by extrapolating the pressure in dilute solution to $X_{\text{solute}} = 1$.

bloodstream is greater than the ordinary solubility because a large proportion of the gas is present as the buffer pair.

An important physiological application of the principle involved in Henry's law is to the "bends" experienced by divers who surface rapidly from appreciable depths. During submersion, the air used for breathing is at high pressure to prevent lung collapse. According to Henry's law, the concentration of gas in solution in the blood and tissues will be considerably larger than at the surface and will increase with the depth of the dive. Thus, for example, the tension of nitrogen, which is especially dangerous because of its generally high solubility in tissue, will be considerably larger than at the surface. The problem is magnified in the case of N_2 by the fact that this gas dissolves slowly. The concentration thus builds up with increasing submersion time. There are additional physiological problems associated with high concentrations of nitrogen in the body during diving, and the gas is often replaced by helium in breathing mixtures. Regardless of the gas involved, however, there are serious consequences of a rapid ascent to the surface. If the diver does not come to the surface slowly enough, the gas tension decreases traumatically, and the dissolved gases come out of solution rapidly enough to form bubbles that can block small blood vessels. In addition, there can be a sudden expansion of air in the lungs, causing serious membrane damage and related effects. Decompression sickness is avoided by a gradual change in depth so that gas tensions are slowly reduced.

9-4 COLLIGATIVE PROPERTIES I—NONELECTROLYTE SOLUTIONS

Just as chemical equilibrium in molecular and ionic systems has been treated separately, it also is useful to examine the properties of solutions containing nonvolatile solutes under the same two headings. Again the same general principles apply to both cases, but the evaluation of concentration terms requires a somewhat different approach for each type of solute.

For a dilute solution the vapor pressure of the solvent depends on the *concentration* of solvent and, hence, on that of the solute. The determining factor, therefore, is not the nature of solute particles but rather their *number* in a given amount of solution. The distinction between molecular and ionic solutes is that in the latter case both the percent dissociation and the alterations in the effective number of particles because of interionic forces must be considered. These effects alter the concentration terms to be used in Raoult's law and the equations obtained from it.

Alterations in the properties of a liquid owing to the presence of a solute are directly related to the change in vapor pressure that accompanies the formation of a solution. This phenomenon is examined initially. The treatment of the other properties then can be based on the relations derived for the vapor pressure.

9-4-I   Vapor Pressure Lowering

When the solute is nonvolatile, the vapor pressure of the solution results entirely from the solvent. Raoult's law in the form of Eq. (9-1) thus becomes

$$P_{soln} = P_2^{\circ} X_2 \tag{9-12}$$

where the subscript 2 designates the solvent. When a solute is present, X_2 is less than unity and the vapor pressure is reduced. For convenience in dealing with dilute solutions, the concentration of the solution is expressed in terms of X_1, the mole fraction of solute. For a binary mixture, Eq. (9-3) applies so that

$$X_2 = 1 - X_1 \tag{9-3'}$$

Substituting this expression for X_2 into Eq. (9-12), we get

$$P_{soln} = P_2^{\circ}(1 - X_1)$$

$$P_2^{\circ} - P_{soln} = P_2^{\circ} X_1$$

The difference between the pressure exerted by the pure solvent and that of the solution is the *vapor pressure lowering* ΔP, given by the relation

$$\Delta P = P_2^{\circ} X_1 \tag{9-13}$$

In dilute solution mole fraction can be replaced by molality, but the proportionality factor is no longer P_2°. In this form Raoult's law for the vapor pressure lowering owing to a nonvolatile nonelectrolyte is

$$\Delta P = K_{VP} m \tag{9-14}$$

where K_{VP} is the *molal vapor pressure lowering constant* of the solvent. The expression for K_{VP} can be obtained directly by converting from mole fraction to molality as follows:

$$X_1 = \frac{n_1}{n_1 + n_2} \approx \frac{n_1}{n_2}$$

where the approximation is acceptable in dilute solution where the number of moles of solute may be considered to be negligible with respect to n_2. Equating each n with the weight in grams divided by the molecular weight gives

$$X_1 = \frac{W_1/M_1}{W_2} M_2$$

The ratio $(W_1/M_1)/W_2$ is the number of moles of solute per gram of solvent or $m/1000$. Making this substitution results in the following equation for the mole fraction of solute,

$$X_1 = \frac{m M_2}{1{,}000} \tag{9-15}$$

Using this expression for X_1 in Eq. (9-13), we get

$$\Delta P = \frac{M_2 P_2^\circ}{1,000} m \qquad (9\text{-}16)$$

Comparison of Eqs. (9-14) and (9-16) shows that K_{VP} is given by

$$K_{VP} = \frac{M_2 P_2^\circ}{1,000} \qquad (9\text{-}17)$$

In using these equations we must keep in mind that since vapor pressure is temperature dependent, the temperature at which a tabulated K_{VP} is applicable must be known. Furthermore, the equation is restricted in general to dilute solutions, as required by the several assumptions involved in modifying Raoult's law to the particularly convenient from of Eq. (9-14).

EXAMPLE 9-3.   Calculate the molal vapor pressure lowering constant for ethyl alcohol at 35 °C where the vapor pressure is 100 torr. Hence, calculate the vapor pressure of a solution of 12.0 gm of a solute of 300 mol. wt. in 100 gm of C_2H_5OH at this temperature.

$$K_{VP} = \frac{M_2 P_2^\circ}{1,000} = \frac{46.1 \times 100}{1,000} = 4.61 \text{ torr/molal solution}$$

$$m = \frac{12.0}{300} \times \frac{1,000}{100} = 0.40$$

$$\Delta P = K_{VP} m = 4.61 \times 0.40 = 1.85 \text{ torr}$$

$$P_{C_2H_5OH} = 100 - 1.9 = \underline{98.1 \text{ torr}}$$

9-4-2 Boiling Point Elevation

The most direct effect of lowering of the vapor pressure is that the temperature required to have $P_{soln} = P_{atm}$ at a particular atmospheric pressure must be higher than the value for the pure solvent. The situation is illustrated graphically in Fig. 9-8.

At some applied pressure P°, the boiling point elevation ΔT_b can be calculated by utilizing Raoult's law and the Clausius-Clapeyron equation. At temperature T_b, the boiling point of the pure solvent, its pressure is P°. The solution will have pressure P at this temperature but will boil only when its pressure has been raised to P°, that is, at the elevated boiling point T_b'. The two vapor pressures and temperatures are related by the Clausius–Clapeyron equation, which written explicitly for this situation is

$$\ln \frac{P^\circ}{P} = \left(\frac{\Delta H_v}{R}\right) \frac{T_b' - T_b}{T_b T_b'}$$

The term $T_b' - T_b$ is ΔT_b, and since the boiling point will not be changed

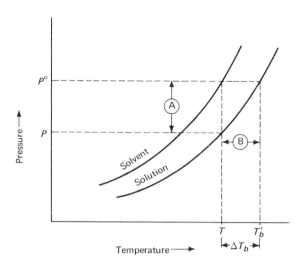

FIG. 9-8 *The boiling point elevation. The graph is an enlargement of a relatively small portion of the general pressure-temperature diagram for the system in question.*

by a very large amount, $T_b T_b' \approx (T_b)^2$. Retaining the natural logarithm in the expression (rather than converting to common logs), and inverting for convenience in making subsequent manipulations, we obtain

$$-\ln \frac{P}{P^\circ} = \frac{\Delta H_v \Delta T_b}{R(T_b)^2}$$

The final step is to utilize Raoult's law to express the pressure ratio terms of solute mole fraction. Here the law can be written as

$$\frac{P}{P^\circ} = (1 - X_1)$$

and natural logarithms can be taken to give

$$-\ln \frac{P}{P^\circ} = -\ln(1 - X_1) = X_1$$

The last equality is obtained by expressing the logarithmic term as a series wherein the higher terms can be neglected.

Replacing the $-\ln(P/P^\circ)$ term in the modified Clausius-Clapeyron equation by X_1, we get

$$X_1 = \frac{\Delta H_v \Delta T_b}{R(T_b)^2}$$

TABLE 9-2. K_B AND K_F VALUES FOR SOME COMMON SOLVENTS[a]

Solvent	T_b	K_B, °C/unit m	T_f	K_F, °C/unit m
Acetone	56.2	1.71	—	—
Benzene	80.1	2.53	5.45	4.90
Bromobenzene	155	6.26	—	—
Nitrobenzene	210.8	5.24	5.7	7.0
Diphenyl	—	—	70	8.0
Ethylene dibromide	—	—	10.1	11.8
Naphthalene	—	—	80.22	6.8
Carbon tetrachloride	76.75	5.03	—	—
Water	100	0.512	0.0	1.85

[a] Applicable at the normal freezing and boiling point of the solvent. Data from *Handbook of Chemistry and Physics*, 45th ed. (Cleveland: Chemical Rubber Co., 1964 p. D88). Reprinted with permission.

or

$$\Delta T_b = \frac{RT_b^2}{\Delta H_v} X_1 \tag{9-18}$$

For convenience, but again only in dilute solution, mole fraction may be replaced by molality, utilizing the conversion relation obtained above, Eq. (9-15), to obtain

$$\Delta T_b = \left(\frac{RT_b^2}{\Delta H_v} \frac{M_2}{1,000} \right) m \tag{9-19}$$

The terms in parentheses are known as the *molal boiling point elevation constant* K_B. This quantity gives the boiling point elevation exhibited by a 1 molal solution. If K_B is known or can be calculated at a temperature of interest, the working equation is just

$$\Delta T_b = K_B m$$

It is apparent that the applicability of the equation is strictly limited to dilute solutions, since otherwise virtually all of the assumptions made are invalid.

In terms of vapor pressure diagram of Fig. 9-9, the derivation of this equation amounts to using Raoult's law to determine the distance A and then the Clausius-Clapeyron equation to move along the vapor pressure curve a distance B on the temperature axis. Some values for K_B for a few common solvents are listed in Table 9-2.

9-4-3 Freezing Point Depression

As the temperature of a solution is decreased we move into the region of the pressure-temperature diagram around the triple point. The relevant vapor pressure and solid-liquid equilibrium curves for an aqueous solution are shown in Fig. 9-9. The lowered vapor pressure curve for the solution intersects the curve giving the vapor pressure of the solid solvent at a lower temperature than the same curve for the pure solvent. Thus the triple point is lowered from T_{tp} to T'_{tp}, and the liquid-solid equilibrium line is shifted to the left, so that at any particular applied pressure the freezing point is lowered from T_f to T'_f. Assuming that it is solid solvent that separates out at the freezing point, that is, that a solid solution is not formed, the solid vapor pressure curve remains unaltered. Similarly, the new solid-liquid curve should rise from the reduced triple point temperature and pressure with the same slope. Thus to a very good approximation the magnitude of the two temperature differences $\Delta T_f (= T_f - T'_f)$ and $\Delta T_{tp} (= T_{tp} - T'_{tp})$ will be the same. Therefore, the freezing point depression can be calculated from the alteration of the triple point temperature, using the same approach as in the derivation of the expression for ΔT_b. Here, however, the Clausius-Clapeyron equation is applied twice: using it to obtain vapor pressure differences for the

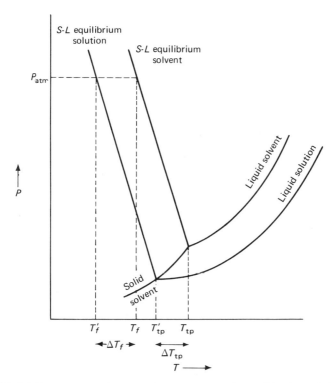

FIG. 9-9 *Freezing point depression for an aqueous solution. The slope of the solid-liquid equilibrium line has been greatly magnified, compared to the vapor pressure curves, to clarify the diagram.*

solid and liquid phases. Raoult's law gives the separation of the two liquid vapor pressure curves on the vertical pressure axis. The two Clausius-Clapeyron equations—one for the solid using ΔH_{sub} and one for the liquid with ΔH_v—are solved simultaneously to obtain the pressure and temperature at which the solid and liquid phase have the same vapor pressure. The resulting equation, stated in terms of molalities using Eq. (9-15), is

$$\Delta T_f = \left(\frac{RT_f^2 M_2}{\Delta H_{fusion}} \right) m \tag{9-21}$$

where M_2 is the molecular weight of the solute, T is the freezing point of the pure solvent, and ΔH_{fusion} replaces the term $(\Delta H_{sub} - \Delta H_{vap})$ from the solution of the two Clausius-Clapeyron equations. In Eq. (9-21) the collection of terms in parentheses is known as the *molal freezing point* depression constant K_F. With this substitution the equation becomes simply

$$\Delta T_f = K_F m \tag{9-22}$$

K_F values for a few solvents have been included in Table 9-2.

Freezing point depression measurements are widely used as a method of determining molecular weights. The values obtained are only approximate since there are usually small but finite deviations from Raoult's law, even in relatively dilute solutions, and the additional assumptions in the derivation of the ΔT_f equation may not be strictly valid. As is evident from an examination of Table 9-2, the value of K_F is larger than that of K_B. Thus the freezing point determination is preferable to boiling point measurement since a solution of given molality will give a larger ΔT in the former measurement.

EXAMPLE 9-4. Calculate the approximate molecular weight of a solid non-electrolyte solute that gives a freezing point depression of 0.210 °C when 3.0 gm is dissolved in 100 gm of *t*-butyl alcohol ($K_F = 8.37$ °C/unit m).

$$m = \frac{\Delta T_f}{K_F} = \frac{0.210}{8.37} = 0.251 \text{ mole}/1,000 \text{ gm of solvent}$$

$$\text{Weight concentration} = 3.0 \times \frac{1,000}{100} = 30 \text{ gm}/1,000 \text{ gm of solvent}$$

$$\text{Approximate molecular weight} = \frac{30 \text{ gm}}{0.251 \text{ mole}} = \underline{120}$$

9-4-4   Osmosis

One of the examples used in discussing the concept of randomness was the situation where a solute is introduced into a solvent and where one observed a spontaneous diffusion of solute particles to all parts of the system. It is instructive to reexamine such a process now from the point of view of

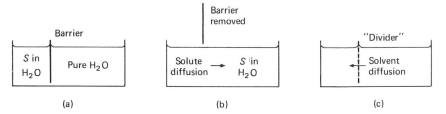

FIG. 9-10 *Movement of solution components in response to a concentration gradient.*

the change in free energy involved in this diffusion. Consider a container such as that illustrated in Fig. 9-10a, in which a solute S in aqueous solution is separated from a section containing pure water by an impenetrable barrier. When the barrier is removed, solute diffusion throughout the whole volume of the system occurs, as indicated in Fig. 9-10b. The process can be written as

$$S(c_1) \rightarrow S(c_2)$$

This spontaneous process must be accompanied by a decrease in free energy. It is instructive to examine ΔG in the specific case where $C_1 = 0.1$ mole/liter and $C_2 = 0.01$ mole/liter. The free energy change for an alteration in solute concentration is given by

$$\Delta G = 2.303 \, RT \log \frac{C_2}{C_1} \qquad (3\text{-}14)$$

For the process in question, at 298 °K

$$\Delta G = 2.303 \times 1.987 \times 298 \times \log \frac{0.01}{0.1}$$

$$\Delta G = -1{,}364 \text{ cal}$$

This result is consistent with the observed spontaneous diffusion process.

This particular example illustrates the general principle, particularly important in biological systems, that dissolved matter diffuses in response to a concentration gradient to a region of lower concentration. Thus, when for one reason or another solute particles are required to move *against* a concentration gradient, the associated ΔG is positive and work therefore must be performed to have the process occur.

Returning to Fig. 9-10, suppose now, as shown in diagram (c), that a device is inserted that temporarily can be referred to as a "divider" that will permit free passage of water molecules but not solute particles. This intrusion of course, does not alter the negative ΔG associated with a decrease in solute concentration, but it does prevent the driving force from being manifested by solute diffusion. The latter, however, is not the only possibility. The concentration of solute also can be reduced by diffusion of water *into the solution*, and thus the decrease in free energy would be brought about by *solvent*, rather than solute, diffusion. The movement is manifested physically by a rise in the water level in the solution portion of the container. This continues until the hydrostatic pressure exerted on the divider is of sufficient magnitude to prevent any further mass transfer. Exactly the same considerations apply if the two compartments contain not a solution and pure solvent but two solutions of different concentrations. In the latter case the solvent would move from the solution of lower concentration to that of higher concentration.

The dividing device that acts in this manner is known as a *semipermeable membrane*. The solvent diffusion process, which gives rise to an *osmotic pressure*, is known as *osmosis*. The formal definitions of these terms is worth examining, particularly since some processes that are similar but not identical to the true osmosis will be discussed later.

SEMIPERMEABLE MEMBRANE: A MEMBRANE THAT PERMITS PASSAGE OF SOLVENT PARTICLES BUT NOT SOLUTE PARTICLES.

OSMOSIS: DIFFUSION OF A SOLVENT SPECIES THROUGH A SEMIPERMEABLE MEMBRANE FROM A REGION OF LOW SOLUTE CONCENTRATION TO ONE OF HIGH SOLUTE CONCENTRATION.

OSMOTIC PRESSURE: THAT PRESSURE WHICH MUST BE APPLIED TO THE REGION OF HIGHER SOLUTE CONCENTRATION TO PREVENT OSMOSIS.

There are a variety of substances that can function as semipermeable membranes, for example, cellophane and certain animal membranes. The latter are frequently involved in selective permeability in which only certain types of solute particles are retained. This phenomenon gives rise to pseudo-osmotic effects where the observed solvent diffusion must be related to the concentration of *nondiffusible solute* rather than total solution concentration. Effects of this kind are particularly important in macromolecular systems and and are dealt with in detail in Chapter 10. The present discussion is confined to true osmosis, resulting from quantitative retention by the membrane of all species except solvent molecules.

A variety of theories have been proposed to explain semipermeability, such as molecular filtration or distillation of the solvent across the pores of the membrane. It is unlikely, however, that there is a unique general mechanism, particularly in the case of selective permeability.

To obtain the quantitative relation between the osmotic pressure Π and the concentration of solute, it is necessary to consider Eq. (3-14) written in terms of natural logarithms and using mole fractions for activities. In this form the equation is

$$\Delta G = RT \ln X_2 \qquad (9\text{-}23)$$

Here the relation has been written for the change from the initial state where $X = 1.0$ (the pure solvent) to the solution where the mole fraction is X_2.

The subscript 2 specifies the mole fraction as that of the solvent. The dependence on solute concentration can be included explicitly by substituting $(1 - X_1)$ for X_2. As discussed in the derivation of Eq. (9-18), if the solution is dilute, $\ln(1 - X_1)$ can be replaced by $-X_1$, so that Eq. (9-23) can be written as

$$\Delta G_{\text{dilution}} = -RTX_1 \tag{9-24}$$

The equation gives the free energy change associated with the dilution of the solvent by introduction of a solute at mole fraction X_1.

The osmotic pressure has been defined as that pressure required to prevent osmosis. From the point of view of the free energy criterion for equilibrium, the pressure must change the free energy of the solvent in the solution to an extent such that its molar free energy on each side of the membrane is the same. The dilution effect causes a decrease in solvent free energy. Thus the ΔG associated with the alteration of applied pressure must be positive. We therefore can regard the osmotic pressure thermodynamically as the pressure that must be applied to the solution to increase the free energy of the solvent by an amount that just compensates for the decrease owing to dilution.

An expression was obtained previously for the free energy change in an infinitesimal process,

$$dG = V\,dP - S\,dT \tag{8-6}$$

By applying this to the present case, the change in free energy with pressure in a constant temperature process is given by $dG/dP = V$. The implication of $dT = 0$ here is that the solvent has the same temperature before and after the application of pressure. Since the volume is practically independent of pressure, the expression can be stated in terms of finite changes:

$$\Delta G/\Delta P = V$$

$$\Delta G = V_{\text{solvent}}\,\Delta P$$

where V_{solvent} is the molar volume. ΔP represents the amount by which the pressure must be increased (beyond that acting on the liquid phase on both sides of the membrane) to give the required ΔG. The ΔG appropriate to this situation will be that produced by the osmotic pressure Π, and thus

$$\Delta G_{\text{pressure}} = V_{\text{solvent}}\,\Pi \tag{9-25}$$

If the solvent is to have the same molar free energy on both sides of the membrane, the total effect of dilution and pressure increase must be zero.

$$\Delta G_{\text{dilution}} + \Delta G_{\text{pressure}} = 0$$

Making the necessary substitutions from Eqs. (9-24) and (9-25), we get

$$-RTX_1 + V_{\text{solvent}}\,\Pi = 0$$

$$\Pi = \frac{RT}{V_{\text{solvent}}}\,X_1 \qquad (9\text{-}26)$$

As usual, for the working equation a more convenient measure of concentration is desired. Here, however, the expression involves the molar volume of the solvent, and thus molarity is more suitable than molality. Since

$$X_1 = \frac{n_1}{n_1 + n_2} \approx \frac{n_1}{n_2}$$

Eq. (9-26) can be written as

$$\Pi = RT\,\frac{n_1}{V_{\text{solvent}}\,n_2}$$

where the term $V_{\text{solvent}}\,n_2$ represents the total volume. Thus, if the solution is dilute, $V_{\text{solvent}}\,n_2 \approx V_{\text{soln(tot)}}$ and n_2/V_{soln} is the molarity C. The simple expression for osmotic pressure therefore is

$$\Pi = C\,RT \qquad (9\text{-}27)$$

In addition to yielding the desired relation between osmotic pressure and concentration, this derivation furnishes an excellent example of how the concept of a free energy change and the criterion of $\Delta G = 0$ at equilibrium can be applied to a physical situation of considerable interest.

The analogy between the osmotic pressure equation in the form $\Pi V = nRT$ and the ideal gas law is interesting. It has been the subject of some discussion in terms of visualizing the osmotic pressure as the force per unit area on the semipermeable membrane counteracting the force exerted by the solute particles dispersed "gaslike" in the solvent. This does not appear to be a particularly profitable viewpoint, and the interpretation in terms of free energy effect is to be preferred on a number of grounds.

Substitution of numerical values in Eq. (9-27) reveals the magnitude of osmotic effects. For example, at 300 °K, a 1-molar solution in contact with pure solvent has an osmotic pressure of 24.6 atm.

$$\Pi = CRT = 1\,\frac{\text{mole}}{\text{liter}} \times 0.0821\,\frac{\text{liter atm}}{\text{deg mole}} \times 300\,\text{deg}$$

$$\Pi = 24.6\,\text{atm}$$

The large value of Π for dilute solutions makes measurement of this colligative property a useful method of molecular weight determination when concentrated solutions cannot be employed. This is especially valuable for high molecular weight compounds where solutions are restricted to rather small maximum practical concentrations. The osmotic pressure exhibited by dilute solutions is of sufficient magnitude to be determined readily with minimum experimental error. This application is discussed in Chapter 10.

9-5 COLLIGATIVE PROPERTIES II—ELECTROLYTE SOLUTES

As discussed previously, the magnitude of colligative properties depends in a general way on the number of solute particles present per unit weight or volume of solvent. In solutions of electrolytes the equations developed for molecular systems must be modified to account for the fact that the solute exists partially or completely in ionic form in solution. In addition, there is the influence of interionic forces that will reduce the effective number of particles, that is, the activity. Therefore, both the extent of dissociation of a given species and the activity coefficients of the ions so formed must be considered.

As a result of ion formation or proton transfer reactions in acid or base solutions, the magnitude of the colligative properties is larger than that predicted by, for example,

$$\Delta T_f = K_F m \qquad (9\text{-}22)$$

The change is usually expressed in terms of the *mole number i* (sometimes referred to as the *van't Hoff i factor*), defined by

$$i = \frac{\Delta T_f}{K_F m} \qquad (9\text{-}28)$$

or the analogous relation for the other colligative properties. For a strong electrolyte such as NaCl, i should be equal to 2.0 when activities and concentrations are the same within experimental error. Values of the mole number for this and other similar solutes approach the theoretical value predicted on the basis of the stoichiometry of the ion-forming reaction only in very dilute solutions.

Solutions of weak electrolytes can be handled rather simply. The extent of dissociation, and therefore the concentration of ions, normally will be small, and activity coefficients of unity can be assumed without introducing serious error.

Consider a general weak electrolyte,

$$A_x B_y \rightleftharpoons xA^{+y} + yB^{-x}$$

If m represents the molality of $A_x B_y$ and α, the degree of dissociation, the concentrations of the various species at equilibrium are

$$[A^{+y}] = x\alpha m \qquad [B^{-x}] = y\alpha m \qquad [A_x B_y] = m - \alpha m$$

The total solute concentration is

$$(1 - \alpha + x\alpha + y\alpha)m = [1 + \alpha(x + y - 1)]m = im$$

Since the term in brackets represents the factor by which the molality m must be multiplied to obtain the total concentration of all species, it can be identified as the mole number i. Conventionally the sum $(x + y)$, the number of moles of ions formed per mole of weak electrolyte dissociating, is designated as v, and i is written as

$$i = 1 + \alpha(v - 1) \qquad (9\text{-}29)$$

To evaluate any colligative property, then, we use the equation applicable to molecular solutes but replace the molality by the product of m and i, the latter being defined by Eq. (9-29). The freezing point depression equation, for example, takes the form

$$\Delta T_f = [1 + \alpha(v - 1)]mK_F \qquad (9\text{-}30)$$

Analogous expressions can be written for the other colligative properties.

For electrolytes that dissociate completely in solution, the situation is less straightforward. As indicated previously, in very dilute solutions where activities and concentrations are the same within experimental error, i will be equal simply to the number of moles of ions formed per mole of strong electrolyte dissociating. As the concentration increases, unit activities coefficients are no longer appropriate, and the evaluation of the magnitude of colligative properties becomes somewhat more complex. As would be expected from the general behavior of γ values, the difference between $i_{\text{theoretical}}$ (that is, the value in dilute solution) and i_{observed} increased with concentration and with the magnitude of the charge on the dissociated ions. For example, i for NaCl decreases from 1.97 in 10^{-3} molal solution to 1.81 in 1.0 molal solution; at a concentration of 0.01 molal, i is 1.94 for NaCl but 1.53 for $MgSO_4$.

In very concentrated solutions the considerations on which activity coefficients such as those obtained from Fig. 5-3 are based are no longer applicable. In fact, ion activity coefficients *greater* than unity are observed. For example, the mean activity coefficient in a 4.0 molal solution of LiCl is 1.46.

9-6 DISTRIBUTION BETWEEN TWO SOLVENTS

The final solution situation examined here is the case where a solute is dissolved simultaneously in two immiscible solvents. This is a three-component, two-phase system.

The aspect of this kind of system, which is of greatest interest, is the quantitative description of the distribution of the solute between two solvents. The empirical facts are embodied in the *Nernst distribution law*,

$$K_D = \frac{C_2}{C_1} \tag{9-31}$$

The subscripts refer to the concentration in the two solvents, and C is the solute concentration. K_D is known as the *distribution coefficient* or *partition coefficient*. The constant value of the concentration ratio at a particular temperature, independent of the absolute value of the concentration, can be rationalized by considering the system as being characterized by the physically reversible process

$$A(\text{in solvent}_1) \rightleftharpoons A(\text{in solvent}_2)$$

Invoking the usual free energy relation, ΔG for the process is given by

$$\Delta G = \Delta G_{\text{dist}}^{\circ} + 2.303\, RT \log \frac{C_2}{C_1}$$

At equilibrium $\Delta G = 0$ and thus

$$\Delta G_{\text{dist}}^{\circ} = -2.303\, RT \log \left(\frac{C_2}{C_1}\right)_{\text{eq}} = -2.303\, RT \log K_D \tag{9-32}$$

As with other equilibria, $\Delta G_{\text{dist}}^{\circ}$ is a fundamental property of the system, independent of concentration in a particular solution. Thus only one value of C_2/C_1 satisfies the indicated equality: the equilibrium constant K_D for the intersolvent distribution process.

Although K_D is invariant for a particular system at constant temperature, Eq. (9-31) is appropriate only if the concentrations C_2 and C_1 are valid representations of the activities—thus the general restriction to dilute solutions. In addition, the distribution process itself must be as described by the stoichiometric equation above. Consequently, if the solute existed, for example, as a hydrogen-bonded dimer in one solvent but not in the other, the equation representing the process would be

$$2\,A(\text{solvent}_1) \rightleftharpoons A_2(\text{solvent}_2)$$

whence the distribution equation would take the form

$$K_D = \frac{C_2}{2(C_1)^2}$$

where C_1 and C_2 are the concentrations of A in each solvent. The factor 2 in the denominator arises because the concentration of the dimer A_2 in solvent$_2$ is $\frac{1}{2}$ the concentration of A in that solvent.

The value of K_D in general is a direct indication of the relative solubilities in the two solvents. Thus, for example, K_D for I_2 in water–carbon tetrachloride is 131 (CCl_4 as solvent$_2$), indicating that iodine is 131 times more soluble in CCl_4 than in H_2O.

A practical application of the distribution phenomenon is to the process of extraction. For example, a compound that is only slightly soluble in water but highly soluble in ether can be extracted from aqueous solution containing water soluble impurities by shaking the aqueous solution with an amount of ether. The compound, being more soluble in the latter solvent, will attain a high concentration in that layer that can then be removed. This cycle is repeated several times with additional quantities of fresh ether. After a number of these sequences, the compound is quantitatively removed from the aqueous layer. The ether layers then are combined, and the pure compound is obtained by evaporation of the solvent.

In the technique of *countercurrent extraction* a series of such cycles is performed simultaneously. An apparatus consisting of a number of interconnecting vessels is arranged so that in operation one of the two solution layers in each vessel can be removed selectively and transferred to the next vessel. In this way a series of distribution equilibria, analogous to the series of vapor-liquid fractionations in fractional distillation, are made to occur. This method is particularly valuable for the separation of biological materials whose solubility in several solvents may be very similar. One of the components in the mixture to be separated concentrates in the solution, moving from vessel to vessel, while the other is partially removed during its residence time in each vessel.

PROBLEMS

9-1. Calculate the freezing point depression for an aqueous 0.1 molal solution of benzoic acid (C_6H_5COOH).

9-2. Using the Clausius-Clapeyron equation in the form of Eq. (8-10), derive an expression for the temperature dependence of K_{VP}.

9-3. The vapor pressure of pure methanol (CH_3OH) is 100 torr at 21 °C. Calculate the weight of a nonvolatile solute of 125 mol. wt. that must be added to 1 liter of methanol (density $= 0.796$ gm/cm^3) to reduce the vapor pressure to 98.35 torr.

9-4. At 42 °C the vapor pressure of n-C_7H_{16} is 100 torr and that of n-C_6H_{14} is 320 torr. Assuming that a mixture of these two liquids behaves ideally, calculate the partial pressure of each component, the total pressure, and mole fraction of hexane in the liquid, and the vapor phase for a mixture of 50 gm of each liquid.

9-5. One liter of water is exposed to a partial pressure of 600 torr of N_2O at 20°C. Calculate the volume, measured at 1 atm and 0 °C, of gas that dissolves.

9-6. Calculate the approximate freezing point depression exhibited by an 0.1 percent by weight aqueous solution of $MgCl_2$.

9-7. What is the molecular weight of a nonvolatile, nonelectrolyte solute that lowers the freezing point of nitrobenzene to 5.61 °C when 1.00 gm of solute is added to 100 gm of the solvent.

9-8. Calculate the free energy of formation of liquid water at 298 °K and 75 atm.

9-9. 1.75 gm of a nonvolatile, nonelectrolyte solute, with mol. wt. of 164, is dissolved in 100 gm of water. Calculate the freezing point depression, the boiling point elevation, and the vapor pressure of the solution at 38 °C where the vapor pressure of pure water is 49.7 torr.

9-10. An organic compound, immiscible with water, has a molecular weight of 176. If the compound is steam distilled at $P_{atm} = 735$ torr at a temperature where the vapor pressure of water is 479 torr, calculate the weight of the compound in 100 gm of distillate.

9-11. Calculate the osmotic pressure at 25 °C of an aqueous solution containing 0.01 percent by weight of a nonvolatile, nonelectrolyte solute of molecular weight 347.

9-12. K_D for acetylsalicylic acid in ethyl ether and water as solvent is 4.7 ($C_2 = C$ in either solution). Fifty grams of an aqueous solution of the acid containing 0.15 gm of the acid is shaken with 50 ml of ether, the ether layer is removed, and an additional 50 ml of ether is added. What weight of acid remains in the water layer at the end of the second extraction?

9-13. Calculate the molal boiling point elevation constant for ethylene glycol, making use of the data in Table 8-2.

9-14. Using the data given in Problem 8-5, determine the molal freezing point depression constant for naphthalene.

Physical Chemistry of

Macromolecular Systems

10

10-1 INTRODUCTION

The physical and chemical systems discussed in previous chapters have all been assemblies of atoms, molecules, or ions of rather small dimensions. Thus, for example, the equilibrium between acetic acid and acetate and hydronium ions in aqueous solution has been examined. From another point of view, however, systems have been encountered where concern is with the properties of matter in what might be termed its "bulk" form. The heat capacity of a liquid, for example, can be used to determine the amount of heat absorbed over a given temperature range without reference to the micro-scopic composition of the liquid.

Of increasing interest in modern physical chemistry is the study of systems that exhibit the properties of both microscopic and bulk matter. These systems classically were known as *colloids*. They were so classified on the basis of rather ill-defined size limits of several thousand angstrom units (1 Å = 10^{-8} cm) within the range of 10 to 10,000 Å. Current activity is increasingly directed toward the type of system where the "particles" are organic molecules of very high molecular weight, *macromolecules*. In classical colloid systems, on the other hand, the particles are usually not individual molecules but rather assemblies of atoms, ions, or molecules. An example of this type is the formation of *micelles* by fatty acids, in the RCOO⁻ form, in solution.

Many properties of macromolecules derive from the effects of their large surface area. For very large particles, such as those in ordinary suspensions, the surface area per unit total volume is relatively small. For example, an assembly of spheres, each about 1 mm in diameter, has a surface are on the order of magnitude of 100 cm^2/cm^3, while the area per unit volume is considerably larger, about 10^5 cm^2/cm^3, for particles with diameters around 1,000 Å. Particulate matter of ordinary atomic and molecule dimensions exhibit even larger surface areas, but in these systems other effects are considerably more important.

Perhaps the most striking general characteristic of macromolecular systems is the virtual coexistence of microscopic and macroscopic properties. To use a protein "solution" as an example, the particulate phase is minute enough so that the system behaves much like a true solution in that the particles do not settle under normal gravitational attraction. Under very high gravity forces in an ultracentrifuge, however, sedimentation can be observed. The particles are small enough to exhibit molecular-like random motion, the *Brownian movement* yet they are of sufficient size to scatter physically a beam of light incident on the system. In the study of macromolecules the general techniques of physical chemistry often can be employed. Their application to microscopic systems has been discussed, but some additional experimental and theoretical factors must be considered. Many macromolecular systems are of immense biological importance. There is no sharp dividing line between physical chemistry and molecular biology in this area.

A complete survey of the interesting and important subject of macromolecular chemistry is beyond the scope of this book. The present chapter is only a very brief introductory survey. The initial discussion deals with the nature of macromolecules and some of the general principles associated with surfaces as they may be applicable to "particulate" systems. Subsequently, several important properties of macromolecular solutions, particularly as they apply to molecular weight determinations, are discussed.

10-2 GENERAL PROPERTIES OF MACROMOLECULES

10-2-1 Nature of Macromolecules

Molecules having very large molecular weights are usually *polymers*. They are constructed from a large number of one or more fundamental units, *monomers*, which occur repeatedly in the molecule. These units are connected through covalent (electron-sharing) bonds. Protein molecules, as discussed in Sec. 4.7.2, are polymers. The units of molecular construction are the various amino acids joined together by peptide linkages, giving macromolecules with molecular weights in the 10^3-to-10^5 range. The number of possible sequences of amino acids with various R groups

$$
\begin{array}{ccc}
\text{H} & \text{R} & \text{O} \\
| & | & \| \\
-\text{N}-\text{C}-\text{C}- \\
& | \\
& \text{H}
\end{array}
$$

that can possibly occur gives rise to a very large number of complex varieties. The elucidation of the sequence of amino acids and the structure of the proteins they comprise is an area of intense interest in current research. Considerable progress has been made in a number of different areas.

The formation of protein molecules through the peptide linkage is an example of the general mechanism of polymer formation known as *condensation polymerization*. The bond linking the monomers is formed following removal of groups (frequently H and OH to form water) from two monomers. Numerous synthetic polymers, notably nylon, are of this type. Others may be formed by *addition polymerization*, in which monomers are added successively through attack on monomers by a reactive site in the partially formed polymer.

The carbohydrates, whose general formula is $C_x(H_2O)_y$, also form polymers, for example, starch and cellulose, which are classified as *polysaccharides*. Both starch and cellulose are polymers of *glucose*,

the units being connected by ether linkages through elimination of H_2O between two monomers.

Another natural polymer of interest is polyisoprene, a linear polymer of isoprene,

Perhaps the most interesting of all macromolecules are the *nucleic acids*. Very elegantly presented discussions of our present state of knowledge are widely available, and it is appropriate here to do no more than outline the salient features of these molecules, which are at the very heart of life processes.

Nucleic acids are constructed from monomer units known as *nucleotides*. Each monomer is a combination of a carbohydrate, either D-ribose (ribonucleic acids, RNA) or D-2-deoxyribose (deoxyribonucleic acids, DNA),

D-Ribose   D-2-Deoxyribose

a phosphate group, and a base. In DNA there are four such bases:

Thymine   Cytosine

Adenine   Guanine

The polymer is formed by bonding of the sugars to adjacent monomers, through the phosphate groups. The structure of DNA proposed by J. D. Watson and F. H. C. Crick consists of a double helix—each strand has the sugar–phosphate "backbone" with the bases attached to the sugars. The two intercoiled chains are held together by hydrogen bonds between pairs of bases.

10-2-2 Macromolecular Weights

Unlike ordinary molecules to which a rigorously defined and unique molecular weight can be assigned, in macromolecular systems there is often a distribution of molecular weights because of the possible variations in the number of monomers in a given species.

For the case of a specific molecule, however, a definite macromolecular weight can be obtained. Thus a molecular weight determination for a well-defined protein can provide valuable data in establishing its composition.

In the more general situation there will be a distribution of molecular weights. It is necessary, therefore, to distinguish between two types of molecular weight data—the *number average molecular weight* M_n and the *weight average molecular weight* M_w. The nature of the molecular weight evaluated depends on the property of the macromolecular system made use of in the analysis.

The number average is defined as the sample weight divided by the number of moles present. M_n values result from measurements, such as osmotic pressure, that essentially count the number of particles present. The statistical significance of M_n is more apparent if the "weight per number of moles" definition is stated more formally as

$$M_n = \frac{\sum_i n_i M_i}{\sum_i n_i} = \frac{n_1 M_1 + n_2 M_2 + \cdots}{n_1 + n_2 + \cdots} \tag{10-1}$$

where n_i represents the number of moles of molecule i, which has a molecular weight equal to M_i.

Molecular weights also can be determined by methods that utilize properties dependent on the mass m_i of polymer present having each of the molecular weights M_i in the distribution. Thus the molecular weight obtained is a weight average,

$$M_w = \frac{\sum_i m_i M_i}{\sum_i m_i}$$

The value of m_i is the product of the number of moles n_i and the molecular weight M_i so that

$$M_w = \frac{\sum^i n_i M_i^2}{\sum_i n_i M_i} \tag{10-2}$$

The relation between the two types of molecular weight is illustrated in Fig. 10-1 for a typical distribution. M_w is invariably larger than M_n since the weight average favors macromolecules of higher molecular weight. The ratio M_w/M_n is a useful indication of the breadth of the distribution—a value of unity being associated with a unique molecular weight—while larger values are indicative of a wide range of molecular weights.

10-2-3 Surface Effects

It has been noted that surfaces are important in determining the properties of colloidal suspensions. As a preliminary to the discussion of some properties of macromolecular systems, it is useful to examine here some of the physical chemistry of surfaces.

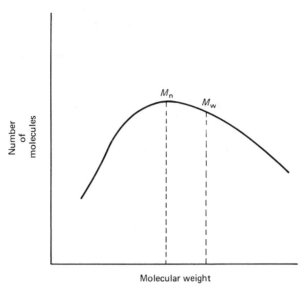

FIG. 10-1 Relation of the number and weight average molecular weights in a distribution of macromolecular weights.

A brief survey of the fundamental properties of surfaces provides an invaluable background to the understanding of some of the unique characteristics of macromolecules. This is also an important area of physical chemistry in its own right.

The concept of a *surface energy* is perhaps easiest to comprehend in terms of a liquid surface. In the bulk of the liquid the molecules are acted upon by intermolecular forces that on the average are spherically symmetrical. At the surface, on the other hand, these forces act only in the direction tending to pull the molecules back into the truly homogeneous interior of the liquid. Thus liquid surfaces have a tendency to contract. We therefore can envisage an energy, the *surface tension*, which offers resistance to any expansion of the surface. The change in free energy associated with such a process is proportional to the change in surface area. The proportionality factor is the surface tension γ. Written for an infinitesimal area change, the expression is

$$dG = \gamma \, dA \qquad (10\text{-}3)$$

In effect the surface tension is the energy per unit surface area. Considering energy as the product of force and distance, γ has the dimensions of a force per unit length, usually dynes per centimeter. At 20 °C pure water has a surface tension of 73 dynes/cm, whereas carbon tetrachloride, a nonpolar molecule with weaker intermolecular forces, has $\gamma = 27$ dynes/cm.

The effect on surface tension because of the presence of a solute is quite variable. In one type of behavior particularly pertinent to the present discussion, solutes may have the effect of decreasing the surface tension by becoming concentrated at the surface. Long-chain fatty acids behave in this way. The polar carboxylic acid group is readily soluble and is pictured as being immersed in the bulk of the solution, while the hydrocarbon chain portion (insoluble of itself) more or less protrudes through the surface. In a general way when "solute" particles become concentrated at a surface we can think of the process as *adsorption*. This phenomenon is described quantitatively by an equation known as the *Gibbs' Adsorption Isotherm*:

$$d\gamma = -\frac{RT\Gamma\, dX}{X} \tag{10-4}$$

which gives the change in surface tension $d\gamma$ when the mole fraction of solute X is changed by an amount dX. Γ represents the concentration of solute adsorbed at the surface.

Another type of adsorption whose characteristics provide some additional insight into surface phenomenon is a gas adsorbing on a solid surface. The general relation applicable to this situation was developed by I. Langmuir, who considered that when a system, wherein adsorption processes occur, comes to equilibrium, the rates of adsorption and desorption are identical. It is considered that adsorption can occur only in a monomolecular layer and only at some number of *active sites*, and θ is taken as the fraction of active sites on which adsorption has occurred. The rate of adsorption will be proportional to the pressure and the fraction of sites unoccupied, $(1 - \theta)$. The desorption rate is proportional simply to θ. Writing the proportionality factors as k_a and k_d, respectively, and invoking the equality of these rates at equilibrium, we get

$$k_a P(1 - \theta) = k_d \theta$$

whence

$$\theta = \frac{k_a P}{k_a P + k_d} \tag{10-5}$$

If one measures at a particular pressure the amount x of gas adsorbed per unit weight of adsorbent, θ is given by the ratio x/x', where x' is the amount required to saturate the surface completely. Using this ratio for θ allows Eq. (10-5) to be written in the form

$$\frac{P}{x} = \frac{k_d/k_a}{x'} + \frac{1}{x'}P \tag{10-6}$$

and thus a plot of P/x vs. P should be a straight line from which the various parameters can be calculated.

The final case of general adsorption phenomena that should be mentioned is adsorption from solution. This process can be described either by the *Langmuir Isotherm*, Eq. (10-6), replacing pressure by the concentration of the absorbing species, or by an empirical relation obtained by H. Freundlich,

$$x = kC^{1/n} \qquad (10\text{-}7)$$

where x has the same significance as in Eq. (10-6) and n is an empirical constant. In situations where this equation is applicable a plot of $\log x$ vs. $\log c$ would yield a straight line of slope $1/n$ and an intercept of $\log k$. Other, more elaborate equations have been developed to correlate data for multilayer adsorption.

In macromolecular systems the most important surface phenomenon is the adsorption of ions from solution. In general, particulate matter of colloidal dimensions presents a charged surface to the surroundings, either by adsorption or by virtue of ionized functional groups on the macromolecule (for example, $-NH_2^+$ or $-COO^-$ groups on a protein). Ions will bind on the surface—which may be somewhat ill-defined for a particular macromolecule— through electrostatic attraction. These forces arise from portions of the molecule having appreciable localized dipole moments or even from essentially nonpolar species through induced polarization. These phenomena lead to the important concept of the *electrical double layer* that is envisaged as surrounding macromolecules in solution.

The formation of the double layer is thought to occur in the manner illustrated in Fig. 10-2. A layer of charged species is fixed at the surface, overlaid with a second layer of oppositely charged ions that also is immobile. In the bulk of the solution near the particle is a diffuse layer of *gegenions* that are free to move. A potential exists from the edge of the outer fixed layer to any given point in the diffuse layer. This is known as the *zeta potential* ζ. If only a small cross-sectional area of surface is considered so that curvature effects can be ignored, the system can be treated as a simple condenser. On this basis the zeta potential is given by

$$\zeta = \frac{4\pi Q d}{D} \qquad (10\text{-}8)$$

where Q is the charge per unit area, d is the distance into the diffuse layer, and D is the dielectric constant* of the medium.

It is apparent that the magnitude of the zeta potential depends on the fundamental tendency of the colloidal-sized particles to present a charged surface layer. Thus for macromolecules where charges are produced at the

---

* The dielectric constant is defined as the ratio of the capacitance of a parallel condenser with the substance between the plates to the capacitance in a vacuum, that is, C/C_0.

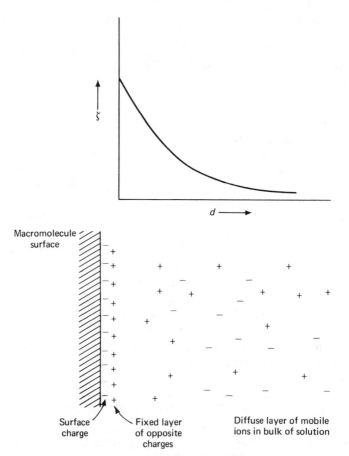

FIG. 10-2 *Formation of the electrical double layer and the magnitude of the zeta potential as a function of the distance from the fixed charged surface.*

"surface" only by proton transfer reactions, the extent of the process and hence the zeta potential will be dependent on solution pH. At the isoelectric point, defined in Sec. 4-7-2, there is no net charge on the protein and the zeta potential will be zero.

The *electrokinetic properties* of a macromolecular system are a direct result of the existence of the zeta potential. The most important of these properties is exhibited in the process of *electrophoresis*. If a system of colloidal particles is subjected to an electric field by means of oppositely charged electrodes inserted into the disperse medium, the particles of the disperse phase move in response to the electrostatic attraction toward the electrode of opposite charge. This motion is opposed by the viscosity of the liquid. When a steady state flow is attained, the *electrophoretic mobility U* can be evaluated by equating the two forces to obtain the expression

$$U = \frac{\zeta E D}{4\pi\eta} \qquad (10\text{-}9)$$

where D is again the dielectric constant, E is the magnitude of the applied field, and η is the viscosity.*

Experiments in electrophoresis frequently involve a specially designed U-tube apparatus, fitted with removable horizontal sections. Through manipulation of these sections a well-defined boundary can be established between the aqueous suspension of the macromolecule, for example, a protein, and a buffer solution. Careful pH control is required because of the sensitivity of the zeta potential to this parameter. When an electric field is applied across the tube, the particles move with different velocities, depending on their zeta potential, and with suitable optical instrumentation the moving boundaries between the suspension and the buffer solution can be followed and characteristic electrophoretic patterns can be observed.

Under suitable conditions the various proteins are concentrated in the several demountable sections, and thus physical separation can be effected. As mentioned previously, if the pH is adjusted so as to coincide with the isoelectric point of a particular protein in the mixture, that component will not exhibit electrophoretic mobility. In this connection the distinction noted in Sec. 4-7-2 between the isoelectric and iso-ionic points is of some significance. The pH at which the net charge is zero usually will not correspond exactly to that predicted from acid–base equilibrium calculations for equal numbers of cationic and anionic groups on the protein because of the presence of adsorbed ions.

Three other electrokinetic phenomena can be explained in terms of the zeta potential. In *electro-osmosis* the electrophoretic movement of the particles is prevented by the insertion of a semipermeable membrane. The system responds in exactly the same manner as in the case where movement of solute in response to a concentration gradient is prevented. There is a net flow of solvent in a direction opposite to that taken by the macromolecules in the absence of the membrane. When a colloidal suspension is forced through a capillary tube, a potential develops between the ends of the tube; this is known as the *streaming potential.* Similarly, when a colloidal system undergoes sedimentation, the movement of the particulate phase creates a charge separation, and a resultant *sedimentation potential* can be observed. This latter phenomenon is known as the *Dorn effect.*

---

* Viscosity is the resistance to liquid flow. For two planes of area A within the liquid, the force F required to move one of these planes with respect to the other at constant velocity $\mathbf{v}$ is proportional to the velocity, the area A, and the reciprocal of the interplanar distance d. Thus $F = \eta\, \mathbf{v}A/d$, where η is the proportionality factor, the viscosity in centipoise, grams $\times\ 10^{-2}$/cm sec.

10-3 PROPERTIES OF SOLUTIONS OF MACROMOLECULES

Strictly speaking, since colloidal systems are two-phase, the term macromolecular suspension should be used. However, that designation is more often employed for classical colloids. The behavior of these systems yields considerable information about the size and shape of macromolecules and, in particular, about their molecular weight. The discussion here is not intended to be comprehensive. A few of the more frequently encountered properties will be discussed to at least indicate how the unique behavior of macromolecular systems can be exploited to obtain information about their physical properties.

10-3-1 Light Scattering

When a beam of light is passed through a suspension of particles of colloidal dimensions, a measurable fraction of the light is scattered randomly. The ability to attenuate transmitted intensity through this mechanism is the most familiar property of macromolecules and colloids. The examples of the phenomenon observable in everyday experience are numerous, perhaps the best example being rays of sunlight made visible as they pass through dust or smoke.

In this and the next section there is a departure from the general approach used previously in that no effort is made to derive the equations used. The details of the development of many of these relations is beyond the scope of this book. It is nevertheless useful to examine the interrelation between the principal equations. These relations are used to describe light scattering and to obtain information about macromolecular systems, particularly molecular weights, from experimental observations.

The scattering of light by a solution of macromolecules is illustrated schematically in Fig. 10-3. The fundamental approach to relating the amount

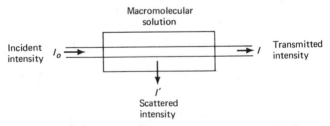

FIG. 10-3 *Light scattering by a macromolecular solution. The amount of light scattered in all directions,* $I_0 - I$, *is represented schematically for one direction only as* I'.

of scattering to the physical properties of the species responsible for the effect is in terms of the attenuation of the intensity. When a light beam is passed through *any* substance, the intensity may be attenuated by absorption of the electromagnetic radiation. The type of excitation produced in the absorber depends on the wavelengths of the radiation. Absorption in the infrared region, for example, produces vibrational excitation. The decrease in intensity owing to absorption is related to the absorption coefficient ϵ, by Beer's law, $I = I_0 e^{-\epsilon c l}$ where c is the concentration of the absorber and l is the path length through which the light travels. Reductions in intensity through the *scattering* mechanism that is characteristic of macromolecules may be treated formally in the same way. The turbidity τ, a quantity analogous to the absorption coefficient, is defined by the equation

$$I = I_0 e^{-\tau l} \tag{10-10}$$

where I is the transmitted intensity, I_0 is the incident intensity, and l is the path length.

The fraction of the intensity that is scattered is generally extremely small. Experimentally, therefore, it is considerably more convenient and accurate to measure the amount of light scattered rather than to attempt to determine the very small decrease in intensity as the light beam passes through the macromolecular solution. Furthermore, since τ is small, Eq. (10-10) can be written in the form

$$\tau l = \frac{I_0 - I}{I_0} \tag{10-11}$$

where $I_0 - I$ is the total amount of light scattered. It is not practical to measure $I_0 - I$ directly by observing the *total* scattered light intensity. The intensity scattered through any particular angle, however, can be determined without difficulty. The angular dependence of scattered light from an unpolarized beam is given by an equation due to Lord Rayleigh,

$$\frac{I_\theta}{I_0} = \frac{8\pi^4 n\alpha^2}{\lambda^4 d^2} (1 + \cos^2 \theta) \tag{10-12}$$

where I_θ is the intensity at a distance d of light scattered through the angle θ from a particular incident intensity I_0 of wavelength λ, n is the number of particles per cubic centimeter, and α is the polarizability* of the particle.

An interesting aspect of this equation is the dependence on the reciprocal of the fourth power of the wavelength that results in shorter wavelengths being scattered appreciably more than longer wavelengths.

* The polarizability is defined as the dipole moment induced by the oscillating electric field of the incident radiation per unit field strength. It can be related in this type of system to the values of the refractive index of the pure solvent compared to that of the solution.

It is for this reason that the sky appears blue during the day since the short wavelength solar radiation (blue) is scattered by particulate matter in the atmosphere, more than the longer (red) wavelengths.

Returning to Eq. (10-12), if the d^2 term in the denominator is brought to the left-hand side, the result is the ratio $I_\theta d^2/I_0$, known as the *Rayleigh Ratio*, R_θ. This quantity can be determined experimentally by measuring the scattered intensity at some convenient angle to the incident beam. It represents a *fraction* of the total scattered intensity, $I_0 - I$, and thus, as indicated by Eq. (10-11), it should be directly related to the turbidity. The relevant equation is

$$\tau = \frac{(16\pi/3)R_\theta}{1 + \cos^2\theta} \tag{10-13}$$

The final step in obtaining the equations necessary to translate light-scattering data into molecular weight information is to modify the equation for the Rayleigh Ratio. On the basis of fundamental light scattering theory, the polarizability α can be related to the refractive indices of the solvent n_r° and the solute n_r. The number of particles per cubic centimeter and the weight of the sample is then used to calculate the molecular weight of the particles. These manipulations yield the following expression for the turbidity:

$$\tau = \frac{32\pi^3 n_r^\circ(n_r - n_r^\circ)CM}{3\lambda^4 C^2 N}$$

where M is the molecular weight, N is Avogadro's number, and C is the concentration in grams per cubic centimeter. All terms except M and C (as it appears in the numerator only) are collected in a "constant," conventionally designed as H, in which case the relation is simply

$$\tau = HCM \tag{10-14}$$

H can be evaluated by taking refractive index measurements on the solution and solvent; τ is calculated from Eq. (10-13) by use of the value of R_θ determined by measuring the intensity of scattered light at some convenient angle θ. Finally, from the value of H, τ, and C the molecular weight is evaluated by use of Eq. (10-14). Since the polarizability depends on molecular size, the value obtained for M in this way is a weight average molecular weight M_w, defined by Eq. (10-2).

These relations are valid only for relatively dilute solutions. For higher concentrations Eq. (10-14) is rearranged and expanded to include a solvent dependent parameter B, as

$$\frac{HC}{\tau} = \frac{1}{M} + 2BC \tag{10-15}$$

When the size of the particles involved exceeds about 0.05λ, further modifications are required to account for interference of light scattered from different sites on the same macromolecule. The resulting equations are of the same general form as Eq. (10-15), and the data obtained from these equations are usually subjected to graphical analysis employing a double extrapolation to $\theta = 0$ and $C = 0$.

10-3-2   Sedimentation

Macromolecular solutions do not undergo sedimentation to an appreciable extent under the influence of normal gravitational attraction. The solute particles are massive enough, however, to respond to very large gravitational forces. This property is the basis of the operation of the *ultracentrifuge* developed by T. Svedberg. The macromolecular solution is placed in a sample tube rotated at extremely high speeds. The centrifugal forces developed are several orders of magnitude greater than normal gravitational attraction. An additional feature of importance is that the instrument is designed so that a light beam can be passed through the cell as it rotates. Thus changes in the optical properties of the solution can be followed as the centrifugation proceeds. Optical measurements can be made on ultraviolet or visible light absorption, refractive index, and interferometric determinations.

When the solution is subjected to the extremely large centrifugal forces, the macromolecules undergo a net movement to the "bottom" of the sample tube. The motion is retarded by the frictional resistance of the solvent. Thus the velocity of sedimentation of the solute species is determined by the balancing of the two opposing forces resulting from the viscosity and the imposed gravitational field. Consequently, a concentration gradient is established in the tube. The rate of sedimentation of macromolecules is usually expressed in terms of the *sedimentation constant s*, determined by the equation

$$s = \frac{dx/dt}{\omega^2 x} \tag{10-16}$$

where x is the distance of the solute species from the center of rotation and ω is the angular velocity.

Three methods are available to relate the characteristics of the sedimentation process to the weight average molecular weight of the colloidal particles being investigated.

By equating expressions for the centrifugal force and the frictional resistance in terms of the diffusion coefficient D and density data, the following expression is obtained, which gives the relation between M_w and the sedimentation constant:

$$M_w = \frac{sRT}{D(1 - \rho/\rho')} \tag{10-17}$$

where ρ and ρ' are the densities of the solvent and solute, respectively. By this *velocity method*, M_w therefore can be determined by measuring experimentally the rate at which sedimentation occurs, provided values of the other parameters are known.

The concept of movement of solution components in response to a concentration gradient was examined previously. In the ultracentrifuge tube, as the sedimentation proceeds, the concentration gradient increases in favor of higher concentrations of macromolecular solute in the outer portions of the rotating tube. There then is the additional possibility of the reverse process—diffusion of the solute back into the regions of cell where its concentration is lower. As outlined in the discussion of osmosis in Chapter 9, the free energy change for diffusion from high to low concentration is negative and can be calculated from

$$\Delta G = 2.303\ RT \log \frac{C_2}{C_1} \tag{3-14}$$

If the centrifuge is operated at reduced speed so that the rate of movement owing to centrifugal force is equal to that resulting from the concentration gradient effect, there will be no net movement. The molecular weight can be determined from an expression based on the equality of the two forces under these conditions, and observed values of the concentration. The frictional resistance is essentially the same for movement in both directions and thus a value for the diffusion coefficient, D, is not required. The technique is known as the *equilibrium method*.

10-3-3   Osmotic Pressure-Donnan Membrane Equilibrium

It is appropriate that the final quantitative discussion deal with a situation to which the free energy criterion can be applied to obtain the equation describing this phenomenon.

In Sec. 9-4-4 the following expression was derived for the relation between the osmotic pressure and solute concentration,

$$\Pi = CRT \tag{9-27}$$

Osmotic pressure is really the only colligative property which can be used effectively to determine the concentration in macromolecular solutions, and hence the molecular weight of the solute. The very large molecular weights involved make it physically impossible to obtain a solution of sufficient molar concentration to produce accurately measureable freezing or boiling point alterations. For example, a 1 percent solution in cyclohexane of a macromolecule of molecular weight of the order of 10^5 would produce a freezing point depression of only about 2×10^{-3} °C. More concentrated solutions cannot be used because of the tremendous increase in viscosity and serious departures from Raoult's law arising from physical interaction between the solute particles.

As seen in the earlier calculations based on Eq. (9-27), however, very small concentrations of solute give rise to readily measurable osmotic pressures. For example, again taking $M = 10^5$, a solution containing 100 gm of solute per liter, and $C = 1 \times 10^{-3}$ molar, Eq. (9-27) indicates that at 300 °C the osmotic pressure would be 0.0246 atm, or 18.70 torr.

Since the process of osmosis depends on the number of solute particles present, the molecular weight of macromolecules obtained from these measurements is a number average M_n, as defined by Eq. (10-1).

A serious complicating factor associated with osmotic pressure of macromolecules is the effect of the ions that may be present. Many selectively permeable membranes that prevent the passage of macromolecular solutes will not hinder the movement of water *or* of ions. The observed osmotic pressure may not be *directly* related to the ordinary concentration or to the concentration corrected by activity coefficients. Such pseudo-osmotic processes have considerable biological importance.

It would be expected from the previous discussion of osmosis that if a membrane through which only certain particles can pass separates two solutions of different solute concentrations, osmosis will occur. For selectively permeable membranes that pass certain solutes in additon to the solvent, however, the resulting osmotic pressure cannot be determined via Eq. (9-27). Only some fraction of the total solute concentration represented by C in that equation cannot move and, therefore, solvent transfer is necessary to satisfy the free energy driving force for dilution. In addition, if ions are involved, the maintainence of electroneutrality on both sides of the membrane will introduce a further restriction. The actual pressure arising from solvent flow through the selectively permeable membrane, in response to the concentration gradient, is called the *tone* of the solution. This quantity is the effective osmotic pressure with respect to a particular membrane. By definition every solution has some osmotic pressure in the sense that a pressure could be observed when the solution is separated from pure solvent by a true semipermeable membrane. Tone and osmotic pressure may or may not be the same, depending on the nature of the solute and membrane. The tone can be defined only with respect to a membrane whose permeability is specified. Two solutions with the same total concentration of solute are *iso-osmotic*. They will exert the same osmotic pressure. Solutions with the same concentration of solute *nondiffusible through a specific membrane* are termed *isotonic*. When two solutions have a different tone, the one with the larger concentration of nondiffusible solute, and thus higher tone, is termed *hypertonic*, the other is termed *hypotonic*.

An unusual effect of considerable importance arises in solutions of strong electrolytes when the membrane passes only some of the ions. In a system containing proteins, for example, the membrane might permit transfer of the small metal ions such as Na^+, which may be associated with the protein, but retain the macromolecular negatively charged species. The distribution of solute on both sides of the membrane when the system comes to equilibrium is governed by the *Donnan Membrane Equilibrium* condition.

In this situation two requirements must be satisfied—the free energy driving force for dilution because of the concentration gradient and the simultaneous maintenance of electroneutrality in the two solutions.

To examine the approach to the problem, consider a simple system where a membrane permeable to small ions separates two solutions of sodium chloride. The concentrations in the two solutions are designated by the subscripts 1 and 2.

Solution 1	*Membrane*	*Solution* 2
$[Na^+]_1$	·	$[Na^+]_2$
$[Cl^-]_1$	·	$[Cl^-]_2$

It is assumed for simplicity that the solution is dilute enough to employ concentrations rather than activities. In this case an expression in the form of Eq. (3-14), giving the difference in free energy across the membrane, can be written for each ion. For Na^+ this takes the form of

$$\Delta G_{Na^+} = 2.303 \, RT \log \frac{[Na^+]_2}{[Na^+]_1} \tag{10-18}$$

If the initial concentrations are such that $[NaCl]_1 > [NaCl]_2$, the equation indicates that ΔG is negative. Thus sodium ions will move spontaneously from solution 1 to solution 2. Similarly, the Cl^- ions move in the same direction in response to the negative ΔG *and* to maintain electroneutrality. At equilibrium the total free energy difference across the membrane must be zero;

$$\Delta G_{Na^+} + \Delta G_{Cl^-} = 0 \tag{10-19}$$

Substitution in this expression for ΔG_{Na^+} from Eq. (10-18) and for ΔG_{Cl^-} from the analogous expression for chloride ions gives

$$2.303 \, RT \log \frac{[Na^+]_2}{[Na^+]_1} = -2.303 \, RT \log \frac{[Cl^-]_2}{[Cl^-]_1}$$

whence

$$[Na^+]_1[Cl^-]_1 = [Na^+]_2[Cl^-]_2 \tag{10-20}$$

This is the Donnan Membrane Equilibrium condition for the system. Here, of course, in the absence of a nondiffusible ion, one could write intuitively that the concentration of sodium chloride must be the same in both solutions at equilibrium.

The situation becomes more complex, however, and more interesting when one solution contains NaCl as before but the other is a strong electrolyte NaP (where P^- is a macro-ion, such as a protein, to which the membrane is impermeable). To circumvent the distraction of algebraic complications it is assumed that the solution volumes are identical and that a pressure equal to the resulting tone is applied to the appropriate solution to prevent solute and solvent diffusion. Schematically the situation can be represented as follows:

Solution 1	*Membrane*	*Solution 2*
$[Na^+]_1$	·	$[Na^+]_2$
$[P^-]_1$	·	$[Cl^-]_2$

Since both the Na^+ and Cl^- ions can diffuse through the membrane, their final concentrations must satisfy the equilibrium condition of Eq. (10-20). The initial concentrations can be represented as

$$C_1 = [Na^+]_1 = [P^-]_1$$
$$C_2 = [Na^+]_2 = [Cl^-]_2$$

Since solution 1 contains no Cl^- ions there will be a diffusion of this species across the membrane to an extent of x moles/liter. Since each solution must remain electroneutral, the same quantity of Na^+ transfers in the same direction. The equilibrium situation therefore can be described as follows:

Solution 1: $[Na^+]_1 = (C_1 + x)$
$[P^-]_1 = C_1$ (nondiffusible)
$[Cl^-]_1 = x$

Solution 2: $[Na^+]_2 = (C_2 - x)$
$[Cl^-]_2 = (C_2 - x)$

Substitution of the concentration terms for diffusible ions into Eq. (10-20) gives

$$(C_1 + x)(x) = (C_2 - x)(C_2 - x)$$

and solving for x, we get

$$x = \frac{(C_2)^2}{C_1 + 2C_2} \tag{10-21}$$

From this equation the final concentrations can be computed for various initial distributions. In this situation the concentration of NaCl is *not* the same on both sides of the membrane at equilibrium because of the presence of the nondiffusible ion. Furthermore, the true osmotic pressure of the two solutions at equilibrium will be different because of difference in total concentration arising not because the membrane is impermeable but because ion movement is restricted by the requirement that electroneutrality be maintained. Other more complex relationships can be derived for a variety of situations in which one or more nondiffusible ions, which may be multi-valent, are in both solutions.

There is one type of Donnan Membrane Equilibrium that essentially gives rise to a chemical reaction. An electrolyte such as NaP may be placed in one solution, in contact through a selectively permeable membrane with pure water. In the migration of sodium ions to the solvent portion, where they are initially absent, an electrostatic imbalance is created. This is counteracted by proton transfer in the solvent. The reaction produces hydroxyl and hydronium ions that migrate so that Na^+ ions moving from solution 1 are replaced by H_3O^+ and balanced in solution 2 by OH^-. The result therefore is a change in solution pH as the system comes to equilibrium.

One practical application of the principle of the Donnan Membrane Equilibrium is to the technique known as *dialysis*. Proteins can be freed from all except chemically bonded ions by placing their solution in contact with water through a selectively permeable membrane. In attempting to satisfy the equilibrium condition, the ions move from the protein solution into the water or buffer solution with nondiffusible ions, which is more often used because of the pH changes outlined above.

The biological implications of the Donnan Membrane Equilibrium are of considerable significance. Although the mechanism of ion transport is complex, it is certain that nondiffusible ions are involved to some extent through this mechanism. From the opposite point of view, movement of ions against a concentration gradient, or in a direction opposite to that required by the equilibrium condition (so-called *active transport*), necessitates the expenditure of energy. Thus ΔG for processes of this type is positive, and its magnitude indicates the amount of useful work that must be performed in causing the process to occur.

10-3-4 Other Properties Used in Determination of Macromolecular Weights

Measurements of the *viscosity* of macromolecular solutions, carried out usually by determination of flow rates through capillaries or the fall of a weight through the solution, have been used to evaluate relative viscosities and to obtain information about the size of the particles. These data also can be used to determine the number of average molecular weights. These are relative values that can be converted to absolute M_n data by comparing the viscosities of solutions containing macromolecules of known and unknown molecular weights.

When colloidal suspensions flow through a tube at high speed the particles tend to become aligned and exhibit characteristic optical properties, the *flow birefringence*, which can be observed when the system is viewed in polarized light. This phenomenon can be related quantitatively to the weight-average molecular weight.

The rate of *diffusion* of macromolecules in solution is appreciably slower that ordinary solutes, and by refractive index measurements in a suitably designed diffusion cell the concentration gradient arising from the diffusion process can be observed and M_w values can be obtained.

Finally, chemical analysis (*end-group analysis*) of the reactive functional groups in a polymer sometimes can be employed to obtain a value for the absolute number average molecular weight.

Use of Logarithms

Logarithms are employed for general computational purposes, in equations involving log functions, and for the calculation of pH and similarly defined (as $-\log x$) quantities.

The general definition of a logarithm may be stated as follows: if $y = a^x$, then $\log_a y = x$ and $\mathrm{antilog}_a\, x = y$.

The constant a is referred to as the *base* of the logarithm. *Common logarithms* have the base 10, whence if $\log_{10} y = x$, $y = 10^x$. If no base is indicated, one assumes that $\log \equiv \log_{10}$. For example,

$$100 = 10^2 \qquad \log 100 = 2 \qquad \mathrm{antilog}\, 2 = 100$$

Most numbers for which logarithms are required cannot be expressed as even powers of ten. The number 273, for example, is greater than 10^2 but less than 10^3; its logarithm, evaluated by the methods outlined below, is 2.4362.

$$273 = 10^{2.4632} \qquad \log 273 = 2.4632 \qquad \mathrm{antilog}\, 2.4632 = 273$$

The digit before the decimal point is called the *characteristic*, and the part after the decimal, the *mantissa*.

The other logarithm base most frequently encountered is the number $2.718\ldots$, symbolized by e. Operations of calculus frequently lead to functions of e, the base of *natural logarithms*. Conventionally, $\log_e$ is written as

324

ln, that is, if $y = e^x$, $\ln y = x$. Although tables of natural logarithms are available, expressions involving ln terms usually are converted for increased convenience to common logarithms by use of the conversion factor

$$\ln y = 2.303 \log y$$

The conversion may be used during the numerical calculation or within the derivation of the equation, often including the 2.303 term with other numerical constants. The two logarithmic systems may be compared for the number 17.50 as follows:

$$\log 17.50 = 1.2430 \qquad \ln 17.50 = 2.8622$$
$$17.50 = 10^{1.2430} \qquad 17.5 = e^{2.8622} = (2.718 \cdots)^{2.822}$$

Since logarithms are exponents, the usual rules of algebraic manipulation for such quantities apply. The normal calculations required can be summarized as follows.

Let $\log A = x$ and $\log B = x$:

Multiplication $A \times B$ *Division A/B*

$$\log AB = \log A + \log B = (x + y) \qquad \log A/B = \log A - \log B = (x - y)$$
$$AB = \text{antilog}(x + y) \qquad\qquad A/B = \text{antilog}(x - y)$$

Powers A^n   *Roots $\sqrt[n]{A}$*

$$\log A^n = n \log A = nx \qquad \log \sqrt[n]{A} = \frac{1}{n} \log A = \frac{x}{n}$$

$$A^n = \text{antilog}(nx) \qquad\qquad \sqrt[n]{A} = \text{antilog} \frac{x}{n}$$

To find the logarithm of a number:

1. Write the number in exponential form with *one* digit before the decimal point.
2. Look up the mantissa in a log table. Ignore the position of the decimal point in the original number in this step. The mantissa obtained from the table is *always* a positive number.
3. The exponent written in 1 is the characteristic of the log and may be positive or negative.

To find the antilogarithm:

1. Ensure that the decimal part of the number is *positive*.
2. Look up the decimal part *only* in the table of antilogs, and write this number with the decimal point after the first digit.
3. The digit before the decimal point in the number whose antilog is being evaluated becomes the power of ten by which the number obtained in B is to be multiplied.

Find the log of 273: write as exponential, 2.73×10^2
find mantissa, 0.4362
write characteristic, 2
$\log 273 = 2.4362$

Find the antilog of 2.4362: decimal part is positive
antilog 0.4362 from table $= 2.73$
characteristic, 2 written as 10^2
antilog $2.4362 = 2.73 \times 10^2$

Find the log of 0.00372: 3.72×10^{-3}
log of $3.72 = 0.5705$
characteristic $= -3$

When negative characteristics occur one of two methods can be used:

1. The log is $-3.0000 + 0.5705$ and thus can be written simply as the sum of these -2.4295. This mode of expression is required for functions involving logarithmic terms, for example, pH.

2. An alternative for computational purposes is the use of the *bar notation*, for this case $\bar{3}.5705$. It is important that it be clearly understood that this representation means $-3.000 + 0.5705$.

Find the antilog of -2.4295: decimal part *not* positive, change to bar notation, $\bar{3}.5705$
antilog of $+0.5705 = 3.72$
characteristic $= -3$
antilog $= 3.72 \times 10^{-3}$

Find antilog of $\bar{3}.5705$: decimal part positive second and third step as above

NUMERICAL CALCULATION EXAMPLES

1. 13.79×0.0832 　　$\log 13.79 = 1.1396$
　　　　　　　　　　　　　$+\log 0.0832 = \bar{2}.9201$
　　　　　　　　　　　　　　　　　　　$\overline{0.0597}$ 　　antilog $= \underline{1.147}$

2. $\dfrac{3.572}{127.6}$ 　　　$\log 3.572 = 0.5529$
　　　　　　　　　　$-\log 127.6 = 2.1045$
　　　　　　　　　　　　　　　$\overline{\bar{2}.4484}$ 　　antilog $= \underline{2.808 \times 10^{-2}}$

3. $(17.66)^2$ 　　　$\log 17.66 = 1.2470$
　　　　　　　　　　　　$\times \quad 2$
　　　　　　　　　　　$\overline{2.4940}$ 　　antilog $= \underline{3.119 \times 10^2}$

4. $\sqrt[3]{29.34}$ 　　　$\log 29.34 = 1.4675$
　　　　　　　　　　　$\times \quad (1/3)$
　　　　　　　　　　　$\overline{0.4891}$ 　　antilog $= \underline{3.084}$

5. $\sqrt{1.327 \times 10^{-13}} = \sqrt{13.27 \times 10^{-14}} = \sqrt{13.27} \times 10^{-7}$

To find $\sqrt{13.27}$: log 13.27 = 1.1229

$$\frac{\times \quad \frac{1}{2}}{0.5614} \qquad \text{antilog} = 3.642 = \sqrt{13.27}$$

$\sqrt{13.27} \times 10^{-7} = \underline{3.642 \times 10^{-7}}$

6. Find the value of $-\log (1.73 \times 10^{-5})$

$= -(\log 1.73 + \log 10^{-5}) = -(0.238 - 5.000) = -(-4.762) = \underline{+4.672}$

Graphical Analysis

Many of the laws of physical chemistry express in some suitable quantitative form how one parameter of a system changes as a second variable is altered. Frequently it is very useful to present this variation in graphical form. Conversely, if the experimental data are plotted as a graph, it may be possible to deduce from the shape of the plot the applicable law. Some more specific applications of graphical analysis are indicated in the discussion below.

A variety of graphical data presentations are encountered in physical chemistry, and it might seem that graphs can be plotted in almost any manner. However, there are a number of basic conventions adhered to regardless of the particular style of presentation. These can be summarized in terms of the typical graph presented in Fig. A2-1. Along the horizontal x axis we take values of the *independent variable*, the parameter that can be varied at will, with positive values to the right and negative values to the left of zero. On the vertical y axis are the values of the *dependent variable*, the parameter whose value depends on the magnitude of x. Here positive values are taken as increasing vertically from zero, and negative values below zero, on the y axis. The point at which the axes cross, where both x and y are zero, is called the *origin*. The y values that occur when x has various values are represented as points, circled so that their position is apparent when the graph is completed by drawing a smooth curve through the plotted points.

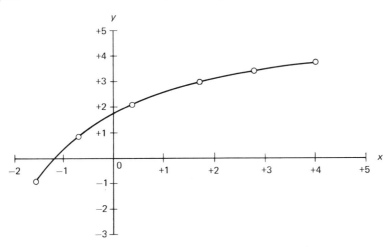

FIG. A2-1.

Many graphical data presentations are confined to positive values of *x* and *y*, and the graph therefore is just the upper right-hand quadrant of Fig. A2-1. Sometimes the choice of the dependent and independent variable is somewhat arbitrary but usually can be made in a straightforward manner from the information available.

It is apparent that before any graph can be plotted there must be some suitable set of data available. One of the chief advantages of graphical presentation is that we can see immediately the relationship between the variables. Consider, for example, the following sets of data:

	$X = 1.00$	2.00	3.00	4.00	5.00
Case A	$Y = 1.58$	3.41	5.24	7.07	8.90
Case B	$Y = 1.58$	3.00	4.05	4.85	5.49

It is obvious that *Y* increases with *X*, but only when the data are plotted, as shown in Fig. A2-2, do we see that in Case A, *Y* is directly proportional to *X* giving the straight line shown.

The *straight line graph* is the simplest type of plot, and in many ways it is the most useful. One advantage of a linear relation that can be made use of without reference to the algebraic interpretation we will examine presently is that values can be obtained of the variables at points other than those used in plotting the graph but in the same range, a technique known as *interpolation* —or beyond the measured or calculated range, as *extrapolation*. The two methods are illustrated in Fig. A2-3, where, for example, the value of *y* = 4.4 when *x* = 6 is obtained by interpolation and *y* = 1.5 when *x* = 0 is obtained by extrapolation. Some caution is required in extrapolation since

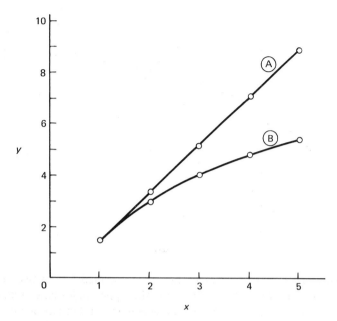

FIG. A2-2.

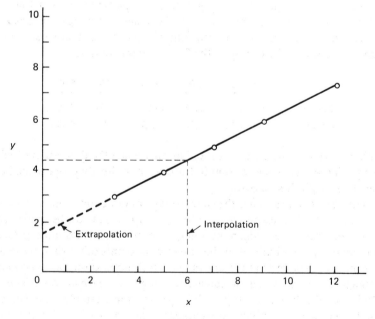

FIG. A2-3.

the data may deviate from linearity beyond the range of points actually plotted.

Any straight line plot can be represented by the simple equation

$$Y = mX + c \qquad \text{(A2-1)}$$

where Y and X represent general dependent and independent variables and not necessarily a single quantity; that is, Y could be $y^2/3$, $\log K + 2.0/a - y$, etc.

Once the value of the constants m and c have been determined, it then is possible to interpolate and extrapolate without further reference to the graph. Thus, if any given set of data result in a straight line plot, the applicable equation, of the form of Eq. (A2-1), can be deduced from the graph in the following manner:

The constant m is known as the *slope of the line*, and it is calculated from the ratio of the change in Y to the change in X between any two points on the line. Consider any two points and denote the X and Y values defining these points as $(X_1 Y_1)$ and $(X_2 Y_2)$. Since both fall on the line, they must satisfy Eq. (A2-1) and, therefore,

$$Y_1 = mX_1 + c \qquad \text{(A2-2)}$$

$$Y_2 = mX_2 + c \qquad \text{(A2-3)}$$

If Eq. (A2-2) is subtracted from (A2-3), the result is

$$(Y_2 - Y_1) = m(X_2 - X_1) \qquad \text{(A2-4)}$$

and rearrangement gives an expression from which m can be calculated directly:

$$\text{slope} = m = \frac{Y_2 - Y_1}{X_2 - X_1} = \frac{\Delta Y}{\Delta X} \qquad \text{(A2-5)}$$

The slope m can be positive or negative, depending on the relative magnitudes of the X and Y values. If the graph is plotted in the conventional manner with increasing values to the right of and above the origin, the m will be positive if Y increases left to right or negative if Y decreases in this direction. In all cases care must be taken not to anticipate the sign of m but rather to determine its value and sign from the defining equation.

The constant c is the *intercept*. It is evident from Eq. (A2-1) that $c = y$ when x is zero and, therefore, c is actually the point at which the straight line intercepts the y axis. The value of c therefore is obtained directly from the graph, extrapolating the curve to $x = 0$ if necessary.

To illustrate the extraction of these constants from a straight line graph, consider the following set of data:

$$r = 2 \quad 4 \quad 6 \quad 8 \quad 10$$
$$t = 5 \quad 8 \quad 11 \quad 14 \quad 17$$

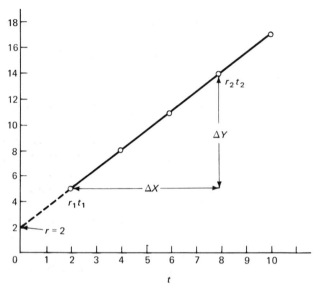

FIG. A2-4.

where the dependent variable Y is r and the independent variable is t. The points are plotted in Fig. A2-4. Taking $r = 2$ and $t = 5$ as $X_1 Y_1$ and $r = 8$ and $t = 14$ as $X_2 Y_2$, we have

$$\frac{\Delta Y}{\Delta X} = \frac{t_2 - t_1}{r_2 - r_1} = \frac{14 - 5}{8 - 2} = m = 1.5$$

By extrapolation the intercept c is 2. Therefore, the equation for this straight line is

$$r = 1.5t + 2.0$$

Having this equation available makes it possible for us to calculate a value of r for any value of t of interest. For example, if $t = 24$, $r = 1.5 \times 24 + 2 = 38$.

    The advantages of the straight line plot are so considerable that even if the initial direct plot of available data is a curve, an effort is almost invariably made to convert the data to some linear relationship.

    Consider the following set of data:

$x =$	2.00	4.00	6.00	8.00	10.0
$y =$	4.50	2.25	1.50	1.13	0.90
$1/x =$	0.500	0.250	0.166	0.125	0.100

Taking the first two lines, values of x and y give the plot shown in Fig. A2-5a. It is apparent that there is some kind of reciprocal relationship between x and y. Using this curve, the next step is for us to calculate at each point the value of $1/x$. The results are given in the third line of data above. Figure A2-5b

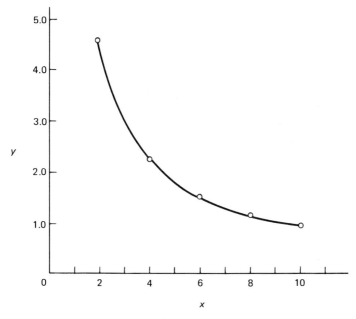

FIG. A2-5a.

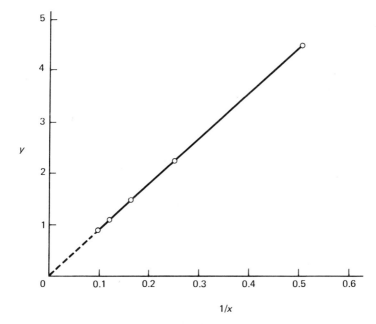

FIG. A2-5b.

shows that a plot of y vs. $1/x$ is a straight line. The intercept is zero, and the slope is $(0.500 - 0.100)/(4.50 - 0.90) = 9$. Substitution into Eq. (A2-1) for m and c gives the simple equation $y = 9/x$. Thus a set of data that give a rather intractable curve in the initial plot have been shown by graphical analysis to be related by the very simple equation obtained.

One other common transformation, applied to relations that do not give a straight line directly, is to convert to logarithms. The most frequently encountered equation is the exponential law

$$Y = ke^{nx} \qquad (A2\text{-}6)$$

where k and n are constant. Taking natural logarithms of both sides, we get

$$\ln Y = nX \ln e + \ln k = nX + \ln k$$

Converting to common logarithms, we get

$$2.303 \log Y = nX + 2.303 \log k$$

$$\log Y = \frac{n}{2.303} X + \log k \qquad (A2\text{-}7)$$

Comparison of this expression with Eq. (A2-1) shows that a plot of $\log Y$ vs. X is linear and that the slope is $n/2.303$. The intercept is equal to $\log k$.

Although functions requiring graphical treatment that have been encountered in this book are readily convertible to linear form, other more complex curves are met in physical chemistry. These relations may not be capable of being transformed to the simple form of Eq. (A2-1). Using computer techniques, however, it is generally possible to find a suitable equation for almost any set of meaningful data, although the equation may be complex and contain a large number of numerical constants. No matter how complex such a procedure becomes, much the same ideas as outlined here are still applied.

Bibliography

Albert, A., and E. P. Sergeant, *Ionization Constants of Acids and Bases*, Wiley, New York, 1962.

Barrow, G. M., *Physical Chemistry*, 2nd ed., McGraw-Hill, New York, 1961.

Butler, J. N., *Ionic Equilibrium*, Addison-Wesley, Reading, Massachusetts, 1964.

Castellan, G. W., *Physical Chemistry*, Addison-Wesley, Reading, Massachusetts, 1964.

Cheldelin, V. H., and R. W. Newburg, *The Chemistry of Some Life Processes*, Reinhold, New York, 1964.

Daniels, F., and R. A. Alberty, *Physical Chemistry*, 3rd ed., Wiley, New York, 1966.

Dixon, M., and E. C. Webb, *Enzymes*, 2nd ed., Academic Press, New York, 1964.

Frost, A. A., and R. G. Pearson, *Kinetics and Mechanism*, 2nd ed., Wiley, New York, 1961.

Glasstone, S., *Textbook of Physical Chemistry*, Van Nostrand, Princeton, New Jersey, 1946.

Hamill, W., R. R. Williams, and C. MacKay, *Principles of Physical Chemistry*, 2nd ed., Prentice-Hall, Englewood Cliffs, New Jersey, 1966.

Klotz, J. M., *Energy Changes in Biochemical Reactions*, Academic Press, New York, 1967.

335

Laidler, K. J., *Chemical Kinetics*, 2nd ed., McGraw-Hill, New York, 1965.

Latimer, W. M., *Oxidation Potentials*, 2nd ed., Prentice-Hall, Englewood Cliffs, New Jersey, 1952.

Lehninger, A. L., *Bioenergetics*, W. A. Benjamin, New York, 1965.

Lewis, G. N., and M. Randall, *Thermodynamics and the Free Energy of Chemical Substances*, McGraw-Hill, New York, 1923; revised by K. S. Pitzer and L. Brewer, 1961.

Linford, J. H., *An Introduction to Energetics with Applications to Biology*, Butterworths, London, 1966.

Maron, S. H. and C. F. Prutton, *Principles of Physical Chemistry*, 4th ed., McGraw-Hill, New York, 1965.

Moore, W. J., *Physical Chemistry*, 2nd ed., Prentice-Hall, Englewood Cliffs, New Jersey, 1962.

Nash, L. K., *Elements of Chemical Thermodynamics*, Addison-Wesley, Reading, Massachusetts, 1962.

Patton, A. R., *Biochemical Energetics and Kinetics*, W. B. Saunders, Philadelphia, 1965.

Tanford, C., *The Physical Chemistry of Macromolecules*, Wiley, New York, 1964.

VanderWerf, C. A., *Acids, Bases, and the Chemistry of the Covalent Bond*, Reinhold, New York, 1961.

Wall, F. T., *Chemical Thermodynamics*, 2nd ed., Freeman, San Francisco, 1965.

Index

Absolute entropy, 55–58
Absorption coefficient, 315
Acids and bases:
equilibria in biological systems, 142–150
polyprotic:
hydronium ion concentration in solutions of, 119–122
titration of, 140–141
strength, *definition*, 104–105
strong:
pH in solutions of, 109–110
pH in extremely dilute solutions of, 110
titration of, 133–137
weak:
equilibrium and pH calculations, 111–115
titration of, 137–139
Acid-base indicators, 130–133
Acid-base solutions, normality, *definition*, 134
Acid-base titrations, 129–141
Acid dissociation constants:
definition, 112
table, 112
Acidosis, 142

Activation energy, 203, 207
Active transport, 322
Activities:
of ions, 152–172
of nonideal solutions, 99
of real gases, 97
use of, to obtain general free energy change expression, 69–70
Activity coefficients:
of ions:
calculation of, from Debye-Huckel theory equation, 159
mean, *definition*, 159
nomograph for determination of, 160
relation to free energy change, 158
of solutes, 100
of solvents, 99
Activity quotient, 71
relation to equilibrium constant, 87
Adiabatic calorimeter, 41
ADP, 77
Adsorption, 310–311
Adsorption isotherms, 311
Alkalosis, 142
Amino acids, 145, 306
Ampere, 153

Amphoteric substances:
definition, 105
hydronium ion concentration in solutions of, 117–118
Area under a curve, 7
Arrhenius equation, 203, 207, 211
Arrhenius theory of electrolytes, 156
ATP, 77
Azeotrope, 280, 282

Base dissociation constants:
definition, 113
table, 112
Bases:
Bronsted-Lowry *definition*, 104
in DNA, 307
Beer's law, 315
Bends, 288
Binary liquid mixtures, 275–286 (*see also* Solutions)
Biological redox reactions, *chart*, 252
Blood, ionic equilibrium in, 142–145
Boiling point elevation, 290–292
table of K_B values, 292
Bomb calorimeter, 41
Brownian movement, 305
Buffer solutions, 122–129
capacity, 127
common ion effect in, 122
definition, 124
efficiency, 126
Henderson-Hasselbalch equation for, 123
physiological, bicarbonate system as, 142–144
preparation of, 128–129

Calculus, 9
Calomel electrode, saturated, *diagram*, 247
Caloric value of foods, 42
Calorimeter, 40–42
figure, 41
Calorimetry in biological systems, 42–43
Carbonium ions, 198
Catalysis, 212–220
by enzymes, 213–220 (*see also* Enzyme catalysed reactions)
homogeneous, 212
Cell constant, 154
Clapeyron equation, 261
Clausius-Calpeyron equation, 262

Co-enzyme, 214
Combustion, heat of, 36, 42
Common ion effect:
in buffer solutions, 122
in slightly soluble salt solutions, 168
Complex reactions, 195–201 (*see also* Reactions)
kinetics of, 200–202
Coulomb, 226
Chain carriers, 197
Chain reactions, 196–197
Colligative properties of solution, 288–300
for electrolyte solutes, 299–300
for nonelectrolyte solutes:
boiling point elevation, 290–292
freezing point depression, 292–294
osmosis, 294–299
vapor pressure lowering, 289–290
Colloids, 304
Concentration cell, 238
Conductance, 153–157
equivalent conductance, 153
concentration dependence of, 155–157
measurement of, 154
specific conductance, 153
Conductivity cell, 153–154
diagram, 154
Conjugate acid-base pairs, 105
relation between K values, 115–116
Consolute temperature, 283
Constant boiling mixtures, 280
Constant pressure processes, 16 *ff*
Constant volume processes, 15
Countercurrent extraction, 302
Coupled reactions, 76
Critical point, 266
Critical pressure, 267
Critical solution temperature, 283
Critical temperature, 267

Daniell cell, 229, 233, 236, 239
Debye-Huckel theory, 157–161
use of, to obtain activity coefficients of ions, 159
Dialysis, 322
Dissociation constants of acids and bases, *table*, 112
Distribution (of a solute between two solvents) coefficient, 301
DNA, 307
Donnan membrane equilibrium, 320–322
Double layer, electrical, 311

Electrical double layer, 311
Electrochemistry of biological redox reactions, 251–253
Electrode:
  glass, 250
  standard hydrogen, 241
  types of, *table*, 228
Electrolysis, 153, 226
Electromotive force (emf), 223 *ff*
  application of in:
  concentration dependence of, 235
  pH determinations, 249–251
  potentiometric titrations, 248
  of half-cells, 240–246
  reversible, *definition*, 230
  temperature dependence of, 234
Electron transfer reactions, 222 *ff*
Electro-osmosis, 313
Electrophoresis, 148, 312
Electrophoretic mobility, 312
Energy, 4 *ff*
  of activation, 203, 207
  definition, 4
  in first law of thermodynamics, 13
Energy change:
  in constant volume processes, 15
  defining equation, 4
  as heat of heation at constant volume, 15
  temperature dependence of, 40
Energy unavailable for work at constant temperature, 50
Enthalpy, 16 *ff*
  of activation, 211
  definition, 16
  as an extensive property, 30
  physical significance of, 16
Enthalpy change:
  in chemical reactions (*see* Heat of reaction)
  defining equation, 16
  temperature dependence of, 38–40
Enthalpy of formation (*see* Heat of formation)
Entropy, 47 *ff*
  absolute, 55–58
  calculation of, 56–57
  equation for, 55
  table of values of, 58
  in Third Law of thermodynamics, 55
  variation of, with temperature, 55
  concept of, 47
  relation to randomness, 48

Entropy change:
  calculation of, in:
  chemical reactions, 58 *ff*
  isothermal ideal gas volume change, 53
  phase changes, 52
  temperature change processes, 53–54
  as a criterion for spontaneity, 49, 62
  defining equation for, in finite and infinitesimal processes, 51
  relation to amount of heat absorbed, 52
Entropy of activation, 211
Enzymes, 213–220
Enzyme catalyzed reactions, 213–220
  competitive inhibition, 218
  inhibitors in, 217
  Lineweaver-Burke double reciprocal plots for, 217–219
  maximum rate of, 214, 216
  noncompetitive inhibition, 219
Equilibrium, 81 *ff*
  and free energy, 85 *ff*
  liquid-vapor, 259–265
  in nonideal systems, 96 *ff*
  solid-liquid, 265–266
  solid-vapor, 265
Equilibrium constant, 81 *ff*
  calculations using, 83–85
  for redox processes, 239
  relation to standard free energy change, 86
  temperature dependence of, 93–96
  thermodynamic definition of, 87
  use of, in calculations involving the standard free energy change, 91–93
Equivalent conductance, 153
Exothermic processes, 17
  activation energy in, 208
Expansion of ideal gases, 6 *ff*

Faraday, numerical values of, 226
Faraday's laws, 226
First law of thermodynamics, 4 *ff*
  applications of, 14–26
  equation for, 13
  example calculation utilizing, 14
  nonpredictive character of, 46
  statement of, 13
First order reactions, 183–189 (*see also* Reactions)
Flow birefringence, 323
Fluids, 267

Four center intermediate, 199
Fractional distillation, 280–282
Free energy:
  of activation, 211
  concentration dependence of, 67, 296
  defining equation, 60
  in derivation of Raoult's law, 274–275
  and equilibrium, 85 *ff*
  of formation, 65 (*see also* Standard Free
    Energy of formation)
  pressure dependence of 66–67, 297
  temperature dependence of, 72–74, 260–
    261
Free energy change:
  application to biological systems, 74–78
  calculation of, for chemical reactions,
    70–72
  concentration dependence
    in generalized process, 67
    in redox reactions, 235–240
  as a criterion for spontaneity, 62–63
  in derivation of osmotic pressure equa-
    tion, 297
  in phase equilibria, 63, 259 *ff*
  in physical processes, 66 *ff*
  physical significance of, 61–62
  pressure dependence of, 66–67
  in redox reactions, 222 *ff*
  relation to electromotive force, 233–235
  significance of, compared to standard
    free energy change, *table*, 90
  in solute diffusion processes, 295
  standard: (*see also* Standard free energy
    change)
    calculation of, 64
    calculations involving the equilibrium
      constant, 91–93
    significance of, 90
    temperature dependence, 72–74
    in ultracentrifugation calculations, 318
    in vaporization, 259
Free energy of ions, relative to activity,
  156
Free radical, 198
Freezing point depression, 292–294
  table of K_F values, 292
Frequency factor, 203, 207
Freundlich adsorption isotherm, 311

Galvanic cells, 227 *ff*
  concentration cell, 238

Galvanic cells (*cont.*)
  conventional representations, 230
  diagram of typical, 228
  half-cell emf:
    measurement of, 240
    oxidation potentials, 242–243
    reduction potentials, 243–244
    use of standard hydrogen electrode to
      determine, 241
    measurement of emf, 230
    voltage, 229
Gegenions, 311
Gibbs adsorption isotherm, 310
Gibbs free energy (*see* Free energy)
Glass electrode, 250
Glucose, 44, 306
Glycine, 146–148
Gram equivalent weight, *definition* for
    acids and bases, 133
    redox reagents, 226

Half-cells, conventional symbols for, 228
Half-life, of a chemical reaction, 187
  equations for, 191
Heat:
  of combustion:
    definition, 36
    relation to caloric values of foods, 42
    table, 36
  definition, 11
  of formation (standard), 33 *ff* (*see also*
    Standard heat of formation)
  of phase change, *definition*, 37
  of reaction:
    at constant pressure, 17
    at constant volume, 15
    processes involving gases, 17
    temperature dependence of, 38–40
  of solution, 37
  of vaporization:
    definition, 25
    determination of, 259 *ff*
    first law of thermodynamics calcula-
      tions involving, 26
    table of values for liquids, 264
Heat capacity:
  at constant pressure:
    definition, 27
    table, 28
  at constant volume, 27
  of gases, 27

Heat capacity (*cont.*)
  graph, of typical variation with temperature, 56
  molar, *definition*, 12
  specific, *definition*, 12
  use of, in calculation of absolute entropies, 55–57
Heat content (*see* enthalpy)
Heating curve, 270
Helmholtz free energy, 80
Henderson-Hasselbalch equation:
  application to:
    amino acid systems, 149–150
    buffer solutions, 124 *ff*
  derivation of, 123
Henry's law:
  equation, 287
  table of constants, 287
Hess' law, 29–32
High energy bonds, 76
Hydrogen electrode, standard, 241
Hydrolysis, in acid-base systems, 116–117
Hydronium ion,
  concentration of (*see* pH)
  formation of, 104
Hypertonic solutions, 319
Hypotonic solutions, 319

Ideal gas:
  expansion, 6 *ff*
  isothermal expansion, 20 *ff*
  reversible expansion, 21 *ff*
Ideal gas law, 255
Ideal solutions, 274–278 (*see also* Solutions)
IEP, 148
Immiscible liquids, 284–285
Indicators:
  acid-base:
    pH color change range of, *table*, 133
    proton transfer equilibria in, 130–133
  redox, 249
Inhibitors:
  in chain reactions, 197
  in enzyme catalyzed reactions, 217
Integration, 9 *ff*
Interionic forces, effect on activity, 157–161
Internal energy (*see* Energy)
Ionic solutions, 152–157 (*see also* Solutions)
Ionic strength:
  calculation of, 159
  definition, 158

Ionic strength (*cont.*)
  effect on pH°, 162–163
  effect on solubility of slightly soluble salts, 170
Irreversible processes, *definition*, 18
Isoelectric point, 148
Isolated system, role in predicting spontaneity, 49
Isotherms, 267–268

Kidneys, pH control by, 145
Kinetics of chemical reactions, 174–220
  (*see also* Reactions, Rate equations, Reaction order, reaction rate)
Kinetic theory of gases, 205

Langmuir adsorption isotherm, 310–311
Law of mass action, 82
Light scattering, 314–317
Logarithms, use of, 324–327
Lungs, CO_2 in, effect on blood pH, 145

Macromolecular weights, 307 *ff*
  determination of, by
    light scattering, 316
    osmotic pressure measurements, 318–320
    other methods, 322–323
    sedimentation, 317–318
    number average, 308, 309, 319, 323
    weight average, 308, 309, 316, 323
Macromolecules, 304 *ff*
  light scattering by, 314–317
  nature of, 305
  osmotic pressure in solutions of, 318
  sedimentation of, in solutions, 317–318
Maxwell-Boltzmann distribution, 206
Mechanism of reaction, 175 (*see also* Reactions)
Melting point, pressure dependence of, 265–266
Micelles, 304
Michaelis constant, 216
Michaelis-Menten equation, 216
Molality, 274
Molarity, 274
Molecularity of reactions, 182
Mole fraction, 274
Mole number, 299
Monomers, 305

Nernst distribution law, 301
Nernst equation, 235–239
  derivation of, 235
  forms of, for half-cells, 246
  summary of relations involving, *chart*, 244
Normality, of acid-base solutions, 134
Nucleic acids, 306, 307
Nucleotides, 307

Ohm's law, 153
Organic photochemistry, 199
Osmosis, 294–299
  definition, 296
  role of free energy in, 297
Osmotic pressure:
  application of, to macromolecular weight determinations, 319
  derivation of equation for, 297–298
Oxidation, *definition*, 225
Oxidation number, 225
Oxidation potentials, *table*, 245
Oxidation state, 225
Oxidizing agent, 226

Partially miscible liquids, 282–284
Path dependent functions, 3
Peptide linkages, 146
pH:
  change during titration, 134–141
  colorimetric determination, 130
  defining equation, 107
  electrochemical determination of, 249–250
  relation to solution classification, 108
pH°:
  comparison with pH, *graph*, 162
  definition, 161
  in strong acids and bases, 161
  in weak acid and buffer solutions, 163–166
Phosphates, 77–78
Polarizability, 315
Polyisopropylene, 306
Polymerization, 306
Polymers, 305–306
Polysaccharides, 306
Potentiometer, *diagram*, 231
Potentiometric titration, 248–249

Principle of electroneutrality, 108–109
Processes:
  constant pressure, 16 *ff*
  constant volume, 15
  isothermal, 18
  reversible, *definition*, 18
  types of, thermodynamic, 19
Prosthetic groups, 214
Proteins, 42, 146, 305, 306
PT diagrams, 270

Randomness, relation to entropy, 48–49
Raoult's law, 68, 286 *ff*
  application to ideal binary liquid mixtures, 275–278
  derivation of, 274–275
  deviations from, 278–280
Rate equations:
  tabulation of, 191
  utility of, 175
Rayleigh equation, 315–316
Rayleigh ratio, 316
Reaction:
  amount of, general problems in determining, 179
  chain, 196–197
  consecutive, 195
  first order:
    differential rate law for, 183–184
    exponential rate law for, 188–189
    half-life expression for, 187
    integrated rate law, 185–187
  heat of (*see* Heat of reaction, Enthalpy change)
  molecularity, 182
  order (*see* Reaction order)
  rate equations for, *table*, 191
  second order, 189–190
  side, 196
  zero order, 190
Reaction coordinate, 208–209
Reaction intermediates, 199–200
Reaction mechanisms, 175
  types of, 198
Reaction order:
  defining equation for, 181
  determination of, 191
    empirical fit methods, 193–194
    half-life methods, 193
    integrated rate law method, 192
    mechanical slope determination, 192

Reaction rate:
    collision theory of, 205–209
    definition, 175
    of enzyme catalyzed reactions, 214 *ff*
    Michaelis-Menten treatment of, 205–217
    general concept of, 175–177
    instantaneous, determination of, 178
    temperature dependence of, 202–204
    transition state theory of, 209–212
Real gases, 256–258
    PV product of, *figure*, 256
    van der Waal's equation for, 257–258
    virial equation of state for, 257
Reducing agent, 226
Reduction, *definition*, 225
Reduction potential (standard), *table*, 245
Reference cell, 240
Resistivity, 153
Respiratory quotient, 43
Reversible processes, *definition*, 18
RNA, 307

Salt bridge, 227, 228
Second Law of thermodynamics, 46 *ff*
    statement of, 49
Sedimentation, 317–318
Sedimentation potential, 313
Semipermeable membrane, 296, 298
Solid-liquid equilibrium, 265
Solubility,
    common ion effect on, 168
    general definition, 273
Solubility product:
    determination of, from emf measurements, 239
    of slightly soluble salts, *definition*, 166
    thermodynamic, K_{sp}°
    definition, 169
    table, 171
Solute, *definition*, 273
Solute distribution between two solvents, 300–302
Solution, *definition*, 273
Solutions, 273 *ff*
    fractional distillation, 280–282
    of gases in liquids, 286–288
    ideal:
        application of Raoult's law to, 274 *ff*
        definition, 274
    ideal binary liquid mixtures, 275–278

Solvent, *definition*, 273
Standard cell, 241
Standard enthalpy of formation (*see* Standard heat of formation)
Standard free energy change (see also Free energy):
    calculations involving the equilibrium constant, 91–93
    relation to equilibrium constant value, 86
    significance of, *tabulation*, 90
Standard free energy of formation:
    definition, 65
    table, 65
    use of, to calculate standard free energy change in reactions, 65
Standard heat of formation:
    definition, 33
    table, 34
    use of, to calculate enthalpy change in reactions, 34–35
Standard hydrogen electrode, 241
Standard oxidation and reduction potentials
    experimental determination of, 247
    table, 245
Standard state:
    definition, 32
    tabulation of conditions for, 33
State, *definition*, 2
State functions, 3 *ff*
Steady-state treatment of reaction mechanisms, 200–202
Steam distillation, 284–286
Steric factor, 207
Straight line graphs, algebra of, 329–334
Streaming potential, 313
Substrate (in enzyme catalysis), 213
Surface tension, 309
Surroundings, *definition*, 2
System, types of, *definitions*, 2

Tandem reactions, 76
Tension, of gases in solution, 287
Thermochemical equations, 29 *ff*
    Hess Law in, 29–32
Thermochemistry, 26 *ff*
Thermodynamics:
    comparison with kinetics, 174
    definition, 2
Third law of thermodynamics, 55–58
Thiyl radicals, 198

Titrations:
  acid-base, 129–141
  potentiometric, 248–249
Tone, of solutions, 319
Transition state, 209
Triple point, 268–269
Turbidity, 315

Ultracentrifuge, 305, 317

van der Waals equation, 257
van der Waals constants, *table*, 258
van't Hoff equation, 93, 262
van't Hoff *i*-factor, 299
Vaporization, application of first law to, 24
Vapor pressure:
  of liquids:
    Clapeyron equation, 261
    Clausius-Clapeyron equation, 262

Vapor pressure of liquids (*cont.*)
  temperature dependence of, *figures*,
    260, 263
  of solids, 265
Vapor pressure lowering, 289–290

Water, proton transfer reactions in pure,
  106
Weston cell, 232
Work:
  definition, 5
  electrical, 222, 233
  pressure volume, 5 *ff*
  useful:
    in redox processes, 222
    relation to free energy change, 61

Zeta potential, 311
Zwitterion, 147